ANNIE BESANT

ANNIE BESANT
A BIOGRAPHY

ANNE TAYLOR

Oxford · New York
OXFORD UNIVERSITY PRESS
1992

Oxford University Press, Walton Street, Oxford OX2 6DP
Oxford New York Toronto
Delhi Bombay Calcutta Madras Karachi
Petaling Jaya Singapore Hong Kong Tokyo
Nairobi Dar es Salaam Cape Town
Melbourne Auckland
and associated companies in
Berlin Ibadan

Oxford is a trade mark of Oxford University Press

© Anne Taylor 1992

All rights reserved. No part of this publication may be reproduced,
stored in a retrieval system, or transmitted, in any form or by any means,
electronic, mechanical, photocopying, recording, or otherwise, without
the prior permission of Oxford University Press

British Library Cataloguing in Publication Data
Data available

Library of Congress Cataloging in Publication Data
Taylor, Anne.
Annie Besant: a biography/Anne Taylor.
p. cm.
Includes bibliographical references.
1. Besant, Annie Wood, 1847–1933. 2. Theosophists—Biography.
3. Social reformers—Biography. I. Title.
BP585.B3T38 1992 299'.934'092—dc20 91-24462
ISBN 0-19-211796-3

Set by Pure Tech Corporation, Pondicherry, India
Printed in Great Britain by
Biddles Ltd.
Guildford and King's Lynn

For Tobias

ACKNOWLEDGEMENTS

THE material on which this biography is based could not have been assembled without help and advice from many individuals and organizations in England, India, and the United States.

In England I must record my warm thanks to W. K. Stead for allowing me to see the letters of Annie Besant to his grandfather, the crusading journalist W. T. Stead. I have to thank Mary Lutyens for her kindness in sharing her recollections of Annie Besant, and for introducing me to the correspondence between the latter and Krishnamurti. Extracts from this appear by permission of the Krishnamurti Foundation of America. I am also very grateful to Mrs T. Lilliefelt for her help with this.

The Revd S. Salter gave me much useful information about Cheltenham in the 1870s. Mr David Eyre drew my attention to the Parochial Diary of the Revd Frank Besant, and the Lincolnshire Archives Office gave me permission to quote from it. The former vicar of Sibsey, the Revd David Loake, showed me St Margaret's Church, and Mrs Frost kindly allowed me to see the inside of the former vicarage. The present vicar, the Revd Bill Price, also helped me. My somewhat despairing request for information about the Marryat family brought a swift response from three descendants: Mrs Sewell, Mrs Young, and Mrs Roper.

I have specially to thank Basil Bradlaugh Bonner.

Dr Hugh Gray, Miss Lilian Storey, and Miss Ianthe Hoskins of the Theosophical Society in England gave advice, and help with papers held in London.

My stay at the headquarters of the Theosophical Society in India at Adyar, Madras, was by permission of the President, Dr Radha Burnier, who was kind enough to discuss Annie Besant with me. While at Adyar I had the pleasure of a conversation with Sri Achyut Patwardhan. The late Mr C. R. N. Swamy put his extensive knowledge of the political papers at my disposal. I have also to thank Elizabeth Sterling for her help with this very fragile material.

Travelling alone, I depended very greatly on Oxford University Press, Madras. The reassuring presence and splendid hospitality of the Manager, Mr N. Parthasarathy, will long be remembered by me. At Varanasi, Dr C. V. Agarwal, General Secretary of the Indian Section, showed me the Theosophical Society compound where Annie Besant's house, Shanti Kunj, is lovingly preserved. I have to thank the staff of many libraries and institutes, especially those of the Indian Institute in Oxford, and of the Brotherton Collection at the University of Leeds, which contains a selection of Annie Besant's early pamphlets. I am grateful to the staff of the British Library, the London Library, the Bodleian Library, Sheffield City Library, the Fawcett Library, the National Secular Society, the Bishopsgate Institute, the University of Essex Library, the House of Lords Record Office, the Records Office of University College London, the Library of the Co-operative Society, Manchester, and the Internationaal Instituut voor Sociale Geschiedenis, Amsterdam.

Extracts from Crown copyright material in the Public Record Office appear by permission of the Controller of Her Majesty's Stationery Office. The letter from Annie Besant to Sir Joseph Hooker is reproduced by permission of the Board of Trustees of the Royal Botanic Gardens, Kew. I am grateful to the Warden and Fellows of Nuffield College, Oxford, for permission to inspect the papers of the Fabian Society. Extracts from the papers of Lord Haldane, the earl of Minto, and Patrick Geddes appear by permission of the Trustees of the National Library of Scotland. The British Library gave permission to quote from the Gladstone, Burns, Morris, Shaw, and Blavatsky papers. I am grateful to the Bodleian Library, Oxford, for permission to quote from the papers of Max Müller and the letters of W. T. Stead to Madame O. Novikoff. I have to thank the Revd Dr Marks, Warden of Liddon House, London.

Passages from *Krishnamurti: The Years of Awakening*, © Mary Lutyens (1975), are reproduced by permission of John Murray (Publishers) Limited. I have to thank the Society of Authors on behalf of the Bernard Shaw estate for permission to quote extracts from Shaw's *Diary* and autobiographical fragments. The London School of Economics and Political Science granted permission to quote from Beatrice Webb's *Diary*, published by Virago, and from the *Letters* of Beatrice and Sidney Webb, published by Cambridge University Press. Bryant & May Ltd. kindly allowed me to use extracts from papers concerning the match girls' strike of 1888.

ACKNOWLEDGEMENTS

The Oriental and India Office Collections of the British Library gave permission for extracts from the papers of the Viceroy, Lord Chelmsford.

In the United States, the Joseph Regenstein Library of the University of Chicago was generous with photocopies of the Helen I. Dennis Papers, as, in the case of the Moncure D. Conway Papers, was the Rare Book and Manuscript Library of Columbia University, New York.

Finally, the warmest thanks to the following individuals: Professor Joseph Baylen, Richard Clive, Anna Davin, Raghavan Iyer, Mary Kavanagh, Patrick Melvin, Major A. A. Persse, Dr Edward Royle, Peter Sutcliffe, David Tribe, Colin Ward, and my editor, Judith Luna.

CONTENTS

List of Illustrations	xiii
1. An Alternative Childhood	1
2. The Half-Angelic Being	18
3. A Husband's Authority	28
4. 'You have read too much already'	41
5. Flight	53
6. Hardship	61
7. Iconoclast	74
8. The Cry for Light	83
9. Heroine of Free-Thought	96
10. The Knowlton Trial	107
11. Life-Enhancing Checks	121
12. Besant *v.* Society	128
13. Tutorials	138
14. 'Let the people speak!'	149
15. Social Reform, not Socialism	161
16. Bradlaugh Estranged	172
17. The New Model Leader	188
18. Annie Militant	203
19. 'Superhuman, spiritual realities'	222
20. Gone to Theosophy	241
21. 'Mrs Besant, I suppose?'	259
22. Aryavarta	277
23. 'Bande Mataram'	293
24. Bullets and Brickbats	311
Epilogue	327
Notes	333
Select Bibliography	369
Index	375

ILLUSTRATIONS

Between pages 178–9

1. Annie Besant in the 1870s
   © *British Library*
2. Frank Besant in Sibsey Churchyard with the Churchwarden and Parish Clerk, 1906
   © *St Margaret's Church, Sibsey*
3. Annie Besant in her lecturing dress, 1885
4. Charles Bradlaugh, in his fifties
   © *British Library*
5. The Match Girls' Strike Committee, Herbert Burrows and Annie Besant standing centre
   © *British Library*
6. Helena Petrovna Blavatsky
7. Annie Besant, 1897, in her Indian robes
   Reproduced by permission of the Theosophical Publishing House
8. Annie Besant with Colonel Olcott and Charles Leadbeater, 1890s
9. Charles Webster Leadbeater
   Reproduced by permission of the Theosophical Publishing House
10. Krishnamurti and Nityananda, 1910
    Reproduced by permission of the Theosophical Publishing House
11. Annie Besant and Krishnamurti arriving at Charing Cross Station, 1911
12. Dr Besant in Trafalgar Square, *c.*1925
    Reproduced by permission of the Theosophical Publishing House

CHAPTER 1

AN ALTERNATIVE CHILDHOOD

ANNIE BESANT was born in London on 1 October 1847 at 2 Fish Street Hill, within the sound of Bow Bells.¹ Her parents were staying with friends at the time and her arrival took them by surprise. For Annie London was to be an unsatisfactory birthplace: it spoiled her perception of herself. 'Three quarters of my blood and all my heart was Irish', she wrote.² Her dedication to Irish causes called for a suitably romantic and Irish place of origin.

Her father's name was William Burton Persse Wood and it was through him that she acquired the modicum of English blood she so despised. His mother was Irish but his father came from a long-established Devon farming family. When Annie was a child the name of Wood was famous in the City and in the country at large, due to the exploits of the eldest son, who left Devon at the end of the eighteenth century and made a career in business and in politics. Matthew Wood, Annie's great uncle, was a famous Sheriff and Lord Mayor of London. He was elected to Parliament as Member for the City in 1817, and championed two notable causes: that of Queen Caroline, and that of the profligate duke of Kent. The former was lost, but Alderman Wood was the salvation of the duke; he helped to manage his debts and was the means whereby the family were allowed to return to England for the birth of their only child. For that Matthew was thriftily rewarded with a baronetcy by HRH's daughter Queen Victoria.³ Matthew's sons also had distinguished careers in the Church, the law, the City, and in Parliament. They were alive and flourishing when Annie was born.

But William Wood, her father, came of the younger branch far removed from the sources of power and wealth. He was born in Galway where his father had married an Irish girl. He grew up in the years of dearth when Ireland suffered a series of such poor harvests that famine threatened, especially in the south and west. William studied medicine at Trinity College, Dublin, where he in his turn married an Irish girl. Annie's mother Emily Morris—once more grandly spelt Maurice—came of a large family whose fortunes had decayed with their name. As a child Emily had to be sent away from home to be brought up by a maiden aunt who instilled in her the pride of descent. Emily was fond of telling her daughter that she could trace her ancestors back to an ancient race of Milesian kings who ruled—tempestuously it seems—in France before taking refuge in Ireland.

The whole family was caught up in the tragic exodus after the devastating famine of 1845. Abandoning his career as a doctor, William Wood took Emily to England, where a relative offered him a position as an underwriter in the City.[4] The means it provided were sufficient to enable them to live comfortably in St John's Wood, where they had three children, Henry, Annie, and Alfred. Emily's parents and her sisters went into decline in an altogether less fashionable part of London: south of the river among the interlocking railway lines of Clapham.

According to Annie, her father

was keenly intellectual and splendidly educated; a mathematician and a good classical scholar, thoroughly master of French, German, Italian, Spanish and Portuguese, with a smattering of Hebrew and Gaelic, the treasures of ancient and modern literature were his daily household delight. Nothing pleased him so well as to sit with his wife reading aloud to her as she worked ... Student of philosophy as he was, he was deeply and steadily sceptical; and a very religious relative has told me that he often drove her from the room by his light and playful mockery of the tenets of the Christian faith.[5]

This description comes from the first of a series of reminiscences Annie wrote, beginning in 1884 when she was 37. As she said, these *Autobiographical Sketches* were 'to satisfy friendly questions and to serve in some measure as defense against unfriendly attack'.[6] By that time her notoriety was such that the attackers outnumbered the defenders. Annie could hardly assess the truth of the portrait of her father or add to it from her own experience; he died when she was 5.

William's job left him time to indulge his medical interests which always absorbed him. He would on occasion accompany medical friends on their hospital rounds or help them in the dissecting room. On one such occasion he cut his finger on the breastbone of a person who had died of rapid consumption. The finger became infected and, a day or two later, a surgeon advised him to have it amputated. Others laughed at the suggestion and William, at first inclined to submit, fatally decided to do nothing about it.

The details of his rapid decline and death which followed as a consequence were frequently told to Annie as a child, and she was not spared the horror and distress suffered by her mother. In Annie's eyes Emily figured always as the heroine. In August 1852 William got wet through on top of an omnibus. A severe cold ensued which settled on his chest. One of the most eminent doctors of the day, who was noted for his graceless bedside manner, sounded his lungs and was asked his opinion by Mrs Wood. 'You must keep up his spirits', he told her, 'he is in a galloping consumption; you will not have him with you six weeks longer.'[7] Emily Wood fainted on the spot but, half an hour later, was at her husband's side, 'never to leave it night or day . . . until he was lying with closed eyes, asleep in death'. Annie was lifted on the bed to say goodbye to her dear papa the day before he died. 'I remember being frightened at his eyes', she wrote, 'which looked so large and his voice which sounded so strange, as he made me promise to be a very good girl to darling mama as papa was going right away. I remember insisting that papa should kiss . . . a doll . . . and being removed crying and struggling from the room.' At the death Emily lost her senses. When she recovered she insisted on locking herself in her bedroom for the night; in the morning her hair, 'black, glossy, and abundant', had turned white. So Annie was told.[8]

On the day of the funeral Emily stayed at home, Annie by her side. The child watched as, 'with vacant eyes and fixed pallid face', her mother followed the service stage by stage in her mind, and suddenly, with the words 'it is over', fell back fainting. Afterwards Emily said she had followed the hearse, attended the service, and walked behind the coffin to the grave. Proof of her second sight was afforded a few weeks later when, on going to Kensal Green Cemetery, Emily found the unmarked grave in the wilderness of stones when her companions had searched in vain.

Annie heard this story many times. Her mother had in her a vivid strain of Celtic superstition, she told her readers in 1884.[9] It was displayed again when, a few months after her husband's death, the baby, Alf, sickened. Emily foretold his death and indeed it shortly occurred.

The gloom now hanging over the family intensified when to the widow's dismay it was found that William Wood had left no money to provide for his family. With Henry aged 7, and Annie $5\frac{1}{2}$, Emily was obliged to move to a less prosperous part of London while it was decided how they were to survive. They went to Richmond Terrace, Clapham, where they lived poorly enough but near to Annie's grandfather in Albert Square. His house was a haven for Irish exiles and every Christmas and Easter members of the family gathered under his roof. Three of Annie's aunts lived there permanently with him and Emily found in them companionship and moral support. Annie loved their house; it had a small and rather dingy garden at the back where she could hide with a book instead of being told to run and play. Her grandfather invented things; in particular he had devised a means of fastening the ends of railway track together (a problem that occupied engineers and promoters a great deal at the time). Though the idea was taken up by several companies, he made no money by it. He was a cheerful man in spite of his financial problems, given to Irish songs and tales of his Dublin youth. As far as is known he was the only adult male close to Annie while she was a child.[10]

William Wood's relatives were disposed to help the young widow but Emily's pride kept them at arms's length. She refused the offer by two of Matthew Wood's sons to take charge of Henry. He could be educated at a City school and afterwards enter commerce, they said.

Emily was adamant that on his deathbed her William had repeated his desire that Henry should go into the law and, though she would have preferred him to take holy orders, a lawyer he would become. Henry, therefore, had to go to public school and afterwards to Oxford or Cambridge. Her stand found critics in the Wood family, especially among its female members. Yet, as Annie recorded, they continued to show her mother many kindnesses.[11]

There were ways of using the City connection to send Henry to public school at little cost—Christ's Hospital springs to mind. But Emily insisted on taking a greater share in her son's education than was strictly necessary. She proposed to find a house in

Harrow to board some boys, thereby earning money as well as qualifying Henry for the lower fees charged by the school to the sons of people of the town. Possibly she welcomed the opportunity to stand on her own feet. Annie thought so: 'never dwelt in a delicate body a more resolute mind and will than that of my dear mother.'[12]

And so when Annie was 7 Mrs Wood took lodgings over a grocer's shop in Harrow while she looked for a suitable house. Fortunately she found a boy about the same age as Henry whose parents were glad to place him in her charge and by this means she was able to engage a tutor to prepare the two for school. Annie remembered little more of him than the cork leg he had which stuck straight out behind when they knelt down for family prayers, 'conduct which struck me as irreverent and unbecoming but which I always felt a desire to imitate'.[13] It took nearly a year to find a house, during which time Mrs Wood enlisted the sympathy and support of the Headmaster of Harrow, the celebrated Dr Vaughan. His permission was needed for boys to live in her house: he gave it on the understanding that a master would live there too so that the boys would not lack a house tutor like the others had.

The Woods moved in on Annie's eighth birthday, 1 October 1855. After the constraint and sorrow of the last three years their new life appeared to promise much to all three of them. Mrs Wood had the satisfaction of having accomplished what she had set out to do and was settled among congenial neighbours. In particular she was befriended by the Headmaster's wife, Catharine Vaughan. Henry was launched into a world of chalk and ink, happily indistinguishable among a crowd of scurrying junior boys. Annie who was very close to her mother, reflected her rising spirits and promptly fell in love with the house and garden. They provided the happiest memories of her childhood. The house

was very old and rambling, rose covered in front and ivy covered behind; it stood on the top of Harrow Hill... and had once been the vicarage of the parish but the vicar had left it because it was so far removed from the part of the village where all his work lay. The drawing room opened... into a large garden which sloped down one side of the hill, and was filled with the most delightful old trees, fir and laurel, may, mulberry, hazel, apple, pear and damson, not to mention currant and gooseberry bushes innumerable, and large strawberry beds spreading down the sunny slopes. There was not a tree there that I did not climb and one widespreading Portuguese laurel was my private country house. I had there my bedroom and my sitting room, my study and my larder. The larder was supplied by the fruit trees... and in the study I would sit for hours with some favourite book, Milton's 'Paradise Lost' the chief favourite of all. The birds must often have been startled when from that small swinging form perched on a branch, came in childish tones the 'Thrones, denominations, princedoms, powers' of Milton's stately and sonorous verse. I liked to personify Satan and declaim the grand speeches of the hero-rebel, and many a happy hour did I pass in Milton's heaven and hell with, for companions Satan and the Son, Gabriel and Abdiel.[14]

It was easy to feel oneself halfway to heaven on Harrow Hill. It rises very abruptly from the murky London plain, an ancient church on its summit whose soaring spire can be seen for miles around. Annie's cherished garden bordered on the churchyard; over the rose-covered fence she could watch the prosperous and stately traffic of the parish. She was, almost, a vicarage child.

'At the end of the terrace was a little summer house', Annie wrote, 'and in this a trap door... which swung open and displayed one of the fairest views in England. Sheer from your feet downwards went the hill and then from below stretched the wooded country till your eyes reached the towers of Windsor Castle far away on the horizon.'[15] Like the background of some medieval Book of Hours a distant prospect of enchanted towers was a recurring image in Annie's life.

When she was only 8 it was abruptly snatched away. She was happy and secure at home and in her mother's love. 'She and I scarcely ever left each other; my love for her was an idolatry, hers for me a devotion.'[16] Then Annie was told she was to go away to be educated, coming home only in the holidays.

In her autobiography she represented this departure as coming upon her mother with a shock equal to her own. Emily resisted the suggestion but was persuaded by the argument that Annie could not be left to grow up in a house full of boys—she was already as good at cricket and climbing as her brother—and ought to have the advantage of the offer made. This was that a wealthy and charitable spinster, Miss Ellen Marryat, would take Annie to live with her and bring her up with her niece Amy, who was already in her care. Annie portrayed her first meeting with Miss Marryat as happening by chance, at a neighbour's house into which she had wandered. According to her the neighbour then approached her mother with Miss Marryat's suggestion as to the child, who had instantly attracted her.

That version was to excuse Emily in her daughter's eyes from a premeditated act of betrayal. Probably Mrs Wood did not see it that way; it was, after all, precisely what she herself had been obliged to undergo. It may be that she devoted as much thought to Annie's education as to Henry's, only to conclude that it was beyond her capacity either to teach her herself— Annie was so *very* clever—or to engage a governess.

Perhaps her husband's family intervened as well. There is some slight evidence of a connection between the Woods and Marryats—Joseph, Ellen's father, sat in the Commons at the same time as Matthew Wood and was also a supporter of Queen Caroline. However it came about there was no doubt as to Annie's good fortune: years later she remarked that she owed Ellen Marryat more than she could say, not least the love of knowledge that had remained with her as a spur to study.

Miss Marryat had a perfect genius for teaching, and took in it the greatest delight. From time to time she added another child to the party, sometimes a boy, sometimes a girl . . . She chose her children—as she loved to call us—in very definite fashion. Each must be gently born and gently trained, but in such a position that the education freely given should be a relief and aid to a slender parental purse. She taught us everything herself except music, and for that she had a master, practising us in composition, in recitation, in reading aloud English and French, and later, German, devoting herself to training us in the soundest, most thorough fashion.[17]

Observation and experience were the means by which they learned. Geography, for example, consisted of painting skeleton maps and putting together puzzle maps in which countries and counties were cut out in their proper shape. 'I liked big empires in those days; there was a solid satisfaction in putting down Russia and seeing what a large part of the map was filled up thereby.' The only grammar they learned was Latin because of its perfection and the solid foundation it gave to modern languages. 'Auntie', as she liked to be called, had a great horror of children learning by rote things they did not understand, and then feigning they knew them. ' "What do you mean by that expression, Annie?" she would ask me. After feeble attempts to explain, I would answer. "Indeed Auntie, I know in my own head, but I can't explain." "Then indeed, Annie, you do not know in your own head or you could explain so that I might know in my own head." '

The fact that Ellen Marryat was still alive when Annie published her memoirs probably inhibited her from giving many details of her life. Yet these are important, for the background from which Ellen came played no small part in determining the kind of woman Annie was to become.[18] Miss Marryat was 41 when she took Annie into her charge. She had been accustomed all her life to affluence, social position, and the duty towards others less fortunate than herself that was a leading tenet of the Evangelicals among whose formidable ranks she placed herself. Her father, who had made a fortune planting sugar in the West Indies, was a Member of Parliament, Chairman of the Committee at Lloyd's, and Colonial Agent for Grenada and Trinidad. He died relatively young, but not before he had begotten fifteen children, of whom Ellen was the last but one. His second son was the renowned sailor and author Captain Frederick Marryat, whose brothers and sisters shared his talent; almost all of them wrote in one way or another. Ellen did not marry, but lived at home as a companion to her mother.

As befitted the widow of a man described as a merchant prince, Charlotte Marryat was *très grande dame*. As well as rearing her enormous family she found time to develop a taste for plants and flowers into an expertise that won her medals and election as a Fellow of the Royal Horticultural Society. Wimbledon House, where she and Ellen lived, was famous in London society as an agreeable place to visit; invitations to drive out to the Common were eagerly solicited. The house, a long, low villa embellished with Ionic columns, faced a park with an artificial lake. The interior was crammed with trophies brought back from his voyages by Captain Fred—tusks, antlers, shells, precious stones, weapons, and wooden carvings. Here for thirty years Charlotte, with Ellen at her side, entertained writers, artists, politicians, and minor members of the royal family; among them Sheridan, Washington Irving, Landseer, Millais, Lord Aberdeen, and the duchess of Teck.

Ellen's famous older brother Frederick was a hero to her as he must have been to the tomboyish Annie Wood. After hair-raising adventures in the Royal Navy he wrote a series of highly popular novels and children's books, including *Peter Simple*, *Mr Midshipman Easy*, and *Masterman Ready*. Coming ashore he stood for Parliament in Tower Hamlets where his programme called for an end to the impressment of seamen and the betterment of Spitalfields. Failing to be elected he retired to Norfolk, where he was an enlightened estate owner and a pillar of the local lodge of Freemasons.

He too died young after a frightful illness during which Ellen looked after his young children. When Charlotte died in 1854 Ellen decided to devote her new-found freedom to the service of others. After Wimbledon House was given up she took a place called Fern Hill, near Charmouth in Dorset, where Annie went with her from Harrow. It was a handsome sunny house standing by itself under the rim of the headland that divides Charmouth from Lyme Regis. There were extensive views: on the one hand out to sea and Portland Bill, on the other over fields and the billowing Dorset Hills. Trees and rhododendron bushes offered a splendid place to play. Inside was a finely proportioned drawing-room with a conservatory and a gracefully curving staircase leading to large and airy bedrooms. Charmouth village was a steep walk down from the house whose life was, in any case, self-contained.

'Daily when our lessons were over we had plenty of fun; long walks and rides, rides on a lovely pony who found small children most amusing and on which the coachman taught us to stick firmly ... delightful all day picnics in the lovely country round Charmouth, Auntie our merriest playfellow.'[19] Riding was always Annie's favourite exercise; in later years she clung to the conventional habit worn to ride side-saddle when the rest of her wardrobe had become, to say the least, unorthodox.

However well Miss Marryat provided for the children's recreation, life at Fern Hill was, and was meant to be, deeply serious. Religion was its chief concern and study. Writing in 1884 for an audience of free-thinkers[20] Annie complained that what she called Miss Marryat's 'somewhat Calvinistic teaching'[21] tended to make her morbid, especially as she fretted silently after her mother at the same time. But her account of how that teaching was imparted leads one to suspect that she was not quite fair to Miss Marryat. Although the older woman's Evangelicalism was resolute and enduring she kept a sense of proportion. Annie and Amy Marryat were spared the effect of excessive zeal which shrivelled the enthusiasm of so many Victorian children brought up too harshly under its austere influence.

Though nothing but the Bible might be read on Sunday, Ellen enlivened the day by getting the children to sing hymns and telling them stories of missionaries like Robert Moffat and his son-in-law David Livingstone, whose adventures among the savages and wild beasts sufficiently excited Annie. They repeated passages learned by heart from the Bible and, when they were older, taught in the village Sunday school. These lessons had to be carefully prepared on Saturday, 'for we were always taught that work given to the poor should be work that cost something to the giver'.[22] Personal self-denial for the good of others was the first important lesson Annie learned, and it was a principle by which she stood for the rest of her life.

She was, however, far from sound on the central issue—salvation through grace. She had no sense of sin; the acknowledgement of guilt, repentance, prayer directed at forgiveness meant nothing to her.

> It was a subject of some distress to me that I could never look back to an hour of conversion; when others gave their experiences, and spoke of the sudden change they had felt, I used to be sadly conscious that no such change had occurred in me, and I felt that my dreamy longings were very poor things ... and used dolefully to wonder if I were 'saved'.[23]

She confessed that she was often praised for her piety when emulation and vanity were more to the point, as when she learned the Epistle of St James more to distinguish herself for her good memory than for its message: 'the sonorous cadences of many parts of the Old and New Testament pleased my ear and I took a dreamy pleasure in repeating them aloud.' The thought that God and Auntie noticed how nicely she prayed quickly dispelled her unease over her inability to feel as others apparently did.

Hell did not come into her dreams except in the interesting shape it took in *Paradise Lost*—the 'beautiful shadowed archangel'[24] whom Jesus, her ideal prince, would save in the end. Thus Annie came to religion, longing for salvation, innocent of guilt or fear, and powerfully attracted by angels wearing human form.

She grew up at Fern Hill in ignorance of how ordinary people lived. She was well instructed and outwardly confident, but lacked

AN ALTERNATIVE CHILDHOOD

feeling for what made others behave as they did, a fault that was partly due to Miss Marryat. Stern in rectitude herself and iron to the fawning or dishonest, her influence, whether she was feared or loved, was considerable. Besides the Sunday school she founded she started a Bible class, visited the sick, sent food from her table to the poor, and interested herself in finding work for anyone who nerved themselves to ask for it. As a 'centre of beneficence'—Annie's phrase— she performed the role that often fell to solitary gentlewomen of independent means living in the country. She was distinguished by self-reliance and composure, but she was remote.

With no experience of ordinary men and women—*her* judgement as to the fawning and dishonest was destined to let her down—Annie could not even learn through her favourite medium, reading. Plays, novels, and all but religious poetry were forbidden by the Evangelical regime. In any case she was to some extent indifferent; ideas and ideals interested her more than people; the flow of events in her autobiography is rarely checked by analysis of character; most of her cast are mere names—like her brother Henry, for example.

In the spring of 1861 when Annie was 13 Miss Marryat announced her intention of travelling on the Continent. Her first object was to place a nephew suffering from cataract in the hands of a famous eye surgeon at Düsseldorf. Annie and a new protégée, Emma Mann, who had taken the place of Amy Marryat at Fern Hill, were to accompany her. Emma was the daughter of a clergyman who had married a sister of Catherine Vaughan, wife of the Headmaster of Harrow. Emma's uncle was the famous Arthur Penrhyn Stanley, dean of Westminster, who was to be of great significance to Annie later on.

The girls already had a smattering of French, since they were made to converse in it at dinner, and had been studying German, at which Miss Marryat excelled, for some months before their departure. Their goal was Bonn, where they were to spend several months in a boarding house overlooking the Rhine. Bonn was a university town with, just then, a mania for all things English, especially, as it turned out, for pretty English girls. According to Annie, Emma was a plump, rosy, fair-haired, typical English maiden, full of frolic and harmless fun; Annie was a very slight, pale, black-haired girl alternating between wild high spirits and extreme pensiveness.

'Mischievous students would pursue us wherever we went, sentimental Germans with gashed cheeks would whisper complimentary phrases as we passed.' Such behaviour was not proper in Miss Marryat's eyes and the girls were sent home for the holidays rather in disgrace. Not before Annie had acquired romantic pictures in her mind's eye, however—the moon as it silvered the Rhine at the foot of the Drachenfels, or the mist-encircled island where dwelt the lady consecrated for ever by Roland's love.

Ellen Marryat went on to Paris where, two months later, in the winter of 1861-2, the girls joined her and their introduction to European culture began in earnest. They had no respite from lessons; the difference from Charmouth was that on Wednesdays and Saturdays they were encouraged to visit galleries and churches. It was a highly significant experience for Annie; the first encounter she had with the tangible expression of Roman Catholicism, which she found appealed to her much more than Evangelicalism. 'I discovered the sensuous enjoyment that lay in introducing colour and fragrance and pomp into religious services so that the gratification of the aesthetic emotions became dignified with the garb of piety.'[25]

From that time on images, symbols, and ritual assumed a very important place in Annie's life for the stimulus they gave to her spiritual imagination. The picture galleries of the Louvre, crowded with madonnas and saints, the incense-laden air and exquisite music of the churches brought joy to her life and 'a more vivid colour' to her dreams. As far as her family was concerned she was free to choose whatever religious denonimation she wished. Annie found Roman Catholicism very attractive and came close to embracing it more than once. But at this particular time, due perhaps to the incumbent of the English church she attended in Paris, she came under the influence of High Anglicanism, whose form of worship was as warm and brilliant as Evangelicalism was cold and, in Annie's eyes, even crude.

In the spring of 1862 the chaplain of the church in the Rue d'Aguesseau took advantage of a visit to Paris by the bishop of Ohio to arrange a number of confirmations. Annie was a candidate and her mother came over from London to attend the ceremony.

During the preparation Annie was plunged into a deeply emotional state. When later she knelt at the altar rails before the bishop she felt as though his touch was the very touch of the Holy Spirit, the Heavenly Dove. She said she took the vows made in her name at her baptism, to renounce the world, the flesh, and the devil, with a heartiness and sincerity equalled only by her ignorance of the things she so readily resigned. She had never been to a theatre and was resolved not to go, while if anyone dared to ask her to a ball she had made up her mind to suffer for conscience's sake and refuse.

In spite of this there was still much to delight her in Paris. She and Emma and Miss Marryat mingled with the crowds in the Champs-Élysées and the Bois de Boulogne and climbed to the top of any, and every, monument that promised a view over the city. The Empire was in its heyday and Annie recalled seeing Eugénie and the prince imperial driving out with their plumed and jingling escort—a pathetic sight in her memory since by the time she came to describe it Louis Napoleon had been ambushed and killed by Zulus in Africa.

Their stay lasted seven months. On their return to England in the spring of 1862 the seclusion of Fern Hill gave way to the sociable though highly decorous atmosphere of the nearby Georgian watering place of Sidmouth, where Miss Marryat again took a house. The change was part of a carefully thought out process of detaching Annie and Emma from their leading strings. They were trained to work alone and when Annie, typically, complained that Miss Marryat was teaching her so little, she was told that she must not expect to have Auntie as a crutch all her life.[26] This gentle withdrawal of constant supervision and teaching was one of the wisest and kindest things Miss Marryat ever did for her, Annie wrote. The custom of keeping girls in the schoolroom until they 'came out' left them incapable of using their precious new freedom to its best advantage. In the days when there were no university places for women, she wrote in 1884, the opportunity for intellectual growth was often thrown away.

Miss Marryat's supervision was further relaxed in the winter of 1862-3 when she moved up to London. Annie was enrolled in the 'admirable' French classes of the no doubt admirable M. Roche. She was now fluent in the language which, with German, was to serve her well later on. For the rest of the time she was encouraged to visit her mother frequently at Harrow. In the spring Ellen Marryat told her that she had done all she usefully could for her. So, after nearly eight years, when she was $15\frac{1}{2}$[27] Annie was dismissed from the schoolroom, took leave of her teacher, whose emotion at the parting is not recorded, and who does not appear again in her life, and went home to Harrow-on-the-Hill.

For the next three years Annie's time was divided between her mother Emily and her own private world. There was no outward conflict between them: one was domestic and social, the other spiritual and intellectual; Annie was blissfully happy in both. Now she saw her intense love for her mother daily reciprocated; Emily, it seemed, could not do too much for her daughter. Priggishness, to which Annie freely confessed, was thrown aside; she accepted every invitation to parties, croquet, archery contests, and even balls. She said she never needed to think of what she would wear until the time for dressing arrived, when she found all she wanted laid out ready for her. 'No hand but my mother's must dress my hair which, loosened, fell in dense masses nearly to my knees; no hand but hers must fasten dress and deck with flowers.'[28] If Annie sometimes coaxed Emily to let her help by sewing in laces, her mother sent her to her books or to play, 'telling me that her only pleasure in life was caring for her treasure'. Emily had a horror of precocity, loved to see girls bright and gay, innocent of premature love dreams, ignorant of 'evil things'. And so Miss Marryat's prohibition on novel reading and the theatre was continued; concerts alone were encouraged, for both of them loved music. Within those limits Annie was free to continue her education, which she did most eagerly, following for the first time her own inclination which, eventually, conducted her into the strangest of places.

Fashionable poets like Wordsworth and Coleridge were too sentimental, she said; more to her taste were the oriental fantasies of their colleague Robert Southey. Homer and Dante were among the classics available in translation which she read. She devoured Plato and found herself annoyed by the 'insatiable questioning' of Socrates. Annie detested cross-examination, an ordeal which her volatility in public life later brought upon her. The highest praise she could bestow upon a friend was, 'You never cross-examined me.'[29]

Under the influence of the religious awakening she had experienced in Paris she moved even closer to Catholicism, scrupulously observing practices that had descended from the early Christian Church. To the intense displeasure of her mother she began to fast, to use the sign of the Cross, and to go to weekly communion. She even experimented with self-flagellation. She was only prevented from converting to Catholicism then by the existence of the Oxford Movement; its writings, particularly those of Edward Pusey, Henry Liddon, and John Keble, ruled her faith. They proved to Annie's satisfaction that the Church of England was Catholic though non-Roman.

But true to Ellen Marryat's teaching Annie had to find out things for herself, to try to trace the lines of doctrine back to their origin. In so doing she came upon John Henry Parker's edition of the Fathers of the Church, a copy of which was in Harrow School library. 'These strange mystic writers won over me a great fascination', she wrote.[30] What thrilled her was their nearness in time and place to Jesus himself. Among the patristic writers were some—Clement of Rome, Polycarp, Barnabas—who were the pupils of the Apostles themselves, who had lived and worked with him. Annie immersed herself in their lives and works and enquired further; she was particularly attracted by Origen, St John Chrysostom, and Augustine.

Even if her mother had known what she was doing she could hardly have foreseen where it might lead. Only scholarly clerics and ecclesiastical historians normally went so far as to read the Fathers; to read them was to enter a world of unfamiliar thoughts and controversial ideas. They stood at the divide between the ancient and the medieval world, the pagan and the Christian. Among the ideas they reviewed were some that had been condemned by the Church into which Annie was baptized. For the first time she learned about such things as the transmigration of souls, divinity through knowledge, the higher virtue of a celibate life, the reality of evil as opposed to good.

As she read she discovered that the Fathers believed in magic (but as Christians condemned its practice); in the power said to reside in images and idols; in the significance of numbers, incantations, and oracles. Because in order to condemn pagan practices the Fathers had to describe them, Annie's head was filled with tales of Greek gods and heroes, of magi from Chaldea, of brahmins coming from the East to mingle their knowledge with that of the priests of Isis. In the many references to the writings collected under the name of Hermes Trismegistus, she found the description of an ancient world religion, originating in pharaonic Egypt, whose secrets were revealed only to initiates. All this information—exoteric and esoteric (words which never failed to excite her)—was stored in Annie's precise and capacious memory. She could always recall anything she had read; her concentration was remarkable.

Even so, whilst she was acquiring this arcane lore she gave it little credence. She said she believed implicitly that Jesus was supreme and unassailable. Her spiritual life revolved round him. 'I often felt that the very passion of my devotion would draw Him down from His throne in Heaven, present visibly in form as I felt Him invisibly in spirit.'[31] To serve Christ through his Church became a definite ideal of hers. Looking back on her 18-year-old self in 1884 Annie confessed that she was the very stuff of which fanatics were made.

Her mother was not: Emily regarded religious enthusiasm as unbalanced and unbecoming. Her Christianity was more social than spiritual, a faith contented by the 'delightful vagueness' of Dean Stanley, whose services at Westminster Abbey she attended. By contrast Emily's sister, Annie's Aunt Minnie, concurred in Dr Pusey's harsh estimate of Stanley as 'unsound in the faith once delivered to the saints'.[32] At Christmas 1865, which was spent with the Morrises at Clapham, Aunt Minnie and her niece offered their services to the local vicar, who set them to decorate the little mission church which had been established in a poor quarter of the area. Its services were satisfactorily 'high' while those of the parish church were not. Annie and Minnie arranged flowers, and rejoiced at the thought they were serving God. Looking back Annie judged that they had only been amusing themselves.

At Easter 1866 Annie was once again in Clapham. According to her *Autobiographical Sketches*, published in 1884, she made a cross for the high altar of camellias, azaleas, and white geraniums. This choice must have struck her afterwards as unsuitable, for in her *Autobiography*, published in 1893, the cross was of spring flowers: 'dewy primroses, fragrant violets and the yellow bells of the wild daffodil.' These delighted the poor people and the children, many of whom, Annie said, had never seen a flower before (let alone a white geranium).[33]

The preceding week while the church was dark with mourning and the altars were bare she followed the stations of the Cross in her mind. In an attempt to make the events of Holy Week more real she resolved to write a history of those days compiled from the Gospels. She set down in four columns the incidents and the times at which Matthew, Mark, Luke, and John said they had occurred. This was a highly dangerous proceeding: biblical exegesis had often precipitated doubt. Like others before her Annie became uneasy as discrepancies 'leaped' at her from her four columns. Finally doubt at the veracity of the story sprang up 'like a serpent' hissing in her face. She struck it down. 'Quickly I assured myself that these apparent contradictions were necessary as tests of faith, and I forced myself to repeat Tertullian's famous *Credo quia impossible*, till from a wooden recital it became a triumphant affirmation.'[34]

It was at this moment of exultation, of temptation put aside, no doubt strengthening her vocation, that Annie met the man who was to be her earthly husband. He was at the time no more than a humble deacon, but his feet were on the path; within a year he was to experience the laying on of hands, to be ordained a priest, and to assume the style of the Reverend Frank Besant.

CHAPTER 2

THE HALF-ANGELIC BEING

IT is hard to say what prompted Mrs Wood to encourage Frank Besant when she turned away several promising young men who, unbeknown to Annie, proposed themselves as suitors.[1] Harrow, whose reputation had been much enhanced during the fifteen years of Dr Vaughan's headmastership,[2] could now command the pick of Oxford and Cambridge. Annie regarded the young masters as friends, 'friends with whom I liked to talk because they knew more than I did',[3] but they found no place in her elevated day-dreams. Perhaps her mother, thwarted in her desire to see her son Henry a clergyman (he was vaguely 'arty'[4]), welcomed a second chance in the person of a son-in-law. Certainly Emily expected that marriage would bring her rapt and brooding daughter down to earth.

Nothing was placed in her way. Annie saw Frank briefly at Christmas 1865 when he assisted at the service in Clapham mission church. He was there again the following Easter as acting temporary curate. The next step was decisive. Emily invited him to spend a week's holiday in a rented cottage at St Leonards-on-Sea where she and Annie and Minnie Morris went in the early summer of 1866. The two young ones were left unchaperoned—deliberately, of course, for Emily was not careless where such matters were concerned. They were companions in all the walks, drives, and rides, when Annie showed off her prowess on her mare Gypsy Queen, or talked to her heart's content.

An hour or two before he was due to leave Frank asked her to marry him, taking her consent for granted since she had allowed him such full companionship. 'A perfectly fair assumption', Annie conceded, 'but wholly mistaken as regarded myself whose thoughts were in quite other directions.'[5] Startled and upset at the hint that she might have been flirting, Annie smothered an impulse to refuse outright. Frank took her silence for acquiescence. He ordered her to say nothing until he had spoken to her mother on pain of dishonour—the worst disgrace in Annie's romantic code—and departed for the station.

Given the circumstances of their week's holiday Frank had reason to expect that his instruction would be obeyed. He was, moreover, the only male in sight. The natural advantage that conferred on him among a group of women was enhanced by his future status as a clergyman of the Church of England. He was 25, seven years older than his intended bride, a schoolmaster who taught at Stockwell Proprietary Grammar School near Annie's grandfather's house in Clapham. He was not rich, relying on his salary and on the fees he obtained from temporary parochial work in the school holidays. He was not immediately prepossessing either, being shy and rather gloomy.[6]

As Frank's distinguished older brother, Walter Besant, described him in his memoirs, their father William was a studious, vague, and timid man who neglected business in order to read and garden. Their home was in St George's Square, Portsea, the very picturesque eighteenth-century quarter of the naval base at Portsmouth. Their mother Sarah was the family's driving force, a clever, quick-witted, ambitious woman, the daughter of an architect, who saw to it that her enormous family—six boys, four girls, of whom nine grew up—went out into the world beyond the garden wall as well equipped as Church and school could make them.

Religious observance governed their lives; the high point of each week was Sunday, when they attended two church services, each with a sermon lasting more than an hour. Their content reflected the prevailing atmosphere of high morality and strict Evangelicalism. Though Walter came to hate that narrow way of life, which was based on the presumption that he and his had been admitted to the ranks of the chosen few destined to be 'saved', Frank embraced it. Frank was 'serious', a word that in the mouths of Evangelicals was the highest accolade.[7]

The boys went first to a local school kept by three sisters, two of whom at least were satisfactorily 'serious'. Later, as an interesting variation, they were coached in Greek and Latin by a clergyman who was 'high'. After that they went as boarders to London, to Stockwell Proprietary Grammar School.[8] It had been recommended by a college friend of the eldest Besant boy,

William. He was born in 1828, eight years before Walter and twelve before Frank, and set a cracking pace for them to follow, becoming Senior Wrangler at Cambridge and, ultimately, FRS. The school had been founded by a group of charitably minded people in the 1830s. It lay between the Clapham and the Brixton Roads in what in the 1850s, when Frank was there, was a prosperous and leafy suburb. The Common was still wild and surrounded by mansions 'sacred to the memory of Wilberforce, Thornton, and Macaulay'.[9]

The school, which was small, was exclusively Anglican; its Headmaster and most of the staff were in orders. If not to the Church the boys looked to the City for their careers, as underwriters, merchants, stockbrokers—as principals, not clerks. Frank, who followed Walter there in 1856, did well, becoming Head of School. He went on to King's College, London, for three terms, where he sat Cambridge Entrance.

King's was a bastion of orthodoxy, a citadel of the Established Church.[10] All the professors were C. of E. men; the least suspicion of heterodoxy among the staff was visited by dismissal and it was a condition of entry that students sign the Thirty-Nine Articles. 'No reputation, no abilities, no services ... could save the heretic',[11] Walter Besant wrote. Frank, like Walter before him, won a scholarship from King's to Cambridge, in his case to Emmanuel College, where he studied mathematics. He did not do as well as his eldest brother William, but he did not do badly, graduating 28th Wrangler in 1863. Cambridge, too, was narrowly clerical in those days. 'The University ... considered only two professions; the Church which included lectureships, professorships, and fellowships at the Colleges; and the Bar. Schoolmastering was a refuge, not a profession; art was an unknown calling',[12] Walter Besant complained.

On graduating Frank immediately took refuge at his old school, Stockwell Proprietary Grammar, as a mathematics master, while preparing for better things which he hoped would come to him on ordination. He was 'called and chosen to the office of deacon' in December 1865. As a teacher he was methodical and extremely conscientious, distinguished by his concern to bring out qualities latent in his pupils. He was not the most popular of masters; he had no interest at all in games and, through shyness, his manner was abrupt.[13]

His unusual surname was old, as was his family; Frank was proud of both. The origin of the word 'besant' was as romantic as Annie could have wished. After careful philological research, which he found much to his taste, Frank decided that it was derived from a bezant, the coin given to Crusaders passing through Byzantine towns as a mark of their service to Christendom.[14] There was confusion as to how to pronounce it. Some made it rhyme with 'pleasant', others put the stress on the second syllable, as did Frank.

It seems, therefore, that, if hardly the catch of the season, the intended Reverend was a perfectly eligible young man whose angularities a wife and children would shortly smooth and tease away. So Annie's mother thought. She was committed to the marriage from the start, though she made a show of reluctance in parting with her daughter. Emily's disposal of Annie's hand was as coolly thought out as was the earlier entrusting of her to Miss Marryat.

In the belief that her mother knew nothing of Frank's intentions, Annie suffered in silence for the two weeks remaining at St Leonards. It was the first time she had withheld anything from her mother and she found it agonizing. On their return to London Frank repeated his proposal. Annie accepted, out of a desire not to cause pain, she said. When the 'secret' was revealed Emily refused to allow a formal engagement on the grounds that, at $18\frac{1}{2}$, Annie was too young and, more significantly, that she was 'childish'. But Frank was not dismissed as other suitors had been, merely asked to wait.[15]

Such a delay was a useful convention. In an age when marriages were tacitly, if not openly, arranged; when divorce was unthinkable; when many husbands were years older than their brides; prudent parents had recourse to it as a defence against blame if the ultimate disaster, a breakdown in the marriage, occurred. Their daughters could not rely on the device so well; once an understanding was approved by the family it acquired a momentum difficult to arrest; they were not expected, and sometimes not allowed, to change their minds.

Instead their nerves were calmed, their thoughts distracted, and their wardrobes much extended by travel on the Continent. Annie was no exception to this judicious, middle-class rule: she spent the rest of the summer of 1866—the 'last free summer' as she came to regard it[16]—in Switzerland. There she and her companions contemplated the Jungfrau, the Mer de Glace, the Simplon Pass, the Lake of Thun, and other Alpine sights made fashionable by the works of Turner and of Ruskin. Their paintings depicted Nature at its most inhuman and sublime. In the awe-inspiring presence of the Alps Annie's religiosity might have taken even further flight had it not been for one of her companions who, in the absence of her mother, acted as her guardian.

William Prowting Roberts[17] was the first in a succession of remarkable men with whom Annie was fortunate to be associated. He was the most experienced person she had met so far and the most humane. He was not scholarly or spiritually minded like her clerical heroes John Keble and Edward Pusey; he was a down-to-earth lawyer. He enjoyed the admiration and gratitude of thousands of trade unionists, for it was to their cause that he chiefly devoted himself. Annie was a stranger to their world as yet, but her passage into it would be greatly eased by her friendship with Lawyer Roberts.

When he and his second wife and daughters took Annie Wood on their tour of Switzerland in 1866 Roberts was 60. Most of the major events in his career were over. They belonged to the decade of Chartist agitation in the 1840s. Although at the time trade unionists no longer risked prosecution for taking unlawful oaths, they were still subject to harassment by unsympathetic owners and managers. This was particularly the case among mineworkers. Roberts rose to fame arguing cases in the magistrates' courts of Durham and Northumberland; in 1844 his expertise was recognized by the Miners' Association, which paid him a retainer of £1,000 a year to defend their members wherever they were. 'He was the terror of many a Bench', his friend and colleague George Jacob Holyoake declared.[18]

At the same time Roberts, whose main office was at Manchester, became a close associate of the Chartist leader Feargus O'Connor, and legal adviser to his scheme for land banks as a means of resettling the destitute poor. Roberts invested, and ultimately lost, a good deal of his own money in this scheme. Annie heard from Roberts of the stirring days in Manchester when O'Connor roused and manipulated crowds with his burning eloquence, capable, it seemed, of turning political agitation into outright rebellion if he chose.

From Roberts she also learned of the existence of the urban poor, of the conditions under which women and children had been forced to work in the mines; how women toiled naked to the waist with short petticoats to the knee, rough, foul-tongued, brutalized. How babies of 3 and 4 were set to watch a door and, falling asleep, were cursed and kicked awake. Annie's response was the more immediate, since her sense of duty and compassion had already been aroused by Ellen Marryat. But Roberts asked for a more considered commitment. He tutored Annie in radicalism and found her an eager pupil. She had taken no interest in politics until then, but unconsciously reflected the 'decorous Whiggism' which had always surrounded her.[19] She was more than ordinarily susceptible to the prevailing climate of opinion, adjusting her position with an agility that disconcerted more sluggish thinkers.

Where once she had regarded the poor as people in need of education and charitable care, who should always be treated with perfect courtesy as befitted the lady Miss Marryat had trained her to become, Roberts taught her to honour them as the producers of wealth, who had a right to determine their own fate, a right to justice rather than a claim upon charity.

Perfect manners joined to intellectual understanding marked Annie's later dealings with the world. But Ellen Marryat and William Roberts were powerless to teach her judgement and moderation, qualities from whose lack she was to suffer all her life. Certainly she could not learn moderation from Roberts, who was almost as hot and rebellious as O'Connor himself. What the Chartist lawyer taught her was the lesson that ordinary people could right wrongs and change the system if they were armed with knowledge and were sufficiently determined.

The result was that, like most girls of her age, Annie longed for something worthwhile to do, for the prestige of something accomplished. She represented this feeling to herself as the desire for 'sacrifice' in a noble cause. Sacrifice in the sense of resignation, of abandoning something cherished in the service of something greater, was not what she achieved. Whatever she undertook in the course of her life was almost always what she wanted to do at that particular moment; if, later, she perceived that she had suffered as a result she wrote it off as something 'sacrificed'. To act in a manner that was alien to her for another's sake was not her way.

On her return from Switzerland in the autumn of 1866 Annie and Frank were formally engaged. Some time later Annie grew restless and proposed to end it. Mrs Wood would not hear of it; her pride objected to the dishonour of a bargain broken. Annie, who appears to have feared as well as worshipped her mother, was obliged to concede the argument used in Frank's favour (how Frank himself felt no one seems to have considered). This was that she would have more opportunity for doing good as a clergyman's wife than as anything else.[20] And what else was there for Annie except marriage?

Frank's own prestige gained immeasurably from his ordination, which occurred shortly before the wedding. The laying on of hands by the bishop translated him to a position where the laity henceforth owed him allegiance, as they did to his brother clergy who supported him before the altar. But Annie, who longed for the Holy Spirit to descend on *her*, could hardly reconcile herself to the humiliating distance which separated the congregation from those set in authority above them. A priest to her was 'a half angelic being';[21] as Frank Besant's wife her place should surely be on the steps of the altar, a nearer worshipper than those who did not approach at all?

During her engagement Annie's imagination was still preoccupied with the figure of Jesus, to whom she spoke as to a lover.

O most sweet Lord Jesu transfix the affections of my inmost soul with that most joyous and beautiful wound of Thy love ... Let Him kiss me with the kisses of His mouth ... Blessed art Thou O most merciful God who didst vouchsafe to espouse me to the Heavenly Bridegroom ... and hast imparted Thy Body and Blood as a new gift of espousal and the meet consummation of Thy love.[22]

These were some of the prayers she recited daily.

Used in a spiritual, almost mystical, context the words delighted her. Vaguely she contemplated a similar emotion on addressing them to an earthly husband. She was wholly ignorant of their earthly meaning. As far as she knew the consummation she invited took place only in heaven. Even at that late stage her mother did not see fit to enlighten her.

Nor was Emily concerned to make sure that Frank and Annie knew what to expect of each other during the year-long engagement. Frank was teaching, or fulfilling temporary duties in a series of parishes during the holidays. Annie was occupied in helping her mother move away from Harrow; they met infrequently.

The move, which Annie regretted, for it deprived her of the much-loved house, took place in the spring of 1867. There was no longer any need to stay in Harrow. Henry was at Cambridge, having won a scholarship in classics to Clare College in 1866.[23] Mrs Wood could lay down the burden she had carried for so long and which was proving far too much for her. She had done well: she had accumulated enough to pay for Henry as an undergraduate with something left over to buy a house at St Leonards, in fashionable Warrior Square, which she might let if she desired. She and Annie moved in during the summer, a summer saddened by the death of Minnie, Annie's favourite aunt.

The autumn before the marriage was sombre too, mourning prolonged by another tragedy. This one was public, the first in a long series of dramatic incidents at which Annie was a not altogether involuntary witness. In September she and her mother went to stay with the Roberts family near Manchester. Annie's education was resumed. Daily as she drove him in to his office in the city William Roberts examined her in her attitude to people. What did she think of John Bright? She had never thought of him at all. Wasn't he a rough sort of man who went about making rows? Roberts was scathing in his denunciation of her opinion. 'I believe some of you fine ladies would not go to Heaven if you had to rub shoulders with John Bright', he thundered, 'the noblest man God ever gave to the cause of the poor.'[24]

Annie shortly found herself in a position to judge how much of a popular leader Roberts himself was. During her visit he was called upon to organize the defence of five Irishmen on trial for their lives at Manchester. Their cause was desperate, the time one of the grimmest in the unhappy history of Anglo-Irish relations. In the autumn of 1867 Manchester was the focus of an outburst over the shooting of a policeman. The city, with its huge population of immigrant Irish workers, was divided into two bitterly hostile camps; neither side was safe in the other's territory. The crisis had been brewing for a number of years while the Fenian Brotherhood, an organization dedicated to the overthrow of British rule in Ireland, built up its strength on the English mainland.

In 1867 there was a plot to seize Chester and the railway to Holyhead, from which port the nationalists would descend on Dublin itself. On 11 September two Fenian leaders, Colonel Kelly and Captain Deasy, were arrested in Manchester, where they had gone for secret discussions with their local supporters. On 18 September, as they were being conveyed through the streets in a prison van guarded by a policeman who was locked in with them, it was ambushed by armed men. The Fenians shot the lock open and, in so doing, killed the policeman who was inside. Kelly and Deasy escaped, but Condon, Allen, Larkin, Maguire, and O'Brien, who had taken part in the rescue, were caught. The man who fired the shot which killed PC Brett escaped.

Five judges were appointed to try the case; Roberts could not conceal his despair at the one who presided—'the hanging judge'[25] he told his family. He arranged for the Chartist leader Ernest Jones to lead for the defence with Digby Seymour, QC, seconding. Roberts then felt they were not firm enough in challenging the jurors, a dangerously unpopular task under the circumstances. He did it himself and ran the risk of commitment for contempt.

There was no question of his wife and daughters remaining at home during the ordeal. They, with their guest, drove to the court each day. Mrs Roberts was connected with Ireland by descent; one of her ancestors had supported the 1798 uprising. Annie was proud to be with them. It was her first experience of an angry crowd; the streets were barricaded, the soldiers under arms, every approach to the court crowded with surging throngs, she wrote. Once, the carriage was stopped as they passed slowly through the Irish section of the crowd, and fists came through the windows with curses at 'the d . . . d English who were going to see the boys murdered'. The situation was critical but it was transformed by Annie's presence of mind. She reached out to the nearest fist and told the man that the targets of his abuse were Mr Roberts's wife and daughters. 'Roberts, Lawyer Roberts, God bless Roberts', roared the crowd and the carriage was allowed to pass.

The defence, that the prisoners were political offenders and had meant no harm to PC Brett, failed. The consequence was infinitely distressing. Annie watched through an open door as an official laid out five black caps. She experienced the appalling hush as the sentence of death was pronounced, and she endured the sight of Allen's fiancée on her knees pleading for his life.[26] Two of the sentences were commuted; on 23 November Allen, Larkin, and O'Brien were hanged—in public, outside Salford Gaol.

As many others did Annie felt the verdict was unjust. In Ireland, among nationalists everywhere, Allen, Larkin, and O'Brien became known as the Manchester martyrs. Their trial haunted Annie; for the rest of her life she was seized with fresh indignation at each recollection of what had occurred. From that day forward justice was supreme in her eyes; something she sought for others as well as for herself. It was a right that must not be denied.

On 21 December 1867 she was married to Frank Besant in the parish church at Hastings.[27] Their union was the beginning of great unhappiness for both of them. To Annie her wedding night was an outrage: shock promoted disgust and fear. These might diminish in time but her pride never forgave the violation. Frank was equally unfortunate, the victim of her untutored anticipation. The author of their misfortune was undoubtedly Annie's mother who, for reasons of her own, disregarded her daughter's fatal innocence and unrealistic expectation. Years later, when their marriage had become public property, its epitaph was pronounced by W. T. Stead and recorded in his journal the *Review of Reviews*. 'She could not be the Bride of Heaven, and therefore became the bride of Mr Frank Besant. He was hardly an adequate substitute.'[28]

CHAPTER 3

A HUSBAND'S AUTHORITY

THE predicament into which her innocence had plunged both Frank and Annie was not unusual, as she acknowledged later on.[1] Once the honeymoon was over other couples managed to regain their equilibrium. No doubt Frank looked forward with impatience to the time when he and Annie would reach that state. But Christmas spent in Paris was uneasy, as were the days that followed at his father's new house.[2] It was in Southsea, near Portsmouth, a substantial villa with a large garden which had been bought from the proceeds of a wine business William Besant had embarked on late in life, almost by chance, and which had proved a surprising success. If Annie felt strange there among her husband's large family, they were both equally new to Cheltenham where, in January 1868, Frank took up an appointment as an assistant mathematics master at the College.[3]

A fashionable spa, Cheltenham, outwardly at least, was a most agreeable place to live. Though they started off in lodgings the young Besants soon moved into 2 Segrave Villas at the entrance to the Park.[4] Houses in Cheltenham were mostly of good square, Regency proportions set in ample gardens adorned with fine trees. Frank's was large enough for College boys to board, as was the practice at the time, so Annie found herself in much the same position as her mother had occupied at Harrow.[5]

Cheltenham College was a more recent foundation. It was begun in 1841 to fill a gap in the existing public school system. It was divided into three departments. The Classical hardly differed from Harrow or Rugby in subject or method. The Military trained boys for entry into Woolwich and Sandhurst, while the Civil prepared them to face the rigours of the examination for the Indian Civil Service.[6] Mathematics, science, and oriental languages were among the new subjects taught in the Military and Civil where Frank Besant was employed. By the time he arrived the College's connection with the East was well established so, however mundane the Revd Besant's own occupation seemed, he carried home with him a whiff of distant and romantic places.

But his position at the College was a humble one. And, though not much older, the head of his department outshone him academically, having graduated from Cambridge 8th Wrangler to Frank's 28th. These matters were important, as Harrow had taught Annie. Her pride was assailed on this front too, which may explain the stiff resistance she put up against the efforts of the masters' wives to make her one of them. In her *Autobiographical Sketches* she was scornful of their narrow domesticity; they were 'ladies who talked only of babies and servants',[7] troubles which Annie found utterly boring and incessant talk of which made her timid, dull, and depressed—all the things she passionately wished she was not.

This sweeping condemnation of the women who surrounded her served to dramatize what Annie plainly felt as intellectual isolation, but it was less than fair to the ladies of Cheltenham College, some of whom were very advanced. Their husbands and fathers, Frank's colleagues, were liberal in their attitude to women's education (Frank most certainly was not). The Headmaster, Thomas Jex Blake, educated his daughters so thoroughly that one became Mistress of Girton College, the other Principal of Lady Margaret Hall. His sister Sophia Jex Blake was involved in a ferocious struggle to enable herself and other women to qualify in medicine. The redoubtable Mrs Josephine Butler left shortly before the Besants arrived because her husband had been appointed to another teaching post. In later life, as a great compliment, Annie was to be compared to her in zeal and the desire for reform. Surely Mrs Butler's influence on her fellow wives must have left an impression that lasted even after Annie had arrived?[8]

As the impulse which had propelled her into marriage spent its force, everything connected with it turned from white to black—a frequent occurrence in her life. But at Cheltenham there was a special reason why she should have felt at odds with her surroundings. The atmosphere of the College was religious, intensely so, as the previous Headmaster, Dr Barry, had devoted his term of office, just ended, to strengthening the commitment of masters and pupils. But their churchmanship was Low, while Annie's at the time was High.[9]

On this account the gulf between her and the masters' wives was likely to be wide. Her manner towards them may have made it unbridgeable. The soul of courtesy to anyone of lower status than herself,[10] Annie was often brusque with her peers and rivals, ignorant—careless certainly—of the offence she caused.

It was not surprising that while Frank was teaching she found herself with nothing to do. Idleness was a state her active mind could not tolerate: it made her ill. At Cheltenham her depression was made worse by the contrast of her surroundings. The town offered all the amusements a lively girl might be expected to enjoy.[11] Its Assembly Rooms rivalled those of Bath in Georgian elegance, and in the frequency of their use for parties, balls, and concerts. A host of dancing and drawing masters, music teachers, and language tutors clamoured for patronage in the pages of the town's directory. Besides the usual complement of gentry, clergy, and professional men, Cheltenham society was leavened by a considerable number of Anglo-Indians, who made the town their English headquarters. Annie was surrounded by people whose lives reproached her; on the one hand soldiers and administrators enjoying a prosperous retirement; on the other boys and young men destined to succeed them in positions of distinction and responsibility. Many of their pastimes were forbidden to her, out of bounds to 'serious' Evangelicals like Frank. Etiquette forbade a bride going into society unaccompanied by her husband or a female chaperone, so Annie was often left to her own devices.

Nor was she immediately able to satisfy her desire to be of service to the unfortunate; they were out of sight in Cheltenham. As a town it was 'genteel and thin' to quote Lawyer Roberts's friend George Jacob Holyoake.[12] 'As the parlours of some prudent housewives are kept for show and not to sit in, so in Cheltenham numerous houses are kept "to be let" and not to live in.'

Annie forgot that she had ever sat at the feet of Roberts or practised charity at the Clapham mission. In her *Autobiographical Sketches* she drew a picture of herself as a young bride that more closely resembles Dora Copperfield than Annie Besant. Her incompetence in household management was apparently equalled by the sketchy way in which she undertook it.[13] She said she knew nothing of the use of money, having never had an allowance or bought herself so much as a pair of gloves. Worse, she was helpless before the servants, as shyness prevented her from criticizing or administering rebukes. Frank, on the other hand, was particular, precise, methodical, and very much the master of the house.[14] Others have testified to the truth of this portrait of him. But, however much her ladylike incompetence might appeal to the working-class readers of the journal in which her memoirs first appeared,[15] Annie's account of herself in this does not ring true. A pupil of Ellen Marryat would hardly escape a thorough grounding in domestic economy, or the example of how to cope efficiently with problems below stairs. A few years after this Annie was to take to business as if she had been reading balance sheets from infancy.

A childish incapacity may have seemed to her the best role in which to counter Frank's overbearing attitude and preserve herself from dwindling into a wife. 'I, accustomed to freedom, indifferent to home details, impulsive, very hot tempered, and proud as Lucifer, had never had a harsh word spoken to me, never been ordered to do anything, had had my way smoothed for my feet, and never a worry had touched me.'[16] Perhaps in retaliation for the behaviour of her husband which these words imply, Annie tormented him with constant and open fretting for her mother. This reduced him to a state of jealous vexation. His temper worsened under the strain of trying to govern her, which, as a husband and a clergyman, he felt it his duty to do. All too frequently her tactlessness reduced him to impatience and harsh words. Annie was incredulous in response, then she had recourse to a storm of tears; lastly she took refuge in what she described as proud and defiant resistance, 'cold and hard as iron'.[17] At times, and on both sides, their behaviour must have looked like sulking.

In this unhappy situation Annie flew to her books and began to write seriously for the first time. Intellect was a crucial factor in her personality; she was not blessed with intuition. Knowledge was food to her: she was avid for ideas. Not in the least original, her capacity would lie in giving brilliant life to other people's thought. Now, in 1868, she tried to recapture the excitement she had felt on reading Milton and the Fathers of the Church. She decided to write an account of the Black Letter Saints—those who are not significant enough to command a red space in the church calendar, or special services, but whose lives had been sufficiently remarkable to earn them canonization. Once more she immersed herself in tales of heroic suffering and sacrifice, and of men who transcended mortal status. When the book was finished she sent it to Macmillan's. They forwarded it to someone who was preparing a series of church books for children. Much later she had a letter from a brotherhood offering to publish it if she would dedicate it as 'an act of piety'. What became of it she never knew.

Simultaneously she tried her hand at short stories which, she confessed, were of a very flimsy type. But with these she stumbled on a market. They were published in the *Family Herald*, which described itself as a 'Domestic Magazine of Useful Information and Amusement' offering, for Gentlemen, Facts and Philosophy; for Ladies, Hints and Entertainment—what else? In Annie's first story Herbert and Evelyn Merton enjoyed a state of perfect wedded bliss; he idolized her; she loved and revered him. One day when out riding, Evelyn fell from her horse and was paralysed. But a friend persuaded her to dedicate her life to others. After thirty years in a wheelchair ministering to the poor (Annie did not explain how, exactly), she was called to her rest, but not before the author made her say, 'she could look back at a life spent in the blessing of others and could see that from an accident which, as she despairingly thought, made her useless and a burden, dated her real life of faithful work and long service to her fellow creatures'.

Herbert, in the author's eyes, *was* a burden, so much so that he disappeared as soon as the suffering began. Poor Annie! Like Evelyn she had been deprived of her beloved horse, Gypsy Queen, and was condemned to endless years of earthbound existence with a husband she could not write out of the story.

Not the least of the attractions of the *Family Herald* was that it paid its contributors as soon as it accepted work, rather than on publication. 'Sunshine and Shade' brought Annie 30*s.*, the first money she had ever earned. 'In my childish delight and practical religion I went down on my knees and thanked God for sending it to me, and I saw myself earning heaps of golden guineas and becoming quite a support of the household.' But her delightful sense of independence was shattered when Frank appropriated the cheque; she had not understood that in those days before the Married Women's Property Act of 1882[18] all a married woman earned belonged by law to her 'owner'—as she took to calling Frank. 'I did not want the money: I was only so glad to have something of my own to give, and it was rather a shock to learn that it was not really mine at all.'[19] That was mildly put, but the disappointment must have been acute. Annie, by her father's death and the arrangements made by her mother for her education, had been deprived of ownership in many things. As soon as she acquired money and influence she was unfailingly generous in disposing of them.

Frank, on the other hand, was by nature frugal, sparing in approval, and in the length of rein he gave his wife. For instance, when they were engaged, Annie wrote a pamphlet 'Notes on Fasting, by a Layman', which was 'very patristic' in tone. This was published after they were married. Pleased, she wrote a sequel, 'Fasting Communion'. On his copy of the manuscript Frank wrote, 'I would not publish this thinking she ought to be satisfied with publishing of the preceding pamphlet.'[20] His patience with High Anglicanism must have been running out. At Cheltenham he was among like-minded people, who looked on practices like fasting and genuflection, on ritual, robes, and incense, as dangerous affectations. But Annie was, as yet, susceptible to 'bells and smells'.

Annie's energy and facility in writing were already such that besides the tome on the Black Letter Saints and the light short stories she also produced a novel. The *Family Herald* sent this back as 'too political', but offered to publish one of purely domestic interest, if it was up to the same standard.

The pressure on Annie to write, to publish, would soon become irresistible; the gates would open and the works pour forth. But in the summer of 1868 the impulse was checked almost as soon as it had made itself felt when she became pregnant. She was ill for months before the birth; from the symptoms of her condition and because of the constraint exercised by Frank as an expectant father. It brought on an attack of acute depression during which she did not touch her books or pen.

The baby, a boy, who was not called Frank, was born on 16 January 1869 and presented for baptism a few weeks later. In the 'Register of Services and Other Events' in which it was his custom to enter every ecclesiastical duty he performed, and the sum he received for doing it, Frank wrote, on 28 February, 'St Philip and St James, Cheltenham. During afternoon service Arthur Digby Besant and two other infants baptized. No fee.'[21]

St Philip and St James was a vast new edifice built to serve the recently created parish of Leckhampton. Its raw red brick interior, wrought iron screens, and gilded sanctuary lamps pronounced it Anglo-Catholic and, therefore, Annie's choice for the ceremony. It is not known why she called her son Arthur Digby. They were not Wood family names nor Besant family names. Digby, by which name the child was always known, was the Christian name of one of the barristers who defended the Manchester martyrs; it was also the name of the clergyman who married her.[22]

Annie doted on the baby and looked after him herself, as they could not afford a nurse. 'My energy in reading became less feverish when it was done by the side of the baby's cradle', she wrote, in a moment of insight into her own character. Digby's presence almost healed the abiding pain of her mother's 'loss'. Even by the sentimental standards of the *Family Herald* this pining for her mother seems excessive. Mrs Wood was not far away. Annie had a lifelong habit of fixing upon certain persons as indispensable; these she clung to, and the first of them was her mother.

Before the year was out Annie was pregnant with her second child. Mabel Emily Besant was born on 28 August 1870, 'somewhat prematurely', her mother remarked, 'in consequence of a shock'.[23] The shock was alleged to have occurred as the result of Frank striking Annie and shouting at her to leave him and return to her mother. In that case its effect was long delayed, for the affidavit made in 1878, when Annie briefly tried for a judicial separation on the grounds of cruelty, put the incident in February 1870—seven months before the birth.[24]

According to this document the cause of Frank's violent anger was her request to him to limit their family; they could not afford more children, she argued. But in his sworn evidence on the same occasion Frank insisted that she had never made such a proposal during their married life.[25] If he was telling the truth (and it seems likely that he was, given the taboos surrounding the subject of birth control at the time, and Annie's total inexperience), there must have been another reason for his outburst. If Mabel was born less than nine months from the time she was conceived, it is possible that Annie did not know she was pregnant again and simply begged Frank not to force another child on her. That she thought him capable of doing so emerges from between the lines of her *Autobiographical Sketches*, where she described her existence with him as 'degraded by an intolerable sense of bondage', and made pointed reference to his high ideas of a husband's authority and a wife's submission. Frank, she recalled elsewhere, was determined to break her in.[26]

Fear of another pregnancy helped to make Annie's recovery from the birth of Mabel tedious and slow. It was also stormy. When such a highly strung and forceful personality was curbed, and deeply at odds with its apparent destiny, explosions were bound to occur. Frank had no experience of how to cope with hysteria; no wonder he shook her from time to time, as Annie complained. In his evidence he denied her accusations of cruelty and claimed alternatively that her behaviour justified his conduct.[27]

The winter of 1870 brought another catastrophe. It was discovered that Annie's mother had been cheated of all her money by a solicitor she had trusted implicitly over a period of years. The consequences bore very heavily on Emily Wood who was obliged to give up her house in St Leonards-on-Sea and seek refuge in London with her son Henry.[28] He was 25 and had graduated from Cambridge two years previously with a Second in classics. The problem of what occupation Henry should follow had been solved with the help of a most useful relative, William Page Wood, a cousin of his late father. This particular Wood had prospered in the law. On coming into office as Prime Minister for the first time in 1869, Gladstone had created him Baron Hatherley and sent him to the Woolsack. The Lord Chancellor had many offices in his gift; a clerkship in the Patent Office was found for Cousin Henry, as well as a part-time job editing the journal of the Society of Arts. Henry was to be successful in both these occupations and eventually to earn himself a knighthood, but in 1870 he was still at the bottom of the ladder, making ends meet apparently with difficulty. To lighten the burden of her presence in his lodgings Mrs Wood stayed out all day and concealed the fact that she had had nothing to eat. So Annie was told, or believed, to her great worry and dismay.

Fate had an even greater blow in store for her. In April 1871 both her small children fell ill with whooping cough. Digby, whose robust constitution was to sustain him through many vicissitudes inflicted on him by his parents, soon recovered. Mabel was only 8 months old and delicate; she contracted severe bronchitis. Annie retreated with her into a makeshift tent before the fire where steam from a succession of kettles vied with coal smoke for the good or ill of Mabel's lungs. Sitting day and night with the choking baby on her lap Annie was plunged into utter despair. Her only comfort in that dreadful time was the sympathy of their doctor, Lauriston Winterbotham, who became a friend and, later, an ally, in that he was willing to give evidence of the adverse effect of Frank's behaviour on Annie's health.[29]

Once when Mabel's body swelled suddenly as a result of perforation of the lungs, Winterbotham, who arrived in response to a frantic call from Annie, put chloroform on a handkerchief and held it up to Mabel's face to soothe the terrible coughing. Such drastic treatment could do no harm at that stage—he was convinced she could not last. Amazingly she withstood the treatment and its frequent—and dangerous—repetition, since Annie had recourse to the chloroform bottle each time there were signs of another paroxysm.

Mabel recovered though she was frail for years and afterwards developed epileptic fits which her mother blamed on this early illness. Annie suffered gravely from it too; she never regarded Mabel as an ordinary child but watched over her with fierce possessiveness, as a lioness its cub.

Once Mabel was out of danger Annie collapsed and lay a week in bed without moving, then rose, a changed woman at the age of 24, resolved to submit no longer but to overcome somehow her adverse fate. Her sense of justice was outraged by the months of suffering she had undergone and had done nothing to deserve. Her religious past now became the worst enemy of her present. Her intense faith in Christ's constant direction of her affairs, the habit of continual prayer and realization of his presence, were all against her; the shock of betrayal was immense, so great had been her love and trust. The presence of pain and evil in a world made by a good God, the pain falling on the innocent as on Mabel, a sorrow-laden world, a lurid, hopeless hell like that in which Frank believed, drove Annie desperate; instead of believing and trembling like the devils in *Paradise Lost*, she believed and *hated*. 'All the dormant and hitherto unsuspected strength of my nature rose up in rebellion; I did not yet dream of denial but I would no longer kneel.'[30]

What followed inflicted great suffering on Annie but at the same time satisfied a deep instinct in her nature, which was to fight adversity—her own or other people's—deploying all her talent and capacity in attack. Frank was unfortunate in that he had a passive role, which was to damp down the fire, to contain her rebellion. It was a hopeless task. Useless for him to argue, as he apparently did, that a woman's business was to attend to her husband's comforts and to see after her children, and not to break her heart over misery here and hell hereafter.

But Frank was more concerned than Annie gave him credit for. It was he who, recognizing the need for older and wiser heads to cope with Annie, brought a clergyman to see her when she had become totally distraught. This man wrote her a series of letters in the summer and autumn of 1871 which reveal an expert in the rounding up of strayed sheep. He was neither horrified nor sanctimonious, but offered a wise understanding that Annie found inexpressibly soothing. Gently he told her that her doubts were natural considering the ordeal she had just undergone. Like others before her, including himself, she would pass through the dark time and be reconciled to faith. Annie did not identify her new friend, W.D., as she called him in her memoirs, beyond the fact that his living was in Cheltenham. In Crockford's list of clergy for 1870 one man stands out as likely to enlist her trust: Edward Walker, rector of the parish church.[31]

He would have enjoyed an immediate rapport with Annie, in that he had spent years as a curate and vicar working among the immigrant Irish of Manchester and Salford: he understood their plight and, like her, sympathized with their aspirations. In Cheltenham, where he had been rector for over twenty years, he was greatly respected for his wisdom and kindness. As his sermons demonstrate, if anyone was capable of handling the creature at bay that Annie had become, he was.

Walker was all too familiar with the suffering experienced by religiously minded people when doubt assailed them. In apostolic times, he wrote, all the children of God were accepted by him. Now the assurance of faith was a privilege enjoyed only by the advanced believer. The weak had been discouraged by the spread of theological argument, and joy in the Lord had almost been banished from the Church. Congregations were bewildered, and ministers of Christ found themselves chiefly employed in allaying fears and resolving doctrinal scruples.

Addressing himself to what he rightly identified as the source of Annie's crisis of faith, W.D. recommended her to read McLeod Campbell on the Atonement. Published in 1856, this work by a Scottish divine had been greeted with relief by the many whose impulse towards worship had been thwarted by their distaste for the concept of original sin—the idea that man, having fallen from God's favour, can only be restored through punishment vicariously endured by Jesus Christ. Campbell, and another of W.D.'s heroes, F. D. Maurice, directed attention away from the actual penalty for sin towards Christ's spiritual intervention. They held that by identifying himself with the sinner and offering up a perfect confession, he made adequate repentance by which God was satisfied. This explanation reconciled numbers of believers to the doctrine of the Atonement, but it did not comfort Annie Besant.

A true Victorian in her need to comprehend what purpose her own life should serve, and how the universe in general was arranged, she had fixed her immediate attention upon proving the existence, or otherwise, of a just God. She could not accept that her faith must be governed by the circumstance to which W.D. had called her attention. Pride would not allow her to confess to sins she *knew* she had not committed, nor did she desire the vicarious suffering of Christ, especially since she felt he had personally betrayed her. Yet what other path was there to the knowledge and experience of the God she so passionately desired? This dilemma preoccupied her for most of her life: the search for a solution governed her actions for many years, dictating the sudden changes of direction which confused all those who knew her.

Annie Besant knew that spiritual insight, knowledge through feeling, was desirable; she constantly sought it but it was denied. As W.D.'s premiss was founded on this he appealed in vain. Her intellect considered the assurance of faith he had won for himself, and which he had offered her, and rejected it as not proven.

In any case she found support for her doubts about the Atonement and related matters of doctrine in the case of the Revd

Charles Voysey,[32] vicar of Healaugh in the diocese of York, to which the newspapers were giving extraordinary attention. In February that year he had conducted an appeal before the Judicial Committee of the Privy Council against a sentence of deprivation of his living. This had been imposed when Voysey had refused to retract statements which questioned the divinity of Christ while asserting the comprehensive nature of the love of God. Annie found these both relevant and sympathetic.

If that was not sufficient claim upon her attention Voysey was an old boy of Frank's school, Stockwell Proprietary Grammar.[33] His appeal had been presided over by no less a personage than the Lord Chancellor of England, Lord Hatherley, Henry's patron, and her distinguished relative. Annie read all she could about the case until she could read no longer.

Mental anguish on top of ill health and great unhappiness finally produced a nervous collapse. Annie lay helpless for weeks, conscious of nothing but 'raging and unceasing pain in the head',[34] unable to sleep, unable to bear the light. Dr Winterbotham covered her head with ice, and gave her opium. Neither remedy had the slightest effect: she had to endure until the pain wore itself out, as eventually it did.

When Annie was convalescent Winterbotham, who knew his patient, prescribed a tonic in the form of books on science and anatomy which he brought her, sparing time to discuss knotty points of physiology in an effort to distract her mind from the dangerous subject of religion.

As soon as she was recovered, the decision was taken to move away from Cheltenham. Frank was to give up the teaching profession to become vicar of a country parish. It is not known which of the two, husband or wife, proposed this major development, or for what reason. It may have been that Frank felt he needed a change of scene, or that Cheltenham had become intolerable to Annie. In the light of her conduct since they had been there College life may have become awkward for both of them. Or it may simply have been that the opportunity arose, as it had not done previously, for Frank to enter upon his true vocation.

However, it was Annie who took the first practical step, as she had to, since it involved an approach to her relative Lord Hatherley. (He was a near neighbour, having inherited the estate of Hatherley on the outskirts of Cheltenham.)[35] Henry's good fortune at his hands may have prompted Annie to solicit one of the livings which, as Lord Chancellor, were in his gift. Obligingly he offered a choice between two. One was Felton in Northumberland, near to Alnwick. This apparently was seriously considered, for the Revd Frank Besant was listed as its incumbent by *Crockford's* for a few months in 1871.[36]

Life might have been happier for Annie if they had gone there. Felton was in the Marches, whose wildness might have eased her sense of physical oppression and pleased her taste for romantic history. The choice fell instead on a parish of some 1,100 souls on a branch railway line—Sibsey, a village in the flat fen country five miles from Boston in south Lincolnshire.

CHAPTER 4

'YOU HAVE READ TOO MUCH ALREADY'

THE landscape which confronted Annie as she stepped from the train and walked the few hundred yards to Sibsey vicarage was unlike anything she had ever known. The houses of the parish were so widely separated that the church of St Margaret's appeared quite isolated, the trees surrounding the graveyard and the vicarage huddled against an interminable wind. The vast expanse of East and West and Wildmore Fens was all that could be seen until the slope of the Wolds intervened miles to the north. All was bleak monotony: 'an intolerable geometry'[1] of dyke and drain—ruler-straight courses in which the sad-coloured water slid imperceptibly along. The reclaimed land was fertile, rich loam over heavy clay, excellent for pasture and growing corn.

The lord of the manor and the other principal landowner lived away from Sibsey. The inhabitants numbered two butchers, two blacksmiths, two wheelwrights, three publicans, a land surveyor, and several small farmers of whom three were women on their own.[2] The only 'society' available to the young Besants consisted of the neighbouring clergy, whom Annie described as 'Tory and prim to an appalling extent'.[3] The previous incumbent had been at Sibsey since 1826, which may explain why Frank, who read himself in at morning service on 31 December 1871, did not bring his family from Cheltenham until the following March. However dilapidated the vicarage may have been when he took over, once refurbished, it was a pleasant house with high ceilings and tall windows looking south. To Digby's lasting delight the garden was huge, with fruit trees, raspberry canes, and an asparagus bed, while the substantial outbuildings included stables and a carriage house.[4]

One of his favourite haunts when he got older was the tower of St Margaret's Church which had a fine peal of bells. Though it was ancient—part Norman, part Early English—the interior must have been a severe disappointment to the new vicar's wife. Drastic restoration in the 1850s had swept away most traces of its long history: there were no tombs, no shields, no effigies, no banners. All was plain to the point of grimness, impersonal, uncherished, and with a strong smell of damp. A frequent complaint in the Parochial Diary kept by Frank was of newly dug graves too full of water to receive their occupants.

His ecclesiastical duties were not onerous. In those days weekday services were rare, and holy communion was celebrated only once a month. On ordinary Sundays, when the numbers at matins fell below a dozen, Frank declined to preach even though by then the pile in his special sermon cupboard had reached respectable proportions. It was a rule largely determined by the need to save candles. Each year Frank recorded the date in the spring when it was light enough at evensong to see the whole body of the church from his place beside the altar, and the closing in of the dark when autumn came and the candles had to be brought out.[6]

Not much else enlivens the pages of his Parochial Diary. He had to complain of drunkenness in one or other of the churchwardens, and sometimes noted the appearance in the village of a tramp stupefied with opium. In 1872, the first year of his tenure, he conducted thirty-two baptisms, ten marriages, twenty-two churchings, and twenty-eight funerals. Their fees supplemented his stipend, which was £350 a year with 170 acres of glebe. Sunday's collection in church added little to that. '£1/5/1 for the *whole* congregation', he noted despairingly on one occasion.[7]

Annie is barely mentioned in the diary. Once, though, Frank noted her presence at the bedside of a dying woman to whom with her daughter he administered holy communion. By her own account, however, she went out and about in Sibsey far more than she had been able to do in Cheltenham. As with any parish there were certain things she, as the vicar's wife, was called upon to do. It was work of which William Roberts would have wholeheartedly approved. Visiting the sick, Annie discovered that some of the farm labourers lived in conditions of appalling overcrowding. In one cottage she found four generations sleeping in one room; the great grandfather and his wife, the unmarried grandmother, the unmarried mother with her child, and three male lodgers. Rheumatism, ague, and scarlatina were commonplace.[8]

It was the moment when a long period of relative prosperity on the land was about to give way to a succession of disastrous harvests and to the worst agricultural depression of the century. Because of the good years rents were rising steeply—from 25s. to £2 an acre was not uncommon.[9] Like others, the tenant farmers of Sibsey were reluctant to pay out more money in higher wages. For labourers these varied from 2s. 3d. to 2s. 6d. a week. Led by the energetic and devoted Joseph Arch, the National Agricultural Labourers' Union, just established, was asking for a guarantee of 2s. 9d. It had begun to be talked about in the fen country, where the farmers were bitter against it and would not employ a 'union' man. Annie had experience of what that meant when a young married man with two children living in Sibsey was 'sinful enough' to go to a union meeting and foolish enough to talk of it on his return home. No farmer would employ him. He tramped about vainly looking for work and recklessly took to drink. Visiting his cottage Annie found his starving wife ill with fever, a fever-stricken child in her arms, the other dead on the bed as there was no other place for it until the coffin came. 'At night the unhappy driven man, the fever stricken wife with the fever stricken child, the dead child, all lay in one bed.'[10]

Annie pondered the stupidity of the tenant farmers, who could well have paid less rent to their landlords and higher wages to the men who tilled their fields. 'They had only civil words for the burden that crushed them, hard words for the mowers of their harvests and the builders up of their ricks.'[11] Why not make common cause with their friends and win an easy victory over the enemy? Seeing all this, she said she learned some useful lessons. Significantly, the emotion aroused in her at the sight of these poor people was nothing so useless as pity, but a profitable anger. She derived no satisfaction, as others might have done in her position, from her influence as the vicar's wife; she was not in the least inclined to smugness, but eager to find a remedy, to attack the poverty at its source.

As her political education progressed theological strife went on inside her head. Winterbotham's attempt to divert her mind was forgotten: she probed and poked as at a nagging tooth, constantly testing the state of her religious belief, holding it up against the declared position of now one, then another of the leading thinkers of the day.

The works of the so-called Broad Church clerics like F. W. Robertson (who as a curate had fled in dismay from the austere Evangelicalism of Cheltenham), Stopford Brooke, and Arthur Penryhn Stanley attracted her initially by their emphasis on the humanity of Christ, a quality which was not a part of the religious attitude of either Ellen Marryat or Frank Besant. Eventually she found herself disposed to reject their arguments as special pleading, skilful evasions of a solution to her particular difficulty over the righteousness of a God who allowed sin to exist, so that evil was eternal and Satan ruled in hell as Christ did in heaven. Venturing into unknown country she came upon Matthew Arnold and was surprised into paying close attention to what he said. She was struck by his definition of Providence as 'a stream of tendency not ourselves which makes for righteousness'.[12] Righteousness was what Annie earnestly desired—for others, including God; she felt no lack of it herself. The idea of morality as religion took root in her mind.

Her continued preoccupation with theology did not please her husband, whose inclination to be masterful had been greatly encouraged by his preferment. As vicar, possessed of spiritual and temporal power in the parish, Frank became increasingly impatient with opposition, whether it came from parishioners who leaned towards the Wesleyans, or from his own wife.[13] He was often irritated enough to lay hands on Annie. In June they had an argument in which Frank took her by the shoulders, 'and struck her violently, striking her with his knee several times thus causing her to leave his house'.[14]

Flight was almost too easy. The station was round the corner, trains to Sleaford Junction where one changed were fairly frequent: London and her mother could be reached within the day.

Mrs Wood was not the only comfort Annie sought in London. She went to hear the Revd Charles Voysey preaching in St George's Hall. After he had lost his appeal against deprivation he had founded a 'Theistic Church' which was attracting a lot of attention. Its message came as a 'gleam of light' to Annie.[15] After his sermon she poured out her heart to him and was most sympathetically received. He and his wife invited her to visit them at their house in Dulwich and she was often there during the weeks she spent away from home. When Frank discovered her friendship with Voysey, on her return to Sibsey, he was shocked.

Voysey's attitude to faith was one that Evangelicals like Frank found deeply distasteful, and which alarmed even liberals if they were concerned about the stability of the Anglican Church. Voysey challenged guide-lines that had been carefully revised after the furore in 1860 over *Essays and Reviews*, a collection that questioned aspects of theology and faith. In the words of one of the authors, Benjamin Jowett, Voysey's offence was that 'he looked too far over the hedge'.[16]

The affair began in 1865, when Voysey published his collected sermons as *The Sling and the Stone*. The content was heterodox. Voysey attacked the doctrine of original sin and eternal punishment as fundamentally cruel; rejected the divinity of Christ; dismissed the Atonement as an unnecessary and revolting idea; and insisted that the Bible was not the revealed Word of God.[17]

By publishing these opinions Voysey placed himself in opposition to the Thirty-Nine Articles of Doctrine to which, on being ordained, he had been required to subscribe, and which he had a duty not to contradict. That duty bore less heavily on the clergy than it had, for the leaders of the Church had recognized the need to relax the hold of the Articles upon the conscience of the clergy after the crisis over *Essays and Reviews*.

The seven authors of that collection were spared the full rigour of ecclesiastical punishment for their audacity. Two of them were to be deprived of their livings but, when their case went on appeal to the Judicial Committee of the Privy Council, the ruling that they should remain as priests was taken as a vindication of their position. The force of the Articles, more especially that of the doctrine of eternal punishment, was thereby diminished.

In 1865 it was decreed that members of the Church of England need no longer actively promote the Articles if they did not so desire; it was enough that they did not teach in contradiction to them. This arrangement was largely the work of Arthur Stanley.

With Benjamin Jowett, one of the authors of *Essays and Reviews*, the dean was a member of a committee to advise Charles Voysey when his refusal to retract views expressed in *The Sling and the Stone* brought on him sensational publicity and the private wrath of both archbishops.[18] Stanley and other Broad Church members were afraid that the ground they had won over the Articles would be lost if Voysey persisted in his defiance. They urged him to accept a compromise, whereby he would keep his living if he climbed down. But he persisted and, in 1869, was sentenced at York to deprivation. To everyone's dismay, including the Broad Church clerics, he won the right of appeal. Further efforts by Stanley to defuse the situation failed when Voysey insisted on going on. 'I shall be an explosive shell among the Articles and Creeds if I have to burst and die in the attempt', he proclaimed. 'My hope is that by staying in the Church to burst its horrid mental fetters for thousands besides myself', he told a friend.[19]

In the words of a rationalist pamphleteer, 'Pontius' Hatherley conducted the Judicial Committee hearing like a quasi-criminal proceeding.[20] The appellant resembled one of Anthony Trollope's less privileged characters. Voysey was miserably poor; the living of Healaugh produced £100 a year with 20 acres of glebe.[21] He had a wife, eight children, and an ailing mother to support. Nevertheless he was gloriously determined not to waver. So were the members of the Privy Council, now they had been obliged to sit.

Opponents of the privileged position of the Established Church rejoiced at the sight of its discomfiture. According to the same pamphleteer, the Voysey affair brought to a head years of unease within the Church that had been bearable only because different heresies had been distributed through various individuals, 'each of whom had his segment of rationalism in connection with ... an eminent or even courtly following [a reference to Dean Stanley whose wife was a lady in waiting to the Queen], or held it with such dexterity of statement that he could not be made a fair test and remained in the Church as its bait for clever young men'. But all these heresies converged at last in one man. 'The orthodox clergy saw all the Broad Church clerics with one neck, that neck being the Revd Charles Voysey, and the outside world saw that the dignity of the Church depended on whether that neck could be cut off or not.'[22] If it were not the Church would become a laughing stock.

As Stanley and his friends had feared, the Judicial Committee of the Privy Council found Voysey guilty of teaching in contradiction of certain of the Articles, a verdict that effectively restored the full rigour of the doctrine of eternal punishment. Lord Hatherley offered mercy—a week to retract and Voysey could keep his living. He refused, not wishing to appear a hypocrite; nine-tenths of the clergy might be so described, he said.[23]

He was better off outside the Anglican communion. He had fought a very public battle, constantly inviting the interest of the press. He attracted moral support from free-thinkers and Dissenters, and financial help as well. An account of his defence before the Chancellor at York found a lively sale; it had been drafted by a barrister friend of Stanley's who was also an able journalist, Fitzjames Stephen. Funds so raised were augmented by a swift lecture tour of England and Scotland following the failed appeal. They were sufficent to set Voysey up in his Theistic Church, and to provide a home for him and his family.

The attention Annie received at his house in Dulwich naturally encouraged her own pursuit of religious 'truth'. Always susceptible to men who acknowledged her intellect, she promoted Voysey to the rank of hero, and copied him—as she would others—as closely as she could.

It was not a judicious choice. Voysey was ill equipped to take a vulnerable young woman under his wing. In spite of his 44 years there was a schoolboyish air about him. He was given to pranks, like the correspondence he began in defence of the Established Church in the pages of the free-thought journal the *National Reformer*, edited by the Devil Himself, Charles Bradlaugh.[24] He badgered people in the public eye to make compromising statements in his support. One of his targets was Frederick Temple (another of the authors of *Essays and Reviews*), who was made bishop of Exeter in 1869 in spite of bitter opposition to his Broad Church views. Voysey invited him to denounce with him the doctrine of eternal punishment and was understandably rebuffed. 'Temple is a horrid sneak',[25] was his reaction.

At Dulwich Annie was introduced to Thomas Scott, *éminence grise* in the Voysey affair, a role he enjoyed and thought to repeat with her. Scott is something of a puzzle. For fifteen years, beginning in 1860, he published at his own considerable expense a series of pamphlets extolling all kinds of unorthodox views, from rationalism and spiritualism to republicanism. Some of them were by authors distinguished in their own field, like Francis Newman, Vansittart Neale, Edward Maitland, and the American minister at South Place Chapel, Moncure D. Conway, who was one of Voysey's most ardent supporters. A 'sheaf of small white arrows',[26] as the pamphlets were described, went out on the first of each month by post to a wide variety of addresses, including rectories and vicarages. They were never advertised but each one had Scott's address and his seal prominently displayed. Conway's pamphlet on the Voysey appeal which castigated 'Pontius' Hatherley may have reached Annie in this way. Scott received back a huge number of letters from people in all walks of life—a correspondence which, as far as this writer can discover, has almost entirely disappeared. Nor are there many details to be gathered about his life, which appears to have been mysterious even to his contemporaries.

It is known that he was born in 1808, brought up in France as a Roman Catholic, and became a page at the court of Charles X, stuffiest of the Bourbons.[27] Of ample means, he then embarked on a life of adventurous wandering which allegedly included living and hunting with American Indians. Some very unusual people were to be found in the pioneering days of the West, but few camp-fires can have been enlivened by so odd a man as Scott, whose head was filled with pagan myths and Christian stories: his passion was the history of the Bible, to interpret which he somehow acquired a fair knowledge of Hebrew.

Returning to England in the late 1850s, he married a woman young enough to be his daughter and settled at Ramsgate. An English east-coast watering place must have seemed quiet after the far frontier, and Scott hankered after an occupation on which to spend his fortune. When he looked about him, he said he saw his countrymen bound hand and foot by 'metaphysical and priestly exclusiveness'.[28] Though he did not underestimate the difficulties posed by clerical prejudice, he determined to liberate them. 'I entered upon [the venture] single handed and entirely on my own responsibility', he explained in 1876, when forced to retire from publishing by ill health. '[I] resolved in a courteous but uncompromising spirit to do my utmost to bring all my forces to bear upon the errors and superstitions so degrading to man's highest nature, and to follow truth, and truth only, wheresoever it might lead me.'[29]

Though his pamphlets addressed themselves to all sorts of subjects, from the gradual development of beliefs and ceremonies in the time of Solomon, and from phallic worship to Christianity, Scott was really interested in undermining the position of the Church of England by a series of assaults upon aspects of its doctrine. Looking back on his career, he said nothing gave him so much satisfaction as the strides free-thought had made among members of that Church in the fifteen years of his campaign. He was never so happy as when he received from a country vicarage a parcel of manuscript with a plea that he might publish it, anonymously of course. A number of his publications bore the cryptic message 'By a beneficed clergyman'.

Scott's personal crusade was vastly encouraged when he encountered Charles Voysey, who allowed him to publish the first version of *The Sling and the Stone*. Scott became Voysey's patron and his chief financial supporter in the struggle with the Church.[30] Scott paid for the defence and canvassed for subscriptions to a fund for the establishment of the Theistic Church (he called himself a theist). In 1870, in order to be near his protégé, Scott left Ramsgate for Farquhar Road in Upper Norwood where he started a salon. Voysey was the chief attraction, but all shades of heretical opinion were represented. In the summer of 1872 pretty Mrs Annie Besant was introduced into the circle by her new friend and mentor.

It was an exciting experience for her to come into contact for the first time with people who had fashioned their beliefs independently of their upbringing or education. Among them were men like Marcus Kalisch, a Jew who had taken refuge in England from Germany, whose biblical criticism demonstrated the 'Continental' method; John Muir, orientalist and Sanskrit scholar, author of a sympathetic and voluminous account of Hindu religion and culture; and Thomas Lumisden Strange, an ardent theist, who had spent all his working life as a judge in India. Annie was particularly attracted by the socialist philosopher and close friend of George Eliot, Charles Bray, who often brought his sister-in-law Miss Sara Hennell to see the Scotts.

The respect with which free-thinkers treated her stemmed from her connection with one of the first and greatest rationalist works of the century. *An Inquiry Concerning the Origin of Christianity*, by her brother Charles Hennell, was published in 1838. His work marked a departure from the ferocious polemics of eighteenth-century rationalists; firm in its statement of conviction, it was always courteous in its attitude to exploded belief.

Most famous of all, more famous than Voysey, was John Colenso,[31] bishop of Natal, whom Annie met at the Scotts'. He was in England to further the cause of his Zulu converts. Scott's knowledge of Hebrew had proved sufficient for him to see Colenso's series of commentaries on the Pentateuch through the press. Like Voysey Colenso had been haled before the Judicial Committee of the Privy Council for heresies, one of which was his rejection of the doctrine of eternal punishment. In his case, though it had been decreed that he could not be removed from his diocese as his opponents desired, no verdict had been passed upon his various beliefs.

In the climate of opinion which now surrounded Annie, 'truth' was exalted at the expense of compromise. Voysey and Scott were the last persons to urge reconciliation on her as, for example, 'W.D.' had done. She could no longer turn to him: his letters had ceased, a fact which is consistent with the identification of him as Edward Walker, vicar of Cheltenham, who died in 1872.[32]

But Annie remained in desperate need of help. She agonized over her abandonment of belief in Christ as God Incarnate. If one took away the dogma of the Atonement how could one account for so tremendous a miracle as the incarnation of the deity, she wondered? She was familiar from her reading with the idea of avatars in Eastern religion and the fact that incarnation formed a part of faith in primitive times. The way was consequently clear for her to challenge Christianity in this respect also. But she shrank from the thought of placing in the crucible a doctrine so dear from the associations of the past: there was so much that was ennobling and soothing in the idea of a union between man and God, between a perfect man and a divine life. It hurt to recall her instinctive response to the treasures displayed in the churches of Paris. Jesus as God was interwoven with all art and all beauty in religion; to break with the deity of Jesus was to break with music, with painting, and with literature.

There was another cruel dilemma. The moment Annie renounced belief in the deity of Christ she had, in honesty, to renounce Christianity itself. As a clergyman's wife how was she to live if she did that? She made one last appeal to an authority she hoped would save her from catastrophe. She entered into correspondence with Doctor Edward Pusey, who, finding her intractable at a distance, summoned her for an interview.

That the wife of an unknown country parson should approach the Regius Professor of Hebrew at Oxford University on a matter of personal belief might be thought audacious. But it was typical of Annie, who took herself most seriously and trusted others to do the same. More remarkably, Pusey represented everything her new associates had fought for years to counter and destroy. He was the Church's champion, who rode out on every occasion, public and private, to defend the orthodox belief. He was unusually well qualified for the task. Early in his career, having identified the potential danger to faith posed by the new school of biblical criticism in Germany, he mastered the language and the method in order to refute it. When, for different reasons, Newman and Keble withdrew, Pusey sustained the Oxford Movement almost single-handed. After the shock of *Essays and Reviews*, he rallied an army of local clergy in defence of the Thirty-Nine Articles.

Some writers have concluded that Annie's motive in approaching Pusey was not straightforward. Either she went in a Voyseian spirit of provocative enquiry (and her account of their meeting must have diverted the readers of her *Autobiographical Sketches*), or she knew his intransigence would reinforce her doubt—to good effect since, the supposition was, she now perceived her loss of faith as a means of escaping from her marriage.

That seems premature, but it is difficult to be sure. Her opinions were so volatile that any number of considerations could have moved her. But it is possible that, as she claimed, she hoped Pusey would rescue her from an adventure that had gone too far. After all, he represented the dream world she had briefly occupied following her departure from Miss Marryat, where nothing disagreeable ever happened to her. He was the learned patristic scholar, full of the wonders of antiquity. Pusey was responsible for the great Oxford edition of the Lives of the Fathers of the Church in which she had immersed herself at Harrow. By that means he had revealed to her the wonders of the early Christian Church, and he was the arch-exponent of the ritualistic practices in which she had delighted.

There are no letters from Mrs Besant in the correspondence preserved at Pusey House. And his letters to her were stolen from her desk in Sibsey, by whom she did not say.[33] But she did remark that their meeting was only possible because she was with her mother in London at the time, and so free from supervision.

At Oxford she found a short, stout gentleman, dressed in a cassock, looking like a comfortable monk, but with 'keen eyes' that told of force and subtlety enshrined in the impressive head. The learned doctor took the wrong line with her, treating her like a penitent going to confession and seeking the advice of a director. Annie's dangerous pride flared up immediately: she and Dr Pusey glared at each other across an unbridgeable gulf. She was an enquirer struggling after truth: he would not even admit the question of the deity of Jesus. 'You are speaking of your Judge', he retorted when she pressed 'a difficulty'. The mere suggestion of an imperfection in the character of Jesus made him shudder; he 'checked her with a raised hand'. She was blaspheming, the very thought was a terrible sin. When she asked him if he could recommend books that might throw light on the subject, Pusey replied (not unreasonably, one feels), 'No, no, you have read too much already; you must pray, you must pray.' A prolonged tussle followed, Annie resenting the implication that all her suffering had been in vain, that she must come back to the point at which she had started, Pusey insisting that it was her duty to accept and believe the truth as laid down by the Church. She had no right to make terms with God as to what she would believe or not, he told her; she was full of intellectual pride.[34]

That ended the confrontation. Annie thanked him for his courtesy and declared that she now had no choice but to go home and face the difficulties that would arise on her next step, of openly leaving the Church. Pusey was ruffled. She must not speak of her disbelief, he cried, he forbade her to lead into her lost state the souls for whom Christ died.

In Annie's mind Pusey appeared thereafter as an Inquisitor, 'perfectly conscientious, perfectly rigid, perfectly merciless to the heretic',[35] suppressing by force, not by argument, all challenge to the traditions of the Church. The violence of her reaction was due to the nerve his last remark had touched—her responsibility in all this to Frank Besant. She believed in the vows she had taken at her marriage; because her husband was a minister of the Church, through him she was bound to it as well. In the autumn she returned to Sibsey.

CHAPTER 5

FLIGHT

WHATEVER comment the long absence of the vicar's wife had provoked among the villagers, life at Sibsey continued much as usual for a time after Annie's return. But a frontier had been established between her and Frank; it was in evidence at the vicarage. She occupied the drawing-room, whose only unusual feature was that it was piled high with books. He retreated into the study, where he was to spend a large part of the next thirty-five years. He had found a pile of old documents in a cupboard in the vestry and, with nothing better to do, began to sort them out. The task appealed to his love of order, and his sense of proprietorship at Sibsey. It was the beginning of a lifelong interest in the transcription of local records—parish registers and the like.[1]

When winter closed in on the Fens Annie suddenly found herself with more than enough to do. As the result of bad drains and the overcrowded living conditions, typhoid rampaged through the village.[2] In those days skilful nursing was held to make the difference between life and death in most diseases, but especially in the case of typhoid, for which there was no known remedy. Fever-ridden patients had to be constantly watched, given nourishment, and bathed, in an attempt to bring down the temperature. As the illness lasted for three or more weeks, Annie, who watched at bedsides in order to give the exhausted relatives a rest, believed that she earned the gratitude of the whole village. No doubt she did, but there was a difference in how she presented her efforts at nursing and her actual experience.

The villagers were led to assume that her devoted service was evidence of a willingness and capacity to care for others. Annie often drew attention to this womanly quality in herself as if to compensate for her shortcomings as a wife and, in due course, as a mother. Rather, the experience exhilarated her; she derived satisfaction from meeting and confounding the threat of approaching death. It gave her 'a strange and solemn feeling',[3] because this time it was not tempered with fear for the loss of a loved one, as in the case of Mabel with bronchitis. In Sibsey, watching over near strangers, Annie felt an influx of power to counter the crisis in the fever when death lay in the balance.

The goodwill of the village earned in that way was important to Annie; she hoped it would help to dispel adverse reaction to her recent behaviour in church. On her return to Sibsey she had obtained Frank's consent to an arrangement whereby she would attend all those services in which God was represented as a loving Father and Creator, but she would not attend those which celebrated the deity of Jesus Christ.[4] As she argued herself into it, there was a kind of logic in this position. The doctrine of the oneness of God, in which she now placed her hope and faith, 'repelled' the doctrine of the incarnation of part only of the Godhead as Christ. A close study of the Synoptic Gospels, in which she tried to visualize the life portrayed therein, reinforced by rereading Renan's *Vie de Jésus*, convinced her that Christ had denied his own Godhead; that if there was any truth in these Gospels they told the story of a 'struggling, suffering, serving, praying man', and not of a God at all.[5]

Casuistry was all very well in the drawing-room, but it looked absurd when it emerged to public view in church. Annie's theological position allowed her to sit comfortably through matins on three Sundays in the month. When, on the fourth Sunday, after the children who were not old enough to be confirmed had filed out, the congregation composed itself again for holy communion, the vicar's wife also had to leave her pew to avoid witnessing a celebration of the Atonement. The first time this happened was in November 1872. A feeling of deadly sickness nearly overcame her as she rose and, with every eye on her, made her exit. In fact everyone jumped to the conclusion that she had been taken ill, and she was overwhelmed with sympathetic enquiries and calls. To any direct question she answered quietly that honesty forbade her to be a communicant, but, 'as the idea of heresy in a vicar's wife is slow to suggest itself to the bucolic mind',[6] to others she offered no comment. As long as she remained at Sibsey she continued to depart each communion Sunday.

For a pupil of Miss Marryat who had been taught never to draw attention to herself, it was an excruciating ordeal, as well as an affront to a nature that was reserved and shy. But Annie, to her surprise, discovered that if the occasion demanded it she was able to put timidity aside. The next encounter with her unknown self took place in Sibsey church. A 'queer whim' seized her that she would like to know how it felt to preach. Perhaps she suspected that she could do it better than Frank, or than some of the visiting clerics whose sermons she had been obliged to sit through. On the pretext of practising the organ, though she took no one with her to pump the bellows, she locked herself in Sibsey church. (It was always her practice to lock the door when she was alone; she never could bear the thought of being taken unawares.[7])

Mounting the pulpit she delivered a lecture on the Inspiration of the Bible to the rows of empty pews. 'I shall never forget the feeling of power and delight but especially of power—that came upon me as I sent my voice ringing down the aisles', she wrote, 'and the passion in me broke into balanced sentences and never paused for musical cadence or for rhythmical expression.'[8] These came unrehearsed, for she was a natural orator with a voice of great beauty, as many of her hearers testified. One of them was Beatrice Webb, who described it as neither female nor male but 'the voice of a beautiful soul'.[9] Softly spoken in conversation, on a platform her voice was clear enough to be heard at the back of a crowded hall without effort on her part, or strain upon her audience.

And so her wings were partially unfolded before Frank realized how dangerous their 'arrangement' might be for him. Sometime during the winter of 1872–3 he discovered, or was shown (the accounts conflict), a pamphlet Annie was writing called 'What think ye of Christ?'[10] Voysey suggested that she write it, and Thomas Scott promised to publish it, though not with that title, whose crudeness offended his rule of impartial courtesy in matters of theological debate.

In 1878, after the marriage was over, Frank Besant swore an affidavit to the effect that he thought the objection he raised on first seeing it was the end of the affair. He was furious when, a few weeks later, the pamphlet appeared under Scott's seal as *On the Deity of Jesus of Nazareth* with the inscription 'By the wife of a beneficed clergyman'. To heap coals on the fire the introduction was by Charles Voysey. Annie deposed at the same time that her husband had accepted that the pamphlet would be published, stipulating only that her name should not appear.[11] He had objected to 'certain things' in the introduction which had been added without her knowledge.

Frank was very anxious to dissociate himself at all times from the views expressed by Voysey, which his own wife echoed. His affidavit concerning *On the Deity of Jesus of Nazareth* was explicit. 'I never consented to the printing or the publishing of the said pamphlet or any pamphlet or pamphlets contrary to the doctrines of the Christian religion or the tenets of the Church of England.'[12] His Parochial Diary shows the care he took at the time to perform his duties to the letter. In March 1873 there were only five people at one Lenten service but, in spite of his rule, he preached, 'in order that it might not be said hereafter that there was any remisseness [sic] on my part'.[13] His unease was understandable in the light of the Privy Council judgement in the Voysey case.

He had always been prone to exaggerated worry over his finances; now with two small children to support he felt they were seriously overstretched. What would he, or could he, do if Annie's indiscretions caused him to be deposed? The example of his own brother terrified him. Like Frank, on leaving Cambridge, Walter Besant had found a teaching post in a school that liked its masters to be clergymen. But at the crucial moment Walter had been deemed unsound on the Atonement (Walter was never 'serious' like Frank). Denied the privilege of taking orders, Walter had lost his job and had been obliged to spend years in exile, teaching in Mauritius.[14]

Though at her husband's insistence Annie temporarily broke off her correspondence with Charles Voysey, Frank remained extremely apprehensive as to what she might do next. The kind of publicity Voysey sought and attracted was dangerous as well as distasteful to him; he feared what the newspapers would do if they got wind of the interesting information that Scott's unknown author, 'the wife of a beneficed clergyman', was related to the judge who had put an end to Voysey's clerical career. The idea seems to have momentarily unhinged Frank. According to Annie, he threatened to shoot her if she dared mention that he had got his living through Lord Hatherley.[15]

Apparently on the verge of collapse Frank received support from an unexpected quarter—from a male relative, not otherwise identified. Annie blamed the ultimate breakdown of the marriage on this man's advice. If he had not intervened, her implication was, she would have been prepared to sustain the 'arrangement' with Frank indefinitely. The man was sent a copy of her pamphlet *On the Deity of Jesus of Nazareth*. However much he agreed with the sentiments expressed in it, he told Frank, he must warn him of the social and professional dangers of having his name associated with such views, as would befall when its anonymous author was inevitably exposed.[16]

The man has been identified, without proof, as Walter Besant. It is the sort of thing he might have said to his younger brother. Walter's views on the Atonement had resembled those of Annie; by 1873 he was ready to dissociate himself from them. His future now partly rested on his reputation as a writer—of novels, of criticism, of historical works. Editors of journals like *Macmillan's Magazine* and the *Nineteenth Century* in which his work appeared would shy away from any scandal attached to the name of Besant. Nor would it be favourably regarded by the learned members of the Palestine Exploration Fund whose Secretary Walter had become shortly after his return from Mauritius in 1867.[17]

As far as the country people were concerned, it was naïve of Annie to suppose her conduct would escape their notice, far less their censure. Their lack of understanding intensified their disapproval which, she now discovered, was disproportionate to their gratitude. As knowledge of her attitude spread, the atmosphere became intolerable. Once more Annie's health threatened to break down. She took refuge in further defiance. She wrote another pamphlet for Thomas Scott which he published as *According to St John: On the Deity of Jesus of Nazareth, Part II: A Comparison between the Fourth Gospel and the Three Synoptics*, 'By the wife of a Beneficed Clergyman. Edited and Prefaced by the Rev. Charles Voysey.' This Gospel, imbued with mysticism, which had appealed most strongly to her favourite among the patristic writers, Origen, did not receive approval from Annie in her present mood. She denounced it as unauthentic, 'an insult to the Justice, the Supremacy, and the Unity of Almighty God'.[18] In writing this pamphlet her correspondence with the Voyseys was renewed, and she also occupied herself in making a collection of hymns in which the name Jesus did not appear, for the use of the congregation of the Theistic Church.

In July 1873 she left Sibsey for Southsea, taking the children with her. One is led to assume from the account in her autobiography that this was flight—from her husband as well as from the parish. But there is evidence to suggest that it was not so. Frank wrote in his Parochial Diary, 'July 20, 1873. I was compelled to be at Southsea owing to AB's illness there. Reginald Jones, curate of St Jude's, Southsea, exchanged duties with me. I taking duty partly at St Jude's and partly at the temporary iron church in St Jude's District called St Peter's.'[19] It cannot have been easy for Frank, who had been in his own parish for only eighteen months, to ask his bishop for permission to make these arrangements. No doubt he welcomed the temporary escape from tension at Sibsey and hoped that Annie might recover in the sea air, while his parents looked after Digby and Mabel.

Annie did not reveal Frank's presence at Southsea in her *Autobiographical Sketches* and, when the second version was published, omitted all mention of the visit. However, she acknowledged that this was the critical period when the future of the marriage hung in the balance. According to her she did nothing more to strain the relationship; trouble occurred with a second intervention from Frank's relative, urging him to issue an ultimatum. She must either return to Sibsey and take her place at all the services of the Church, including holy communion, or she must leave her family and go to her mother. Conformity or exclusion from home was how Annie interpreted the demand. 'Hypocrisy or expulsion. I chose the latter.'[20] It was the very same choice that had made a hero in her eyes of Charles Voysey.

Expulsion is a telling word in this context; she might have been expected to say 'escape'. That difference, and the bitterness with which she spoke of the intervention of the male relative, are the only indications of regret for the drastic step she took. One wonders if she might have continued as she was, given Frank's complaisance, if the ultimatum had not fatally aroused her pride. Was it possible that, as with her recourse to Dr Pusey, she wanted to pull back from the direction in which she had strayed? Would the occupation she found at Sibsey nursing the sick, the crying need of the farm workers for sympathetic help, the ordinary work of the parish, to say nothing of the demands of her two small children, alert her instinct for self-sacrifice and make her stay?

They did not, but by the time she looked back at her expulsion she bitterly regretted the loss of that comfort, shelter, and assured position in society that was granted to a member of a Victorian family. All Annie's great strength and courage expended on the effort to survive as an independent person was never enough to ease the pain of becoming an outcast from that society.

She did not lack warning of the dreadful consequences of her decision to leave. Her mother, frantic with distress, went down on her knees to beg her daughter not to condemn herself to such a fate, particularly not for the sake of her belief; in Emily's view it was not worth it.[21] But Emily's belief was not like hers, Annie remarked. Her mother's Christianity was wide and vague and loosely held. Emily could not understand the intensity of her daughter's feeling that where she did not believe she could not pretend.

After Southsea, when the children returned home with Frank, Annie went to London to be with her mother at her brother Henry's house in Brompton. There a marked deterioration in her health forced her to consult her doctor. He diagnosed nervous exhaustion and warned her that if she went on as she was the function of her heart would be impaired. The household was further upset when, in September, Frank Besant appeared in their midst without warning. His demeanour was alarming. He shouted and stamped about before delivering himself of the ultimatum as to Annie's attendance at holy communion.[22] Henry Wood decided it was time to do what he could for his sister: his only recourse was to the law, which offered little comfort.

Grounds for divorce at the time were restricted to adultery, desertion, and cruelty. Heresy, which was really the issue between the Besants (so Frank deposed, though he used the word religion[23]), had once been a cause admitted by the ecclesiastical courts, but not since the Reformation. Henry Wood had to try to extricate his sister without clear grounds from a situation in which her personality was deemed to have been absorbed into that of Frank Besant. The law regarded husband and wife as one individual.

A deed of separation was the usual recourse for people in such circumstances, particularly when there were religious scruples against divorce, as was the case with Annie at the time, and with Frank. Since Annie as a married woman had no legal standing her brother and a solicitor had to act as trustees on her behalf.

The deed, which was dated 25 October 1873, was very unusual in that it gave Annie custody of her daughter Mabel, while Digby was to remain in his father's charge. Each child was to spend one month a year with the other parent. The Custody of Infants Act which permitted this arrangement was barely six months old.[24] Before it was passed in April 1873 no father was entitled in law to divest himself of responsibility for any of his children; to try to do so would have made a deed invalid.

The agreement reflected the fierce possessiveness Annie felt for Mabel. Frank's consent to it was not easy to obtain. It seems that her trustees had to use coercion, in that they threatened to pursue an aspect of Frank's behaviour in the marriage which he preferred not to have revealed. (The later charge was one of cruelty.) One quarter of his income—£110 a year—went to Annie to maintain their daughter. For the rest she was only to take with her gifts and personal possessions; all the household goods were to remain at Sibsey.[25]

So did Mabel for a few weeks yet. Annie had nowhere to take her. Her brother Henry offered her a home with him. That was gallant in the circumstances for he had himself just got married. But he found it necessary to stipulate that Annie should drop her friendship with Charles Voysey and Thomas Scott; this she refused, and left his house, and saw little of her brother and his wife from that time on.

CHAPTER 6

HARDSHIP

ANNIE'S difficulties now began in earnest. She was poor. She had left Sibsey with nothing in her pocket and only £9 3s. 4d. to rely on at the end of each month.[1] On that, and what she might earn—for she was determined to find work—Annie hoped to support herself, her mother, and Mabel, while holding up her head in society.

Until she left Frank Annie had had very little experience of society; she had no idea how unkind people could be to those who offended against its rules. She was ignorant of the effect the religious scandal which drove her from Sibsey would have upon her reputation. Naturally it was much discussed, in low tones and not before the servants.

Loss of faith and husband all at once seemed due to something more than misfortune (on which in any case the Victorians looked with unforgiving eyes); it was positively dangerous, threatening to pull down religion and marriage. On these two pillars to which Annie had been tied, good moral order—hence society—was held to depend. Doubt being a condition of the age, there were people ready to excuse her religious scruples—they were common enough—but few who did not judge her harshly on the breakdown of her marriage.

A gallant first attempt at earning money revealed at once how callous the world could be. A Birmingham firm advertised, for a fee, the opportunity of adding to one's income. Annie sent off the money and received a pencil case with the explanation that she was to sell such articles, including cruet stands, to her acquaintances. Annie did not feel equal to pencil cases or cruet stands, and forfeited the fee. Precious shillings spent in employment agencies were also unproductive.[2]

There was almost no one to whom she could turn; many of her friends closed their doors to her. William Roberts, who might have proved her best adviser, had died in 1871. Support and sympathy came to her from Thomas Scott, but the first person to offer practical help was Moncure Daniel Conway.[3] He and his wife Ellen were good friends to Annie Besant at the very beginning of her independent life, and steadfast later on when the going became so rough. As Americans they could withstand the social panic that afflicted the English when confronted by a solitary woman, however undeserved her state; they went so far as to invite her to stay with them until she could plan for her future. Conway did this because he knew, through Walter Besant, an acquaintance, that Walter's brother Frank was not in a position to provide any 'worthwhile' sum of money to his estranged wife. Annie gratefully accepted his hospitality.

It filled him with astonishment that a young man should be willing to part from this beautiful and accomplished wife for the sake of any creed, Conway wrote. 'Her face was beautiful, its delicate oval contours and feminine sensibility were heightened by the subtlety and sincerity that comes of good breeding and culture.'[4] Conway tried to find useful occupation for her. He was preparing a treatise on demons and demonology for which he had obtained a number of obscure German books. Annie was put to sifting them for suitable material.

Conway's sympathy may have been tinged with a certain amount of guilt; he was partly responsible for the disaster that had overtaken Annie. He was the author of the Scott pamphlet that had made a hero of Charles Voysey;[5] he had been the principal advocate of the vicar of Healaugh's ill-fated appeal to the Privy Council, arguing that the burden of proving heresy ought to be placed squarely on the Church. He was a man of remarkable powers: brilliant, conscientious, but not well balanced. 'A very queer specimen of a latter day prophet', was how T. H. Huxley described him.[6]

Huxley's acquaintance sprang from an invitation to lecture at one of the Saturday gatherings regularly held at the institute founded by Conway as an adjunct to the chapel of which he was the minister. It was at South Place, Finsbury, just outside the City. Among other lecturers, each eminent in his own field, were the physicist John Tyndall; the geologist Sir Charles Lyell; the historian Sir John Seeley; and the chief apologist and doyen of the Unitarian sect James Martineau.

On Sundays Conway himself held forth on a subject previously advertised in the Saturday newspapers. These discourses—they were never referred to as sermons—were eclectic. As minister of South Place since 1864 Conway virtually set the agenda and chaired the gatherings of fashionable free-thought in London.[7]

By 1873 his personal allegiance was no longer to any particular religious belief. He conducted enquiries into diverse creeds, including Eastern ones, intending to show that each was capable of revelation. He selected their most valuable aspects to present to his congregation in a series of readings. In the late 1860s Conway's choice of extracts from Hindu and Buddhist literature—from the *Bhagavadgita* for example—was very popular, filling the 1,000 seats in the chapel.[8]

As a healer of spiritual wounds Conway's effectiveness stemmed from his own experience of loss of faith. Like Annie Besant he had once been utterly absorbed by the doctrine and practice of a particular creed—in his case Methodism—only to be forced to abandon it when its dogma failed to pass the test of reason.

Born in 1832 into an aristocratic Virginian family whose members served with distinction as politicians, lawyers, and journalists, Conway embraced Methodism in his adolescence, becoming a minister at the very early age of 20. A year on circuit was enough to show him he could not truthfully subscribe to belief in original sin, everlasting punishment, and vicarious sacrifice; they offended his highly developed critical sense. 'My belief is that I gradually preached myself out of the creeds in trying to prove them by my lawyer-like method',[9] he wrote, describing a process that Annie Besant must have recognized.

Conway did not suffer from his loss of faith as keenly as Annie did; his interest was mercifully more detached. But like her, in his bewilderment he turned to books; in his case a guide to the way forward came from the writings of Ralph Waldo Emerson. In 1852 Conway went north to Massachusetts and became a disciple of the famous sage of Concord. Emerson's library contained rare translations of oriental books in which Conway immersed himself. From 1854 to 1856 he studied at Harvard Divinity School where he embraced his next great enthusiasm—the abolition of slavery.

Returning to his home state he was reckless in proclaiming his new belief. *Ante bellum* Virginia was not the place to express such views: Conway and his family were ostracized and finally driven out. With the onset of the Civil War, however, his eloquence was recognized by the Northern Government. Concerned by the substantial support in England for the Confederacy they sent Conway to London to lobby for the anti-slavery cause. The Virginian won many friends among English abolitionists and came to be well regarded by society at large. He carried letters of introduction to Richard Cobden, John Bright, Robert Browning, Thomas Carlisle, and Alfred Tennyson.

In 1864, when the South Place Chapel faced disintegration upon the death of its long-serving incumbent W. J. Fox, its congregation asked Conway to be his successor. Fox's Unitarianism had been enlivened by his participation in radical politics, and his interest in the arts. Conway was encouraged to continue the tradition.

When Annie first entered it when staying with the Conways, South Place Chapel was barely recognizable as the Unitarian centre of worship it had once been. The sacramental vessels were preserved, but as relics; flowers stood on the former communion table; sherry was set out in the vestry. It was a foretaste of the heterodox 'world beyond the churches'[10] which Annie was to explore during the next ten years.

Much was to happen, however, before that state was reached. Annie's immediate problem was to find a place to live. With the help of Thomas Scott a house to rent was discovered in Colby Road, Gipsy Hill, Upper Norwood, just round the corner from Scott in Farquhar Road.[11] Small though it was, it had to accommodate Annie, her mother Emily, her daughter Mabel, and, in spite of their straitened circumstances, something they did not dream of doing without—a maid. It would not be available for six months, so Annie decided to look for a temporary living-in post; that part of her annuity thus saved would go to buy the rudiments of furniture.

In the meantime Mrs Wood had gone to stay with Annie's grandmother and two of her aunts at Folkestone. The Morrises, being churchgoing folk, knew that the vicar, Mr Woodward, was in the throes of a domestic crisis. It was arranged that Annie should go to him as governess to his children.[12] Mabel was to be allowed to accompany her, and they were to receive their keep. The arrangement was a tribute to the vicar's broadmindedness.

Mabel, aged 3, duly arrived at Folkestone station with a label round her neck, having been dispatched from Sibsey in the sole charge of the guard.

Shortly after her arrival at Mr Woodward's house the crisis intensified when the cook and parlourmaid fell ill.[13] Annie, called to fight on all fronts, had recourse to an unfamiliar set of books; these instructed her in the mysteries of cooking. Broiling and frying were well within her capacity, she found, and making piecrust rather pleasant. Stews were interesting; it was touch and go if they tasted of anything but onions. But kettles and saucepans blistered her hands—scouring in those days was no joke—and her sweeping was bad since she lacked muscle. Then one child fell ill with diphtheria. Mabel was dispatched to her grandmother and Annie thankfully turned nurse. No sooner was her first patient out of danger than the youngest boy contracted scarlet fever. This frightened the household far more than diphtheria; it was known to spread very rapidly.

Annie isolated herself and her patient on the top floor of the house. There, according to the method recommended in Mrs Isabella Beeton's *Book of Household Management*,[14] she cleared away the carpets and curtains. Annie hung sheets over the doors which she kept wet with chloride of lime and, in an atmosphere of ferocious disinfectant, took her meals from a tray left outside on the landing. Cowed by these heroic measures the fever retreated, leaving Annie triumphant; the other children in the house remained unscathed.

As the spring of 1874 advanced she was able to congratulate herself on having achieved the first part of her plan; she could now spare money to furnish Colby Road. Eagerly she looked forward to a resumption of family life, in her own home, and with the two people she loved best—Mabel and her mother. After six miserably hard years she would have peace, and support from a loving companion. She and Emily spent hours discussing the share each would take in Mabel's education.

Fate decided otherwise. Towards the end of April Annie and her mother parted for a while. Emily went to stay with Henry at Brompton while Annie saw to the arrangement of the furniture. She had enough for two rooms only, bought on the instalment plan, but Mrs Scott kindly supplied some extra pieces for her bedroom. After about a week Annie received a telegram from a doctor called to attend on Mrs Wood, who had fallen so gravely ill with heart disease he thought she could not last three days. Annie hastened to her bedside and nursed her night and day with 'a very desperation of tenderness'.[15] Two crises were surmounted; through her mother's will to live and her own tenacious spirit, Annie believed. But when dropsy supervened they had to accept that death was inevitable.

As with William Wood's last illness, circumstances lent themselves to a display of calculated emotion by his widow. Emily used the fact of her own approaching death to try to force her daughter's return to the Church. Though Annie testified that her mother's faith had been undemanding, Emily now begged to receive the sacrament, but only if Annie did so too. 'I would rather be lost with her than saved without her', the dying woman declared.[16]

This was to present Annie with a cruel dilemma at a time when she was already overwhelmed with grief. She had not communicated since she walked out of Sibsey church in November 1872; to do so now would be to acknowledge that she had been wrong to leave; had sacrificed security, position, and Digby, her son, in vain. Nevertheless she bowed to her mother's wish and went in search of someone to perform the rite. It did not occur to her to pretend belief for the occasion; mindful as she was of the priest's spiritual integrity. Though they knew her well first one clergyman refused, then another.

Failure did nothing to persuade Emily to relent. Despairing, Annie resolved to approach the one person who, by repute, offered a slender hope of acceding to her request. She knew he would at least receive her sympathetically. She set out for the deanery at Westminster. There, according to her *Autobiographical Sketches*, she followed the servant upstairs with a sinking heart, was left for a moment alone in the library, and then the dean came in. She said she never in her life felt so uncomfortable as when Arthur Stanley gazed at her, waiting for her to speak. Clumsily she blurted out that she was not a Christian; that her mother was dying, and was fretting to take the sacrament but would not take it unless Annie took it too; that two clergymen had refused to allow her to take it so she had come to him, feeling how great was the intrusion. Stanley came at once to her rescue. Of course he would come to see her mother, and if Annie would not mind talking over her position with him he felt he might see his way to doing as her mother wished.[17]

Having talked to both of them he administered the sacrament. By receiving it Annie made a concession to her mother she had not granted to her husband. Did her mind go back to the time at Sibsey when, still a believer, she had watched Frank do the same for a dying mother and *her* daughter? Stanley told Mrs Wood she should not distress herself over Annie's heresy, since her daughter's heart was set upon truth. The honest search for truth would never be displeasing in God's eyes, Arthur Stanley said. His action sprang from the fact that he regarded charity as supreme, taking precedence over all else. He was an ultra-liberal, the broadest of Broad Churchmen, whose conduct often shocked his stricter colleagues, and for which he was bitterly attacked. It is likely that he would have preferred his intervention to remain private, it was so very controversial.

Annie Besant published her version of it in 1884, three years after the dean's death. By that time she was one of the leaders of free-thought, in open conflict with the Church, delighting to tease and embarrass.[18] Satisfied that it was in keeping with his subject's character, Stanley's first biographer, Rowland Prothero,[19] quoted her version at length, having found no trace of it among the dean's own letters and papers.

That Stanley administered the sacrament to a non-believer was deeply shocking to clergy and laity alike. For example, Henry Liddon, canon of St Paul's, expressed outrage when told the story by Annie's close friend, the journalist, W. T. Stead, on one of their regular walks together along the Embankment.[20] Liddon had been pained by Stanley's boldness in the past. Of a sermon by the dean, delivered from his own pulpit in St Paul's, his diary remarked, 'calculated to unsettle young men, I should fear, very seriously'.[21] But Stanley's response to Annie Besant was, Liddon felt, much worse.

Annie's account drew the dean even further into the open. With views as broad as his, she asked, innocently, on his second visit to her mother, how did Dr Stanley find it possible to remain in communion with the Church of England? It was the kind of provocative question Voysey excelled in posing, but Stanley answered 'gently'. He felt he could be of more service by staying to widen the Church's boundaries from within than if he left it.

As dean he could make Westminster Abbey of a wider service to the nation than would otherwise be possible.[22] Annie remained devoted to Stanley's memory and came to look on him as if he had been her mentor; the second person after Thomas Scott who had stressed the high importance of the quest for 'Truth'.

When Dean Stanley's work was done the doctor gave permission for Annie to take her mother to Colby Road. It was argued that the purer air of Upper Norwood—in those days a hilly, country place—would do the invalid good. What Henry thought is not recorded, but Annie rejoiced at having her mother to herself. Shortly after her arrival Emily suffered another attack. Though she lingered two days she died in the evening of 7 May, aged only 58.[23] Frequently, during her last hours, she expressed regret at leaving Annie all alone, lamenting the course of events that had brought her daughter to this pass. 'My little one... you have always been too religious', Emily declared, 'It has been darling Annie's only fault; she has always been too religious.'[24]

Emily was buried in Kensal Green Cemetery beside her husband and the baby, Alf. Though smothered under the sentimental language of the day—all three were 'sleeping in chill earth damp with the rains of spring'[25]—Annie's grief was bitter. As a young, attractive, unsupported woman without her mother's presence at Colby Road, Annie was desperately vulnerable to malicious gossip.

The next two months were hard. Annie's first care was to see that Mabel, and Mary, the maid, had enough to eat, which often meant going without herself. Mary proved to be worth far more than her keep, a wonderful contriver in domestic matters who kept the house bright and welcoming.

In July Mabel was sent to Sibsey to spend a month with her father in accordance with the Deed of Separation. Frank Besant engaged a Miss Louisa Everitt, who kept a boarding school for girls in Boston, to look after Mabel at the vicarage for that period.[26] The little girl was subjected to inspection to see how the separation had affected her behaviour. All appeared in order; Mabel repeated the Lord's Prayer night and morning and looked forward to being taken to church. But when Digby's turn came to spend a month with his mother he was upset at seeing a half-forgotten figure.[27] That, and his great distress when the moment came to part from her again, was more than Annie could bear.

She tried to lose herself in work. Each time she left Colby Road to travel up to Bloomsbury she pretended she was having 'dinner in town'. Like many an impecunious intellectual before and since, her refuge was the Reading Room of the British Museum. There, under the great blue and gold dome, she was, if permanently hungry, dry at least while studying for the new series of pamphlets commissioned by Thomas Scott. He paid her valuable guineas for them. He and his wife were unfailingly kind; they did not let two days go by without sending to see if she was all right, and frequently invited her to dinner.

Annie's deliberations in the Reading Room produced pamphlets on *Inspiration*, *The Atonement*, *Mediation and Salvation*, *Eternal Torture*, and *The Religious Education of Children*. Unlike her first two efforts these were signed—A. Besant, or Annie Besant. In them she declared herself a theist. She was seeing a good deal of the Conways at the time and acknowledged the Virginian's influence in widening her views on the question of a divine existence.

But she did not achieve the same freedom of thought and approach to theological matters that Conway had taught himself to enjoy. As the fact of her mother's death was borne in upon her her spiritual life received a check. She ceased to pray, ending the habit of a lifetime. Prayer, which had been her means of going forward, now seemed a blasphemous absurdity. Like any mystic Annie needed an act of constant renewal in order to sustain belief; she could not neglect this without losing contact with the object of her faith. When she did not pray she could not feel the presence of God. If she did not feel she found it very hard to believe.

It was typical of Annie that she did not keep these perilous doubts to herself, but aired them at a meeting of the Liberal Social Union, a group she had joined in the hope of alleviating her loneliness. She was surprised and dismayed to find herself cold-shouldered. Scott, on the other hand, was pleased by news of her altered state of mind; he had seen it coming, he said. She asked him if she might analyse her position in a new pamphlet. 'Write away little lady', her memoirs had him say, 'write away.'[28]

In late July she enlisted his aid to advance her career in a new and startling direction. Scott was taken aback by the strength of her desire to become a public lecturer. A quasi-theatrical pursuit, lecturing was all the rage, a lucrative way of making money. But those who occupied the platform were almost exclusively male; female lecturers risked losing their respectability. Those who tried it, like Frances Wright in America some forty years before, exposed themselves to unfavourable publicity, even physical abuse.

There was only one woman in England at the time well known for her appearances on a platform, Mrs Harriet Law, one of the leaders of the National Secular Society. She was a farmer's daughter, in florid middle age, genial, loud voiced, perfectly able to deal with the hecklers and roughs who frequented the gatherings she addressed. Scott disapproved of her opinions, which he described as 'unmitigated atheism'. And it seemed to him that a woman was out of place when lecturing. 'She has her sphere, that of household economy and social morality ... She seems to me to unsex herself when she lectures on abstract themes to poorly educated men.'[29]

No doubt Scott, who had conceived an old man's doting love for Annie, felt she would be incapable of unsexing herself. He asked George Holyoake to arrange for her to be tried out as a lecturer.[30] Holyoake's opinion would be valuable: he was the father of secularism, he had spent years in halls up and down the country listening to speakers who were invariably sincere and almost always inescapably solemn. He would know at once if it was worth Annie's while to develop the talent she had stumbled upon in the pulpit at Sibsey. The former concert room of the Princess's Theatre, which had just been taken over by the Cooperative Institute, was booked for the evening of 25 August. The subject of Annie's début was to be 'The Political Status of Women'. While that should sufficiently test her capability it was safely removed from the swampy ground of religious controversy.

But Scott's careful plans suddenly went awry. In the three weeks that remained before 25 August Annie Besant embraced 'unmitigated atheism'. One encounter with its chief apologist, Charles Bradlaugh, and she was gone beyond Scott's reach. Again it was the Conways who proposed the unthinkable, that a lady should go so far as to enter the Hall of Science, where Bradlaugh presided over meetings of the National Secular Society. In those days nothing terrified the English middle class more than Bradlaugh—'brawling, swaggering Bradlaugh'. He stood for atheism, republicanism, Malthusianism, any one of which posed a dire threat to the peace and good order of the kingdom. His reputation even gave bold Annie pause. 'Mr Bradlaugh is rather a rough sort of speaker, is he not?', she asked the Conways.[31] They told her he was the finest speaker of Saxon English in the country, with the exception of John Bright, and his power over crowds was marvellous. Every aspiring orator must hear him.

With the Conways' advice in mind Annie stopped one day on her way home from the British Museum at the bookshop in High Holborn kept by the veteran publisher of radical works Edward Truelove. There she bought a copy of Bradlaugh's paper, the *National Reformer*, and sat reading it on top of the omnibus taking her to Victoria Station. She was amused to see the look of horror on the face of an old gentleman seeing her, a young woman—respectably dressed in crape—with such a production in her hand. The *National Reformer* was the leading atheistic journal and as such a target of attack throughout the country. As, for example, the vicar of Liversedge told a meeting of ratepayers, it was notorious for pulling the Bible to pieces more than any other paper, 'anything more *fearful*, more *dreadful*',[32] he had never seen.

Annie liked its opinions and was interested in the description of the National Secular Society in that issue. She wrote to enquire if anyone might join. Receiving a courteous reply, on Sunday 2 August she presented herself at the Hall of Science in Old Street. She was surprised by what she found.

The hall was crowded to suffocation and at the very moment announced for the lecture a roar of cheering burst forth, a tall figure passed swiftly up the hall to the platform, and, with a slight bow in answer to the greeting Charles Bradlaugh took his seat. I looked at him with interest ... The grave quiet, stern, strong face, the massive head, the keen eyes, the magnificent breadth and height of forehead—was this the man I had heard described as a blatant agitator, an ignorant demagogue?[33]

Annie had feared that the views she was about to hear might not meet with her approval. She was relieved and greatly impressed therefore when Bradlaugh launched into a discourse on the similarity between the Christ and Krishna 'myths', a subject with which she was already familiar. 'I could test the value of his treatment of it and saw that his knowledge was as sound as his language was splendid.'

In fact she was overwhelmed by the figure on the platform. Eloquence, fire, sarcasm, pathos, passion, all these in turn were exercised by Bradlaugh against Christian superstition. It was the first time Annie had recognized the power that dwells in the spoken word, she said. Afterwards Bradlaugh came down into the audience to hand out certificates of membership. Even before she identified herself Bradlaugh recognized her. She was startled by his prescience, being unaware of the unusual figure she presented among that company.

Their first encounter was 'brief, direct and satisfactory'. Annie, who was determined to hold on to this extraordinary phenomenon, sought a further meeting. This was conceded; she was invited to visit Bradlaugh at his lodging in the East End. At 29 Turner Street, between Whitechapel and the Commercial Road, Bradlaugh disposed of two tiny interconnecting rooms whose windows, at the back of the London Hospital, had to be kept closed against smells in summer and drunken midnight cries.[34] His furniture consisted of a child's bedstead, a nursery washstand, and six ramshackle chairs. Books lined the walls except for a space over the bed where hung his most treasured possession, a painting of a boy organ grinder, asleep in a doorway, watched over by a monkey.

Bradlaugh lived there for 3s. 6d. a week, though he earned at least £1,000 a year from lecturing, and had some capital. He had taken refuge in Turner Street five years before when his business affairs had suffered—as countless others did—from the spectacular collapse of the banking firm of Overend Gurney. At the same time he banished his wife Susannah, who was a dipsomaniac, to live with her father in Sussex. Their daughters Alice, aged 18, and Hypatia, 16, took it in turns to look after their mother and spend a month in London as their father's amanuensis.

Annie recoiled from the lack of light, air, and space at Turner Street. She persuaded Bradlaugh to a second meeting at her house in Gypsy Hill, brushing aside his warning that his presence there would scarcely commend her to her neighbours. As a result of these meetings, at which they discussed Annie's crisis of faith,

Bradlaugh arranged to take the chair at the Co-operative Institute meeting on 25 August.

He also advertised it in the *National Reformer*, albeit cautiously, as a 'discussion' to be 'introduced' by Mrs Besant.[35] Somehow news of this reached Sibsey. Appalled that his name should be so used, and in such a paper, Frank Besant rushed to his solicitor to see if Annie could be prevented from her purpose, or at least obliged to appear under a pseudonym.[36] She could not.

On 25 August she was sick with fright before she mounted the platform at the Princess's Theatre. All went well; Holyoake praised the easy self-possession of her delivery. 'She had then the sweet low voice so charming in a woman',[37] he remembered. But Annie's enthusiasm for her chairman grated on Holyoake; relations between him and Bradlaugh were uneasy at the best of times. In recent years Bradlaugh had taken over the secularist leadership from Holyoake, at whose feet he had once sat. Something Annie said offended Holyoake, and Thomas Scott condoled with him. 'I think you have been very forbearing in your conduct to Mrs Besant', Scott wrote. He was himself out of temper with his protégée. 'Mrs Besant is an ambitious woman and when the fit is on her will do and say any mad thing. She is very young, and in all worldly matters, very foolish.'[38]

CHAPTER 7

ICONOCLAST

'PROBABLY the best speech by a woman to which we have ever listened', the *National Reformer* proclaimed on 30 August in its report of Annie's début. Obviously relieved, it promoted the occasion from a discussion to a full-scale lecture. 'Her first public speech and marvellously successful.'

The eulogist was Bradlaugh who was determined to engage this promising ingénue for his own production. 'Mrs Besant est une femme très intelligente que j'espère de voir sur notre platforme dans l'avenir', he explained in his regular letter to his daughters (in French, because Alice and Hypatia were to have all the advantages he had lacked).[2] 'Bien élevé [sic] et sa santé est assez forte, capable de faire une grande sensation comme auteur et orateur.' And as his anticipation ran ahead of his pen, Bradlaugh went on in his own language, 'If she stays with us poor Mrs Law may say her prayers.'

Bradlaugh's brutal frankness was a measure of his need. Harriet Law had shared many a platform with him over the years. But she must give way to the newcomer, whose exceptional talent, youth, and good looks, combined with the unshakeable air of a gentlewoman, he was convinced would draw the crowds. Annie's ability as a writer, demonstrated in the pamphlets published by Thomas Scott, was an extra advantage. Bradlaugh decided that she must work for him on the *National Reformer*, his chief means of communication with his followers in the National Secular Society.

When Annie first encountered him Bradlaugh was engaged in rebuilding the membership of the Society which he had founded in 1866.[3] In 1871 he had been obliged by overwork, ill health, and pressure of debt to resign the Presidency. Left in other hands the Society had virtually collapsed. Bradlaugh knew that he must return to the United States in the autumn of 1874, to resume a lecture tour which had been cut short in the spring by his hasty return to England for the parliamentary general election, when he had been an unsuccessful candidate for the industrial town of Northampton. Lecturing in America was the only means he had of earning a large sum of money in the shortest possible time.

Unfortunately his printer, publisher, and right-hand man Austin Holyoake had died in the spring. The burden of grief at the loss of a very dear friend was intensified by the strain of office. After a full week's work in London Bradlaugh frequently travelled down on Saturday night to whatever association of free-thinkers had engaged him, spoke on Sunday—often twice—and travelled back the same night. At Turner Street the empty fire grate was invariably filled with the shreds of letters he sat up to answer. The lectures were exhausting in themselves; the audience never guessed how, afterwards, behind the scenes, Bradlaugh would collapse, panting and perspiring, into a chair.

'The Bradlaugh', as he called himself (not always in fun),[4] exuded an enormous sense of purpose. His massive figure lent itself to ceremonial; his natural appurtenances proposed themselves as wigs, seals, chains of office, robes. He was introduced to his American audiences as the first President of the English Republic.[5] He did not claim that office, but he was confident that a Republic would come about—constitutionally—in under twenty years. In 1871 he had publicly predicted that the prince of Wales would never sit upon the throne.

The storm this raised reflected the apprehension he inspired in Queen Victoria's subjects. The Irish journalist T. P. O'Connor (known to many as Tay Pay) vividly conveyed this sense.

A true description of him . . . recalls the portraits of the leaders of the French Revolution. The burning words in which Taine draws his picture of Danton would also fit Bradlaugh . . . Imagine . . . a huge creature, some six feet one or two high, with immense shoulders, broad, strong, high, massive, with arms that, even in clothes looked like those of a great bruiser, conjure up a vision of a man who looked at once like a coal heaver or a pugilist, and a great thinker and a protagonist in the fight of ideas.[6]

Bradlaugh's face was a strange contradiction, O'Connor thought. Long, with huge heavy jaws, like the mask of a Roman emperor, except for the short almost snub little nose that seemed to end unwillingly. The eyes were Bradlaugh's most remarkable feature: large, protuberant, of an expressive grey-blue, they had depths of brilliancy, of passion, of menace that were calculated somehow or other to make you feel as if they could freeze your blood with terror. 'Imagine this man with that heavy jaw, with that long upper lip, with those terrible eyes, master of the guillotine, and you could think of him uttering such a sentence as that of Danton; "Europe has declared war against France. France has answered by throwing the head of a King at the feet of Europe." '[7]

With a caution Thomas Scott must have approved (and may have suggested), before she accepted Bradlaugh's offer of a job, Annie asked advice from Austin Holyoake's elder brother George. 'I said Mr Bradlaugh had considerable capacity and was capable of great usefulness', was Holyoake's somewhat cool assessment.[8] He warned Annie that Bradlaugh turned on those who dared to disagree with him. She was not dismayed; it is hard to imagine what might have dissuaded her from accepting Bradlaugh's offer. The salary—a guinea a week[9]—was the least of it. To be associated with Bradlaugh, both as his employee on the *National Reformer*—he was editor and proprietor—and as a member of the National Secular Society of which he was the President, was to be translated in one bound from deprivation and despair to shining fortune. Now she had within her grasp things she had craved: occupation for her restless intellect, adventure, camaraderie. Above all she recognized in Bradlaugh, as others did, a vaulting ambition that surely would be satisfied.

An inflexible will, ferocious appetite for work, intelligence as quick as hers, and a rampant egotism—these were qualities already tested in Charles Bradlaugh when Annie met him: he radiated force. He had Cromwell's 'berserker' temper, his friend John Mackinnon Robertson wrote;[10] when that took hold of him he would thunder into a crowd, knock heads together, and flail his heavy arms about to the detriment of all who got within their prodigious reach.

Born in 1833, 'the lad', as he was affectionately called by those who had supported him in his early days as a street corner speaker, was probably the most remarkable figure to emerge from the working class in the nineteenth century. By 1874 he had already accomplished much, and almost single-handed. But the great effort of his life was still to come;[11] there would be a part in it for Annie Besant. He was a child of the East End, born in Hoxton, an area of artisans' dwellings and railway junctions lying between Finsbury and Bethnal Green. His father, a solicitor's clerk, was by no means among the poorest of the inhabitants, but Charles, as the eldest of five children, had to go out to work as an errand boy by the age of 12. His education continued by his own effort, and at Sunday school.

In 1848 an incident occurred which led to his embracing free-thought. The Sunday school to which he went was conducted according to the Lancastrian method whereby the brighter pupils passed on lessons learned by rote to their fellows. Bradlaugh was 15 and had become a monitor. He was put to examining the Gospels so that he might pass on his knowledge to members of a confirmation class. Like Annie Besant, Bradlaugh found discrepancies between the versions of Matthew, Mark, Luke, and John and reported them—as no more than interesting—to the superintendent, the Revd Packer. His reaction was to suspend Charles as a monitor. Given the general horror of 'infidelity' that might have been a reasonable response if it had stopped there. But Packer proved a nervous bigot; he forbade the boy to enter church for three months, while complaining to his father that he showed dangerous leanings towards atheism.

With his Sundays free, Bradlaugh took to attending Bonner's Fields, a famous open-air meeting place in Bethnal Green. There the doubts that had arisen in his own mind as the result of deduction from evidence—the process that was to shape his life—were reinforced by listening to the various forms of unbelief put forward by a wide range of speakers. Two years previously an English audience had gained access to the German school of biblical criticism through George Eliot's translation of Strauss's *Life of Jesus*. This gave renewed impetus to an older indigenous stream of free-thought descending from Godwin and Paine through Charles Hennell to George Holyoake. It had been made to run more freely by courageous publishers like Richard Carlile, Henry Hetherington, James Watson, and Edward Truelove. Though these works, and the speeches at Bonner's Fields, were primarily directed against religion, the general public in Queen Victoria's time read a sinister political purpose into them. 'The absurd doctrine that the French Revolution was brought about by religious freethought', John Mackinnon Robertson, himself a free-thinker, wrote, 'was ... devoutly believed by the mass of nominally educated men ... every freethinking book was a menace to the social order.'[12] So Charles's father thought, and forbade his son the house.

Almost penniless he took refuge with Eliza Sharples, who had been the common law wife of Richard Carlile. He had served nine years in prison for selling, among other works, Paine's *The Age of Reason* and *The Rights of Man*, and Shelley's *Queen Mab*. Bradlaugh therefore found himself admitted to the very hearth of infidelity.

He began to lecture for the cause and made his friends among its supporters, especially Austin Holyoake. But his efforts to set up in business did not prosper; his views were against him: as one customer remarked, a smell of fire and brimstone hung about him. In December 1850, harassed by debts, he accepted the Army's bounty as the only way to pay them off. He was posted to Ireland as a private in the 7th Princess Royal's Dragoon Guards.

He had to wear his lecturing clothes on board the ship taking the regiment to Dublin; he had no others. The decayed black suit and silk stovepipe hat made him fair game among his colleagues. They thought him odd when they kicked open his box to find it contained a copy of Euclid, a Greek lexicon, and an Arabic dictionary. Odder still was his alacrity in preaching teetotalism, a hopeless cause in that company. But Bradlaugh never gave up a principle from fear of incurring displeasure, much less ridicule. His companions stopped persecuting him when they found that he would fight, and that he could win. Besides their respect he shortly earned their gratitude when he began to apply Army regulations for his, and their, benefit. He was already aware of the power knowledge conferred; the Army gave him a chance to practise it. A mere private, he won privileges for himself and other ranks through a study of the rule book, and got the better of his officers.

He did not enjoy the kind of soldiering he was called upon to do in Ireland. He had to witness scenes far worse than the horrors of war; when deliberate cruelty was visited upon the people in the name of law. He had to stand by while bailiffs evicted starving families and demolished the mud houses to prevent them creeping back.

In 1855 a legacy enabled Bradlaugh's mother to buy him out of the Army. Back home in London he passed up jobs because of his strong and apparently hopeless attraction for the law. At last he came in the way of Thomas Rogers, a solicitor with an office in Fenchurch Street. When Bradlaugh asked him for a job Rogers could only offer something any other dashing young Guardsman would have rejected out of hand. Bradlaugh jumped at it and became once more an errand boy. The time was, fortunately, ripe. The Common Law Act of 1854 widened the jurisdiction of the courts,[13] leading to an avalanche of litigation. So great was the pressure of work in Rogers's office that Bradlaugh was given duties that properly belonged to a solicitor's clerk. Within nine months he was in charge of the firm's common law business; he was 21.

Over the next seven or eight years court work gave Bradlaugh a remarkable knowledge of the law. Its adversarial nature was eminently suited to his love of battle. Its methods answered to his grasp of detail, while its procedure gratified his need, not only to be right but to be seen to be right. At the same time he resumed his personal crusade against religion, speaking, and writing in Holyoake's paper the *Reasoner*. As his employer, Rogers found himself the target of anonymous letters urging him to condemn his clerk's behaviour out of hours. To his credit all Rogers did was to ask Bradlaugh not to let his propaganda become an injury to his business. Bradlaugh complied by adopting the pseudonym that, for a time, became more famous than his real name—Iconoclast.

Soon crowds were flocking to his speeches. Bradlaugh, with his sonorous tones, dramatic gestures, and authoritative presence, was worth going to hear whether one agreed with him or not. By the late 1850s he had become the acknowledged leader of secularism, replacing G. J. Holyoake who, with his 'thin tin kettle' voice[14] and bland manners, had little effect as a speaker.

Bradlaugh's appearance was calculated to inspire confidence and respect; he always dressed in black, set off by an expansive white shirt front—the mark of a gentleman in that sooty age.[15] It was easy to mistake him for a clergyman. 'In some respects he reminds us of a Puritan who has lost his way', the *Northern Echo* wrote. 'The apostolic zeal, the vehement impatience with false doctrine, the abiding faith in great principles, the assertion of his doctrine in season and out of season, the eagerness with which he seizes any opportunity to proclaim the faith that is in him—are all characteristic of the Puritan.'[16]

Two of the great principles stated by Iconoclast very early in his career had nothing to do with secularism. The first was the need to limit the size of families, the second the right to political democracy. The first—neo-Malthusianism as it was called by the few who could bring themselves to mention it—was put forward by Bradlaugh in a pamphlet published in 1861 called *Jesus, Shelley and Malthus*. In it he called for a recognition of the truth of Malthus's argument, that the world risked running out of food for its ever increasing population, and a discussion of the means of preventing this from happening. His colleagues were alarmed; the subject was as far from 'respectability' as it was possible to go.

The idea of political democracy went down more easily, though the form that Bradlaugh personally advocated still did not. Republicanism was hard for Englishmen to swallow, even though the existing monarchical order was represented by figures variously sinister—in France, Napoleon III: outrageous; in Sicily, Ferdinand II (King Bomba): ridiculous and expensive; in England the Hanoverian uncles of the queen, and her children by Prince Albert. They were the subject of a caustic, sneering attack in Bradlaugh's most notorious work, published in 1871, *The Impeachment of the House of Brunswick*. 'I loathe these small German breast bestarred wanderers, whose only merit is their loving hatred of one another. In their own land they vegetate and wither unnoticed; here we pay them highly to marry and perpetuate a pauper prince race.'

Bradlaugh attached no theory to his concept of democracy; political dogma was as displeasing to him as the religious kind; he was in all things an individualist. 'I am in favour of the establishment of a National Party which shall destroy the system of government by aristocratic families', he wrote in his first attempt to get into Parliament in 1868. The poorest members of the community should have fair play (and no more than that) in their endeavour to become statesmen and leaders.

Democratic progress in England in the 1860s was impeded by the wrangle over the terms of a new Reform Bill. Bradlaugh did his best to keep things on the move, becoming a prominent member of the National Reform League, devoting fees from lectures to the cause, and taking part in public demonstrations in favour of a wider franchise. His presumption in putting himself forward at Northampton, at the election which swept the Liberals into power in 1868, shocked many people who liked to think themselves broadminded. Even Arthur Stanley could not abide the thought of Iconoclast in the House: he used his pulpit in Westminster Abbey to denounce Bradlaugh's candidature.[17]

Bradlaugh had evidence from his lecturing as to how anxious the people were for the extension of the franchise; he was convinced they would flock to secularism once they had had it explained to them. When the National Secular Society was founded in 1866 it seemed likely that he would shortly be proved right. The audiences at his lectures were numbered in thousands; the halls were full to overflowing, the circulation of the *National Reformer* rose. Yet the activists were very few. Bradlaugh stood almost alone at the head of the movement. A small band of enthusiasts, among them the late Austin Holyoake and another lifelong free-thinker, the Secretary of the NSS, Charles Watts, helped to run the paper and manage the Hall in Old Street where Bradlaugh's London lectures took place. Middle-class liberals were nowhere to be seen. 'The Bradlaughites met in Halls of Science, on public commons, under railway arches and at street corners with the design of lacerating the feelings of their opponents by allusions to disgusting Biblical subjects', one hostile observer complained.[18] Female secularists were conspicuous by their absence; all the more reason why Bradlaugh welcomed Annie Besant.

He was quick to instruct her in the responsibilities of her new position. Like him, she was to serve the readers of the *National Reformer*: Bradlaugh took immense pride in putting to use the knowledge of law and other matters he had laboriously acquired. Like his, her education was far superior to that of the great majority of secularists, a fact which imposed a serious obligation on both of them. It was as easy to earn indiscriminate praise from secularists as it was to provoke indiscriminate hatred from Christians, Bradlaugh told Annie. Like him, she must be her own harshest judge; he and she must be sure they knew thoroughly every subject they proposed to teach.[19] Self-examination was the first duty Bradlaugh had commended to his daughters;[20] a rule that governed his actions as well as theirs. It formed the basis of his adamantine self-possession. Bradlaugh sceptical? There was no one more cocksure than he, one adversary cried.[21]

Annie's first appearance in the *National Reformer* was less than a month after she had encountered Charles Bradlaugh—on 30 August 1874. Though weekly journalism was a strenuous and exacting occupation, Annie found she had the necessary gifts. Reading was as natural to her as breathing, and she remembered all she read. Thanks to Miss Marryat her general knowledge was already wide and she knew how to look things up. She had the born journalist's ability to go straight to the heart of a matter, and a facility of exposition a barrister might have envied. 'Though not an original thinker', Hypatia Bradlaugh wrote, 'she had a really wonderful power of absorbing the thoughts of others, of blending them, and of transmuting them into glowing language.'[22]

One thing only Annie lacked which, if she had been able to deploy it, might have turned her into a national heroine—wit. Irony and sarcasm had to serve instead to point up the long paragraphs signed by her pseudonym, Ajax. This was to spare Thomas Scott, for whom she was still writing under her own name, the obloquy of an association with the *National Reformer*. The name occurred to her because a copy of the famous statue was at Crystal Palace near her house in Gypsy Hill. In the *Iliad* Ajax cried for light, though it might have delivered him to his enemies. That cry appealed to Annie's sense of the dramatic; it underlined the portentous nature of the events to which she now addressed herself. She chose the title Daybreak for her regular column, and in the first of them called her readers to attention.

There is a strange interest in standing on a mountain top, watching for the first faint signs of the coming day. Here and there a tiny white cloud warms into soft yellow ... here and there a gleam flashes across the sky ... But a deeper keener interest swells the heart of those who are watching for the rising sun of Liberty: each ray, be it ever so feeble, each tiny gem of colour in the dull, grey sky, brings to the foremost soldiers of the army of Freedom a message of hope ... The series of papers of which this is the first, will weekly point out these signs of the coming day; whether they betoken freedom of thought, or freedom of action they are dear to us as signals of that coming reign of Liberty, when men shall dare to think for themselves in theology, and to act for themselves in politics.[23]

CHAPTER 8

THE CRY FOR LIGHT

PEOPLE complained of ambition in Annie Besant as if it threatened them.[1] Often it did, particularly as she grew older, but while she was still so young the trait that captured their attention was, rather, an uncalculating eagerness for experience, a desire not to disappoint expectation. It was a most attractively dashing quality that proved valuable to the *National Reformer*, where versatility had to compensate for lack of staff. Annie had barely a month to find her touch as a columnist before she was tumbled into the roughest job of all, as a political reporter.

In September one of the two sitting Members of Parliament for Northampton died suddenly. A by-election was called. Bradlaugh, who had expected to be on his way back to America at the beginning of October to resume his lecture tour, had to postpone his passage in order to stand. Charles Watts, who normally sub-edited the *National Reformer*, was called upon to organize Bradlaugh's campaign. Annie Besant was asked to join the party to do the day-to-day reporting for the paper.

So Ajax set the scene in the 4 October issue with a vivid description of the town, not forgetting the odour of leather everywhere perceptible.[2] Annie was shocked at the sight of the small, crowded, ill-built houses where the Northampton cobblers worked, shut in for up to sixteen hours a day to earn no more than 30*s*. a week. Their support for Bradlaugh at the general election, though not sufficient to return him, had shown a heartening increase over his first appearance in 1868. This time he was really hopeful of getting in.[3] A radical, he expected the Whigs to combine with his supporters to defeat their mutual enemies, the Tories. As it turned out he was foolish to suppose they would so readily accept an infidel. After a good deal of politicking the Whigs put up a Liberal candidate to make a three-cornered fight. Not content with that, the bigots among them descended to scurrility, on the grounds that an atheist was capable of whatever scandal they cared to invent. Specifically they charged Bradlaugh with not being legally married and with having deserted his family.

Bradlaugh had been married in church in 1854 because his wife Susannah had wanted it.[4] But he scorned to say so, or to explain the reason why he did not live with her. Ajax reported that Bradlaugh's radical supporters had been driven 'almost to madness' by these taunts;[5] they could not endure to hear slander against their candidate. Sporadic fighting broke out whenever they were mocked by Whig supporters.

On 6 October the Tory won the poll, Bradlaugh coming bottom once again, having collected an extra 113 votes. Annie was sitting with him in the George Hotel, both of them worn out, when the landlord rushed in crying that Bradlaugh's supporters were attacking the Whigs at their headquarters, the Lord Palmerston. Bradlaugh quickly made his way there to rescue those who had so unjustly maligned him. 'He charged alone right through the mob, thrashed one or two and thundered them all back from his rival's windows',[6] Ajax reported. At 9 o'clock that evening he had to leave to catch the boat at Queenstown for America. Left at her post Annie heartily regretted his departure. 'We chose the name Ajax because of his celebrated cry for light, and not with any wish to assume warlike qualifications', she commented ruefully in Daybreak. 'To war with the pen is one thing but stone throwing *n'est pas mon métier*.' Nevertheless that night she went out bravely into the town where the 'Bradlaugh riots' were once more in full swing. The damage was extensive; soldiers had to be called out.

Annie's exertions at Northampton helped to bring on an attack of congestion of the lungs. When she recovered she decided to move into Town to save herself the daily journey by train and omnibus to the *Reformer* offices just off Fleet Street. She arranged to take a furnished apartment in a tall, thin house in Bayswater: 19 Westbourne Park Terrace, which she shared with two other ladies.[8] Mary did not accompany her, but a maid was always on hand to look after Mabel. The house was in a reputable neighbourhood, two minutes from Hyde Park.

In Bradlaugh's absence in America, Charles Watts, who normally supervised the production of the *National Reformer*, became its acting editor. The son of a Wesleyan minister, he had embraced free-thought at the age of 16. Now 41, hard work and above average ability as a lecturer had made him one of its leading figures. By trade a printer, he was also *de facto* publisher to the secularist movement. 'Mrs Besant is exceedingly good altogether',[9] was his verdict that autumn.

Her work was certainly varied. Daybreak was for teasing the opponents of free-thought—be they Anglican celebrities or Nonconformist divines. In her first column Annie used her anonymity to chastise the bishop of Lincoln for his habit of pronouncing in public on all kinds of unrelated subjects. Christopher Wordsworth was a scholar and a disciplinarian whose presence had loomed uncomfortably large during her dispute with Frank Besant. He was an invaluable man, Ajax jeered; few helped forward the cause of free-thought as Bishop Wordsworth did: 'he will not be quiet, he is always in a fuss.'[10]

Next she took Joseph Arch to task for refusing Bradlaugh's offer of help in the strike of the National Agricultural Labourers' Union. Land reform and related matters greatly interested Bradlaugh. Besant concurred: 'the Land laws need a touch but it is the touch of a bigger hand than that of Joseph Arch', she told her readers. In Bradlaugh's absence she put her shoulder to the wheel with a fluent account of the state of the nation's agriculture that took three issues of the *National Reformer* to complete. It drew heavily on the personal experiences of the vicar of Sibsey's wife, though neither the place nor the person was identified.

Praise was given where Ajax decided it was due. 'Dean Stanley often gives us a sign of Daybreak in his church; his courtesy towards his opponents, his growing liberality of sentiment as well as his broad doctrine and deep learning entitle the Broad Church Dean to a respectful hearing at our hands.'[12]

Annie's second appearance on a public platform was arranged by Moncure Conway, who invited her to lecture at his Institute on 'The True Basis of Morality'. A few weeks later, as chairman of a meeting of the Dialectical Society, he called on her again, though she was a very new member, to lead a discussion of a lecture by Dr C. M. Davies, 'The Poles of Religious Opinion'. This was virtually a second audition. A good notice from Davies would greatly assist Annie as a public lecturer. His articles on the world beyond the churches, a phrase he coined, were very widely read. A clergyman turned journalist, he had once been a fervent Tractarian, but reneged on the movement in a series of sensational novels.[13] His articles, however, were sober enough for the *Daily Telegraph* to print. Collected into books with titles like *Heterodox London*, *Unorthodox London*, *Mystic London*, and—when all else was exhausted—*Orthodox London*, they provided glimpses of religious happenings which their readers would not venture to attend themselves.

The Revd Dr Davies was present the next time Annie lectured, on 19 January 1875, at South Place Chapel. It had just been refurbished. Conway presided, sharing the platform with the speaker and a grand piano covered with green baize, behind a red mahogany rostrum, 'somewhat reminiscent of a grocer's counter', Davies remarked.[14]

The occasion signified a new departure for Annie. After much thought and self-analysis (instead of prayer), she had decided to give herself wholly to propagandist work as a free-thinker and a social reformer, 'to use my tongue as well as my pen in the struggle'.[15] Charles Watts signalled her new status in the *National Reformer*. Its announcement of her lecture revealed that she was none other than the celebrated Ajax. The lecture, on 'Civil and Religious Liberty', with reference to the events of the French Revolution, went exceedingly well. Davies waxed enthusiastic in the *Sun* but, to Annie's fury, suppressed her name, having been got at by the ubiquitous Walter Besant.[16] (It does not seem to have occurred to anyone, including Annie, that she might have lectured under her maiden name.)

The decision to nail Besant to the new masthead had one consequence that must have been regretted, at least by Thomas Scott. Her newly publicized connection with the *National Reformer* obliged him to issue the last pamphlet she wrote for him anonymously. When *On the Nature and Existence of God* appeared in 1875, it paid a glowing tribute to Bradlaugh's writing on the same subject. Taking her cue from him Annie argued, not that God did not exist, but that man had not so far discovered any proof of his existence. Yet she had to concede that to worship was a most powerful instinct in man. What could be the explanation? 'If the Deity exists He does not want us to know of His existence', Annie concluded. The reason? That religion was in the hands of meddling priests who produced nothing but mischief and despair. Morality must be detached from religion and formed solely by thought and experience.

The negative aspects of religion had so preoccupied her during the ordeal of the last four years that she had paid scant attention to anything else—science for example. In September 1874 one of her first duties on the *National Reformer* was to make a résumé of the celebrated speech by Professor John Tyndall to the British Association at Belfast. It was a crucial utterance in the long-running popular debate as to which contained the truth of man's origin, science or the biblical version. Annie was struck by the novelty—to her—of Tyndall's exposition of the role of science. The idea that in Nature lay every possibility of intelligence as well as physical life opened up a whole new vista of positive belief which she found infinitely heartening, and on which she prepared to base her developing idea of a new morality.

Thomas Scott sent a copy of *On the Nature and Existence of God* to Gladstone as part of a lengthy discussion they had been having on the efficacy, or otherwise, of prayer. Annie dismissed prayer as arising either from childishness or impertinence. 'Who are we that we should take it upon ourselves to remind Nature of our work or God of His duty?', she wrote. There was only one possible way to pray, she now decided, silently in adoration of the greatness, beauty, and order in the world. Her pamphlet was important, Scott told Gladstone; it forced the discussion of its subject on the attention of those who claimed to be religious guides. 'Every day proves to me that the reading and thinking men and women are now doing for themselves in lightening the darkness that prevails in these subjects is what I had hoped those in authority would have done long ago.'[18] Gladstone, who had sustained the discussion during his tenure of Downing Street without conceding anything to Scott's scepticism, now gently called a halt. Annie's pamphlet was almost the last work Scott published; he was ailing and, like her, his other correspondents were drifting away.

In February, while Bradlaugh was still in America, Annie embarked on her first provincial tour on behalf of the National Secular Society. In Glasgow she found herself for the first time alone in a strange city and had to conquer an impulse to sit down and cry. The temperance hotels she was obliged to stay in for the sake of propriety were grey and grubby—water was provided solely as a substance to drink, she commented bitterly. At Aberdeen she confronted a grimly silent audience of sons of the Covenanters, 'granite like their granite city'. 'I felt I would like to take off my head and throw it at them', Annie wrote, 'if only to break that hard wall.'[19] After twenty dreadful minutes a provocative phrase drew a hiss; Annie retorted; there was a burst of cheering and the granite crumbled. She hugged the memory to herself on the interminable journey back to London in the stuffiness and gloom of a third-class carriage.

On 25 February she stood for the first time as a speaker on the platform Iconoclast had made famous, in the Hall of Science itself. It had been erected by subscription in 1868 and Bradlaugh thought it the most elegant auditorium the party had ever had.[20] The Revd C. M. Davies, on the other hand, told his readers it looked like two railway arches thrown together under a corrugated iron roof.[21] The approach to it was guarded by a hostile encampment where flew the banners of Bradlaugh's implacable foes, the Christian Evidence Society (which, nevertheless, rented the Hall on Tuesdays).

According to a reporter venturing east of Aldgate as if into undiscovered country, the crowd that surged through the narrow entrance passage, where portraits of the speakers were on sale, were dressed in decent black coats and hats—fifty 'billycocks' to one silk 'topper'. These hats, he reassured his readers, were neither too glossy nor disagreeably napless, their wearers resembled elders and deaconesses of a very staid and stiff persuasion; a gathering of Baptists, for example, eagerly distributing leaflets and tracts. Tracts against tracts, so that the Hall of Science became 'an infidel tabernacle expressing all that is horrid and hideous in politics and religion as politics and religion are understood in Mayfair'.[22]

About one-third of the audience were usually women, though that number diminished whenever Mrs Besant spoke. 'Hers is an essentially womanly mind', the same (male) observer wrote, 'but women even when they are Freethinkers prefer to be led by a mind that is masculine.' T. P. O'Connor was scarcely aware of Annie's mind.

What a beautiful and attractive and irresistible creature she was then [he wrote] with her slight but full and well shaped figure, her dark hair, her finely chiselled features, her eyes with their impenetrable depths of emotion and thought, with that short upper lip that seemed always a pout, softening the intensity of the look and making her a mixture at once of soft and warm and intellectual femininity.[23]

The dress she wore to lecture heightened the effect that smote Tay Pay. It was simple, softly sweeping, black with an edge of white around the high collar that framed her face; it was the sort of dress Mary, Queen of Scots might have chosen on going to her execution. And, unusually, Annie's hair was short, a delightful cluster of curls on her erect, small head.

Annie's portrait in her lecturing dress went on sale and could be had in various sizes, singly and in dozens. Her past was made known to the membership, who responded sympathetically. As a young man, the Positivist Malcolm Quin remembered seeing her at her first appearance in Leicester, a town noted for its adherence to free-thought. That was in the early summer of 1875.

'Perhaps she was too unfailingly earnest', Quin wrote, 'and the relieving touches of wit and humour seldom had a place in her discourses.'[24] But her solemn demeanour seemed appropriate to him. The audience had had it impressed on them that Mrs Besant was a martyr for truth and liberty, having won her freedom from domestic and clerical oppression at the price of social proscription and isolation.

It was at that same meeting at Leicester that Annie was confronted by the prospect of a far more painful martyrdom than she had suffered so far, one to be inflicted by unfavourable propaganda which she would be powerless to refute. In the discussion that followed the lecture an irate member of the audience, whose Christian feelings she thought she had wounded, got up to berate her. But instead he accused her of supporting a book that she had never heard of—*The Elements of Social Science*.[25] Annie heard herself accused of copying Bradlaugh in endorsing free love as it was advocated in the book which was notorious as 'the Bible of the brothel'. For once defenceless she could only respond that Bradlaugh, who was famous among secularists for his strait-laced ways, was the last person to approve any such thing.

Upon investigation Annie discovered that she was caught in a dangerous situation created by Bradlaugh. The book, of which the original title was *Physical, Sexual and Natural Religion*, was the work of George Drysdale, a 28-year-old medical student. He had felt a call to instruct the world in the errors of human behaviour with regard to sex. The book was published anonymously by Edward Truelove in 1861, at which time Bradlaugh had reviewed it in the *National Reformer*. He praised its honesty of purpose and commended its exposition of the Revd Malthus's doctrine on population. Apparently he quite overlooked part II, 'Sexual Religion'. In this the Bible was denounced for its pernicious doctrine of mortification of the flesh, and Christianity accused of inhibiting the right development of women. As no organ in the body could be healthy unless it was regularly exercised, sexual abstinence must lead to mental and physical decay. Preventive sexual intercourse was the only hope, young Drysdale claimed, especially for boys and girls, and outside marriage. That institution spoiled the enjoyment of youth, and degraded women.

Bradlaugh's enemies pounced on his failure to condemn these deeply shocking ideas as evidence of looseness in his morals. The issue would not go away, and Bradlaugh's opponents were only too pleased to extend their censure to the woman of uncertain status who, in association with him, was displaying herself on public platforms up and down the country.[26]

Annie may not have recognized the danger to her reputation; she did little to defend it. Her best protection was a kind of wilful innocence arising from impulsiveness, self-absorption, and a lack of judgement where other people were concerned. This appeared at its most remarkable in her dealings with the opposite sex. At the age of 28, having suffered an unkind marriage, and borne two children, in her writing as in her private life Annie proclaimed the same idealized view of love that had betrayed her adolescent self. This was in contrast to the sensuality she conveyed on the platform. Annie appeared to be as ignorant of this effect as, before her engagement, she was of sexually teasing Frank Besant.[27] Now she insisted on praising courtly love represented, for example, by those dreamy youths and maidens in paintings by Pre-Raphaelites like Dante Gabriel Rossetti and Holman Hunt. Their fashionable romanticism afflicted her prose; it was studded with pseudo-medieval expressions like 'methought' and 'yclept', so embarrassing to the modern ear.

Reading between the lines it is clear that Annie desired to attract Charles Bradlaugh in this way. During his absence in

America she composed a series of articles on the French Positivist philosopher Auguste Comte.[28] These drew attention to his unusual relationship with Madame Clothilde de Vaux. Comte's exaggerated worship of his dead companion (whom he had known in life for only a year) struck most observers as unfortunate, if not macabre. Annie defended it with a vehemence which sprang from personal conviction. She compared their passion in its 'purity' and 'reverence' with that of Dante for Beatrice, or Petrarch for his Laura. 'Those who are too base to believe in a true and noble friendship between a man and a woman will alone try to cast any slur on the frank and noble love which bound these two souls', she claimed. When the articles appeared in the *National Reformer* some of her audience inevitably replaced Comte with Bradlaugh.

Comte's private life influenced his achievement to an unusual extent, Annie declared, thereby prolonging the comparison. Inhibited by an unhappy marriage, his genius rejoiced in the appearance of a perfect soul mate. As Comte himself said, without Clothilde de Vaux, he would never have been able to add the career of St Paul to that of Aristotle by founding his universal religion on sound philosophy, after having drawn that philosophy from true science. But Annie, who had investigated Positivism in the despairing period after her mother's death, was not prepared to grant its author so high a place in the history of ideas. She admired Comte's method, his habit of mind which, through the progressive study of phenomena, brought order to the natural world, making of science an organic whole—that was good, that must be studied in order to check the slide to anarchy. She was also beguiled by Comte's view of woman as angel, a being whose spirituality was higher than that of man. But she rejected the Positivist ideal state. Too much emphasis was laid on rules to govern it, and on the governors, who were to be Catholic priests—albeit much reformed. The scheme was noble in scope but childish in detail, grand in aspiration and puerile in its petty direction, Annie declared. 'Positivism is the death of individualism and therefore of all liberty and life.'

It was a conclusion she confidently expected Bradlaugh, the arch-individualist, to applaud. Her thoughts had dwelt on him during his long absence in the United States; left alone she continued the habit of intensive brooding which had distorted her adolescent view of life. When, in March, his return was imminent, she wrote artlessly to Ellen Conway:

this stupid lad of mine started on February 27 as settled, in the 'City of Brooklyn', Inman steamer; now the White Star 'Adriatic' started *on the same day* and reached Queenstown last Monday morning early. Why didn't he sail in that, like a sensible man; everybody knows that the White Star ships go quickest. I feel very provoked about it, and inclined to cut him when he does come: it is ruining my work, for I *can't* get on: I've struggled in vain, and yet here are some four lectures waiting to be prepared. I can't work so it's no use, and I don't know when I've spent three such uncomfortable days as the last. I do want to see him so much, and it is specially aggravating to know that the other ship is in all this time. I shall soon be in *delirium tremens*, or else shall have to turn Xtian.

Yours ever affectionately,
Annie Besant.[29]

Evidently when she and Bradlaugh met it was in private, for Annie was not betrayed. She had overreached herself: Bradlaugh's affections were engaged elsewhere. Besides the fact that he had a wife, his colleagues in the free-thought movement knew that for the past five years he had sustained a close relationship with a lady they called 'the French countess'. Reportedly a beauty on whom time was just beginning to tell,[30] the charm, distinction, and ultra-fashionable appearance of Madame la Vicomtesse de Brimont Brissac much enlivened the social functions of the National Secular Society which she attended on the arm of its President.

They had become acquainted during the Franco-Prussian War, when Mina de Brimont had tried to enlist Bradlaugh's talent for publicity in order to capture the sympathy of the English working class for the French Republican cause. Madame de Brimont wanted him to denounce the Foreign Secretary Earl Granville, for scheming to replace Napoleon III on the French throne once the invading Germans were defeated. This was as preposterous an excursion into secret diplomacy as Madame's accomplice in it—no less a personage than Prince Jerome Napoleon, head of the imperial family while the emperor of the French was in a German prison: the prince known to all as Plon-plon. In spite of his raffish style of life, which Bradlaugh could not approve, Prince Napoleon's genuine concern for democracy, and his adherence to free-thought, led to a long-standing friendship between the two.

Though there was no suggestion of impropriety, Bradlaugh and Mina de Brimont were closer still. When peace was restored in 1871 she persuaded him to send Alice and Hypatia to school in Paris, where she acted as their guardian. He wrote to her often—affectionate letters full of political discussion, family affairs, and comment on the chaotic business transactions of Prince Plonplon.[31] As soon as he returned from America in 1875 he made arrangements to visit her in Paris.

Annie was devastated. On 4 May she wrote again to Ellen Conway from Glasgow, where she had gone on another lecture tour.

Dear Friend,
I am going to avenge myself on you for feeling very poorly and out of heart. My lectures here on Sunday could not have gone better. I doubt if I ever spoke better than I did in the evening.

The only nuisance is that I feel particularly unwell, which is stupid, as feelings can't be allowed to influence one's work. I go on to Aberdeen tomorrow—it takes from 9 to 4 to get there,—and lecture there Thursday and Friday ... I shall hope to see you on Tuesday next ... I think I shall come late and stay with you—if I may, madam—as I very much need a little chat with you. I have serious thoughts of taking a ticket to the Antipodes, and of living among the Kangaroos, and if my present state of mind lasts long, shall certainly do something of the kind. I am not, just now, in charity with any man, or woman either, and should find a malign pleasure in spiking anybody who would give me the chance. Being in this frame of mind on Sunday last, I *did* spike my opponents and gave the bitterest lecture by far that I ever spoke, going in for bitter irony and savage jokes in a way worthy only of 'Bradlaugh'. I scared my audience somewhat and had no opposition.

Altogether, I don't like things. They all go crooked, and I am beyond measure puzzled and ... worried. If there was any [certainty?] of a beyond it would be almost worth while to take the plunge in order to get out of *this*. If there were a God, how one would hate him for making things in such a muddle.[32]

While Madame de Brimont often figured in Bradlaugh's correspondence with his daughters at this time, he did not mention Mrs Besant to them at all in the letters that have survived from 1875. In August, however, he made public a startling intervention on her behalf. On the 22nd of the month the *National Reformer* began its leader column with an announcement headed 'Personal', whose tone was, to say the least, pugnacious. Bradlaugh told his readers that an attempt had been made to deprive Mrs Besant of her daughter Mabel. After strenuous efforts by lawyers, however, Mabel had been reunited with her mother. But Bradlaugh, who was about to leave for another lengthy tour of America, warned against any further attempt at what he did not hesitate to call kidnapping.

> It was decided against Percy Bysshe Shelley that an Atheist father could not be the guardian of his own children [Bradlaugh wrote]. If this law be appealed to and if anyone dares to impose it we shall contest it step by step and while we [Bradlaugh's 'I'] are out of England we know that in case of any attempt to retake the child by force we may safely leave our new advocate to the protection of the stout arms of our friends... So far as the law courts are concerned we have the most complete confidence in Mr George Henry Lewis and we shall fight the case to the House of Lords if need be.[33]

Lewis was Bradlaugh's solicitor. Though he had not yet achieved the kind of fame that came to him as a result of the famous Bravo murder case in 1876, he was already accorded healthy respect within his own profession.[34] George Henry Lewis's participation in a case guaranteed close attention to detail, unorthodox methods, a comprehensive knowledge of criminal law, and a singular tough-mindedness.

Bradlaugh brought him in to deal with the Revd Frank Besant as soon as he detected signs that Mabel's father hoped to vary the terms of the Deed of Separation which gave Annie custody. Arrangements concerning both children were altered in the summer of 1875. Digby, who was $6\frac{1}{2}$, did not visit his mother; according to Frank Besant she did not want to see him; she feared a repetition of the distress he had suffered the previous year. Mabel's visit to her father, on the other hand, was eagerly awaited at Sibsey, where Miss Louisa Everitt had once again gone into residence. But instead of calling at Westbourne Park Terrace, Frank was obliged to collect the 5-year-old from the Holborn offices of Lewis & Lewis.

As a result of the furore over *The Elements of Social Science* and, generally, of his wife's association with Iconoclast, Frank's concern for Mabel's welfare was now acute. His worst fears were realized the first evening of Mabel's visit. At bedtime, when he said 'Good night, God Bless', Mabel asked him why he said it: 'her Mama had told a servant not to say so.'[35] This in contrast to the previous year when Mabel had shown no sign of Annie's unbelief. Deeply perturbed at what he took to be the result of Bradlaugh's influence, Frank Besant approached the trustees of the Deed of Separation to ask them to make sure Mabel continued to be instructed in religion. One of them, Cornelius Neale, did not reply. The other, Annie's brother Henry Wood, referred him to George Lewis. A second letter to Wood was returned unopened.[36]

Frank sent Mabel off to Southsea, to his parents, while he considered what to do next. When Mabel did not appear at Lewis & Lewis on 9 August it was Annie's turn to lose her head. The following day she presented herself at Sibsey vicarage accompanied by Bradlaugh, whose conduct when he saw the Revd Besant was allegedly most offensive. 'He used both violence and threats towards me', Frank deposed, 'and forced his way past me into the hall of my house without stating his business.'[37] When Bradlaugh refused to leave, the police were called and, not without difficulty, he was ejected. Annie remained to demand her daughter. Frank replied she was not there. 'To whom shall I apply?', asked Annie. Frank told her: 'Albert Besant, solicitor, of Southsea'—his younger brother. George Lewis had to issue a threat of court proceedings before, a few days later, Mabel was reunited with her mother.[38]

In September Bradlaugh departed on another American lecture tour; he did not expect to return until the following spring. In his absence Annie brooded on the scene at Sibsey. A victory that appeared conclusive had been won by the forces of light, represented by no less a figure than the leader of free-thought, over the powers of Christian darkness which, in the person of her husband, had grievously oppressed her for so long. Annie's vigorous imagination represented this as signalling another new departure. By his public intervention over Mabel, Bradlaugh had revived her trust in him, had proclaimed himself her champion. Once again she entertained the keenest anticipation of special consideration at his return.

CHAPTER 9

HEROINE OF FREE-THOUGHT

'I DON'T care for life if I can't fill it with work',[1] was a cry that grew familiar to Annie's friends. Each autumn from now on the *National Reformer* was to publish a series of long articles on a particular theme whose preparation used up many of her otherwise idle hours. That year the subject was the French Revolution seen, as far as possible, from the standpoint of the common man. As a historical incident its ability to stir Annie was second only to the death of Christ. Faithful to her method she brooded over the events until she 'lived' them as once she felt she had the Stations of the Cross.[2]

She had two hansom cab loads of books from Bradlaugh's precious library to assist her; they were delivered to 19 Westbourne Park Terrace before he sailed. That he could bear to part with them to her was proof of his high regard; it would console her for the six months of his absence.

For several weeks the news from America was little different from that in previous years; Bradlaugh delivered lectures to enthusiastic crowds, then boarded a train at the last moment which carried him through the night to the next engagement. But in December came a sudden terrible anxiety; Bradlaugh was in a New York hospital in danger of his life. Exposure to the biting cold while in a state of extreme fatigue had brought on pleurisy, which was complicated by typhoid fever. Fortunately, Moncure Conway, who was on a rare visit to his home country, heard of his plight—Bradlaugh was penniless since he could not fulfil his engagements—and proved as staunch a friend to him as, two years before, he had to Annie Besant. Conway spent hours at Bradlaugh's bedside and, as soon as the sick man could stand, lent him money for his passage home. Observing him with professional interest as he lay close to death, Conway decided that Bradlaugh was religious after all; that there was a supernatural origin to his famous thunder.[3]

'I do not try to thank you, the words crowd too much and my eyes fill', Bradlaugh wrote to Conway shortly before his steamer left New York. 'If, as I believe, my life lasts, I will try to live my thanks to you.'[4] His convalescence was marked by fits of sweating, faintness, and shortness of breath. From Newcastle, where he went for his first public appearance after his illness, he wrote to his daughters, 'so weak and the folks cheered so I literally wept on the platform'.[5] He took a long time to recover his strength: some of his friends thought he was never the same after this illness.[6] When CB and AB, as the free-thought party now referred to them, took to the lecturing circuit in the spring of 1876, the newspapers sensed a change. Bradlaugh would have to look to his laurels, they jeered, Mrs Besant was the 'more venomous'.

Annie proudly pleaded guilty to the charge. Her attacks on Christian theology were bitter.

Smarting under the suffering inflicted on myself, and wroth with the cruel pressure continually put on Freethinkers by Christian employers [she wrote], speaking under constant threats of prosecution [for blasphemy] identifying Christianity with the political and social tyrants of Christendom, I used every weapon that history, science, criticism, scholarship, could give me against the Churches.[7]

Confidence, nourished by the crowd's applause, vigorous wellbeing, and a keen enjoyment of her own developing capacity, made Annie reckless. She was too bold even for the impeccably liberal members of South Place Chapel, who had a long tradition of sympathy with radical causes. On his return from the United States Conway had to quell a sharp revolt by some members of his congregation, who urged him to delete Mrs Besant from his list of speakers; he refused.[8]

The opposition at her lectures was increasingly organized; 1876 proved a severe test of her ability to stand up to attack. Two Protestant 'missionaries'—names unknown—took it upon themselves to visit in advance the towns where Annie was advertised to speak. They had two or three set discourses which one or another of them delivered, usually to the lowest element in the place. 'A sort of joint stock slandering company, limited (in brains)', was how Annie described them to her readers.[9] These men founded their attack upon the most extreme suggestions contained in *The Elements of Social Science* concerning free love and birth control. The town was placarded with questions: 'Will Mrs Besant defend etc. etc.?' An open air meeting just before she arrived completed the process of stirring up the population.

At Hoyland near Barnsley, for example, Annie found the hall crowded to excess, and noisy with the effort to suppress the roughs until the right moment arrived. The signal was given after her lecture when one of the 'missionaries' challenged her to debate. Annie refused, on the grounds that to fight with a sweep was to get soot on one's hands. 'Christians' leapt on forms, yelled, roared, made faces, shook their fists, and generally conducted themselves like maniacs. Annie remained throughout the perfect gentlewoman; 'if one only had a glass to show them how silly they looked with their faces twisted and their mouths wide open!'[10] As further rational debate was impossible she departed between a row of jeering faces. Kicks were aimed at her and an attempt was made to overturn her cab, which the driver foiled by whipping up the horse.

Such tumult was heartening to free-thinkers; it showed how rattled their opponents had become. Nevertheless it was a strain for Annie. She preserved a grateful memory from these wanderings of a place where she was received with rare gentleness—in the towns and remote villages of the Tyne coalfield. The miners welcomed Annie not only for her message, to which they listened with grave attention, but for the memory of their advocate, and her old friend, William Prowting Roberts. 'How kind they were to me these genial, cordial miners, how careful of my comfort and motherly the women!'[11] In one pit village her host invited a dozen of his fellows to supper with her. They talked politics. Neither she nor they had the vote, and they were uneducated, but she found them shrewder and more knowledgeable than their so-called social superiors—far better fitted to be citizens, she decided. Although she valued the experience above all else she had to record a deficit from eight days lecturing in Northumberland and Durham of 11*s*.

Annie counted it a financial success if one of her lectures cleared as much as £7 after expenses. The profit came from the more expensive seats: railed off from the rest of the audience these cost 1*s*.[12] It was a matter of anxious calculation which subject would attract the greatest number of shilling hearers. The rest of her income came from the guinea a week for her job on the *National Reformer*, and £9 3s. 4d. a month from Frank Besant for Mabel. She was far from well off. Yet in April that year she wrote to Moncure Conway asking him to guarantee the rent on a house she proposed to take in St John's Wood.[13] Hypatia Bradlaugh was puzzled. 'We were told she was extremely poor; she lived in furnished apartments at Westbourne Park Terrace. But she took and *furnished* a house in Mortimer Crescent. And Annie's clothes, which were in faultless taste, struck the Bradlaugh girls as shockingly expensive.[14]

Oatlands, as the new house was called, had a garden, a stable, and a conservatory which was filled with uncaged birds. All at once Annie acquired a piebald Irish mare called Kathleen, a rough-haired terrier, and a St Bernard puppy whose name was Lion. Riding lessons were arranged for Alice and Hypatia who had to accompany Annie, mounted on Kathleen, down to the office off Fleet Street. It was an excursion they detested: they felt conspicuous waiting outside on horseback, and precarious, for their mounts slid about on the asphalt street. Annie also engaged a woman to teach them German. They were obliged to be with her a great deal, since one or other of them had to accompany their father when he appeared in public with her. She was kind to them for his sake, Hypatia thought, but they had little in common.[15]

She and Alice were capable, high-minded, and hard-working, devoted to the secularist cause and to their father. Though they were ashamed of the reason why their mother Susannah lived apart from Bradlaugh, they loved her; and they respected Mina de Brimont. They did not take to Annie. Hypatia wrote a private account of their relationship after Annie's death. This stated that while she had no feeling of hostility on her own account she could not forget how Annie made her father suffer, how she used him and afterwards repudiated his teachings, and all he stood for. Hypatia portrayed Annie as capricious, histrionic, and impossibly demanding, and Charles Bradlaugh as patient, magnanimous, and forbearing. By her own admission Annie's relations with her and her sister never recovered from their first encounter. Alice and Hypatia had been looking forward to meeting the famous Ajax. But Annie greeted them perfunctorily, waved them to a sofa in the corner of the room, seated herself beside Bradlaugh to whom she talked in whispers, and ignored his embarrassed attempts to draw the girls into their conversation.[16]

The money to support Annie's new kind of life, which she had not enjoyed since her days at Fern Hill, may have come from the Wood side of the family. Annie's grandfather Robert Wood had a brother Benjamin, whose widow Maria long survived him. Known as 'Aunt Ben', she was an immensely wealthy recluse who lived at Eltham near London, where she was looked after by Annie's cousin Katharine (Kitty) O'Shea, mistress of Charles Stuart Parnell. It may be that, for Mabel's sake, Aunt Ben gave Annie money to secure an appropriate place to live. While there is no evidence of that, we do know that Annie hoped to inherit something from Mrs Wood's vast estate.[17]

Nothing could be more delightful than St John's Wood in spring. The streets of small houses had been built with taste and imagination, if not at vast expense. Italianate villas sat next to Victorian Gothic cottages. Almost every front garden had a green acacia bush beside the short path leading to the front door, while fruit trees blossomed at the back. It was where Annie had spent the few unshadowed years of her childhood before her father's death. Whether it was the most suitable place for a young woman separated from her husband, with no visible means of support, to bring up a child in her turn was a question Annie, typically, did not consider, though others did.

'Old Jolyon looked about him with interest for this was a district which no Forsyte entered without open disapproval and secret curiosity',[18] John Galsworthy was to write, describing the raffishness that clung about the Wood. Too many artists and writers lived there, too many kept women, for his respectable contemporaries to feel entirely sure about it. Of course Annie was not unchaperoned; the rules of social conduct made that obligatory. And adult company was essential to her well-being, for she hated to be alone. Her mother's sister Constance Marion Morris came to Oatlands to be with her, and to act as governess to Mabel, who was 6. The arrangement ought to have reassured Mabel's father, as 'Co' was a devout member of the Church of England.[19]

But, to his distress, the name Besant was again publicly associated with that of Bradlaugh when it stood at the head of a

'Monster Petition' against parliamentary grants to royal personages presented to the Commons in June. Annie had taken it upon herself the previous year to superintend the collection of signatures for this latest demonstration of republican feeling against the Crown. There had been doubts within the free-thought party as to the wisdom of her doing this. It would be embarrassing if sufficient signatures did not materialize, and the effort of collecting them would be prodigious. Nor was the moment particularly opportune; Disraeli's skilful coaxing of the queen out of her retirement had begun to restore the popularity of the royal family.

However, Annie's effort vindicated her presumption. On 16 June 1876 the Monster Petition, carefully stuck together to make a mile in length, was rolled around a pole and driven in a carriage, with much ceremony, to the House of Commons. There 'The Petition of the undersigned, Charles Bradlaugh, Annie Besant, Charles Watts and 102,934 others' was introduced into the Chamber by the miners' leaders: Thomas Burt, MP for Morpeth, and Alexander Macdonald, who sat for Stafford.[20] Both had been elected in 1874 and were the first representatives of the working class to win their way to the House. The newspapers gave a flattering amount of serious attention to the incident, which had no effect at all; the Tory majority saw to it that public money continued to be bestowed on members of the queen's huge family.

From his slippery footwork over the Suez Canal shares to his provocative backing for the Turks against the Russians, everything Disraeli did aroused the anger and foreboding of republicans, among whom Annie Besant was proud to count herself. That autumn her anxiety at the threat of war (it broke out between Turkey and Russia a few months later) provoked her into print. *Why did Gladstone Fall from Power?*, Annie's first venture into the realm of foreign affairs, was published by Charles Watts. It delivered a solemn warning to the nation of the dire consequences of Disraeli's policy, and called for Gladstone to replace him. Where Gladstone had ignored her treatise on the nature and existence of the deity, he now praised her exposition of the Eastern Question. He wrote her a letter of splendid courtesy which ended, 'I beg you, Madam, to accept this note as a willing tribute to the ability and force, as well as the integrity and sense of justice with which you have discussed a question of vital interest.'[21]

At Christmas, in its review of 1876, the National Secular Society's *Almanack* declared:

> Too much praise cannot be awarded to Mrs Annie Besant whose extraordinary labours have done much in the present year to extend our cause. Few men could undergo the exertions which appear to this lady to be a labour of love. By night and day, undeterred by the extremes of heat and cold, Mrs Besant has spread her principles from Scotland to Devonshire, wringing even from our opponents a tribute to her ability, her refinement, and her thorough earnestness.

Was there a hint of sarcasm in this? If so, did Annie notice? What did Charles Watts, who wrote it, mean by 'a labour of love'? It was the last time he had anything favourable to say about her, the last time he wrote anything for the *Almanack* or for the *National Reformer* while Charles Bradlaugh was its editor.

The incident which ended the long working relationship of two close friends, Watts and Bradlaugh, arose out of a clear sky on 11 December. Watts received a telegram from Henry Cook, who sold free-thought literature in his Bristol bookshop. Cook wanted to know what to tell the magistrates, who had summoned him for selling an allegedly indecent book—the *Fruits of Philosophy*, published by Watts. 'Fear not, nothing can come of it', Watts replied, 'the book is Certainly Defensible and were I summoned I should defend it . . . It is absurd for the magistrates to try to suppress it. They *Cannot do it*.'[22] But if they tried, Watts advised Cook to ask for an adjournment, when he and his colleagues in London would supply the bookseller with a defence. An affidavit accompanied Watts's letter. This said that Watts, and others before him, had published the book without prosecution for twenty years; it was openly on sale in Edward Truelove's Holborn bookshop; it was not very profitable and the type was old, having been set up some forty years before. In effect—what was all the fuss about? At this stage Watts had not read the book.

It was a treatise on birth control written by an American doctor, Charles Knowlton, and first published in 1832. As author Knowlton had been convicted of indecency by a Massachusetts court, fined, and spent a short time in gaol. The book, however, had not been suppressed and enjoyed a steady sale. Plates for an English edition had belonged to the celebrated free-thought publisher James Watson. At his retirement in 1853 the right to publish had been taken over by George and Austin Holyoake. At

Austin's death, in 1874, Charles Watts succeeded to the business and, at the same time, bought the plates of Knowlton's book from Watson's widow. Charles Watts, therefore, was the person on whom responsibility for the publication of the book in England rested inescapably.

When he made the affidavit Watts did not know that, in putting the Knowlton book on sale in Bristol, Cook had made changes. He had added two illustrations which were discussed in lowered tones, their nature was so shocking. The book's subtitle, *The Private Companion of Young Married Couples*, and the fact that Cook had raised the price from 6*d*. to 1*s*. 8*d*., was taken as a hint to the salaciously minded that they would find their money's worth inside. Watts and his colleagues did not discover until too late that Cook had already served time in prison for offences under the Obscene Publications Act.

Almost immediately Annie Besant was drawn into the affair. Bradlaugh was away from London, so Watts who, like most printers, was disinclined to read the type he handled, asked Annie for her opinion. She was dashing off to lecture so she took the work with her on the train. Arriving at her destination she telegraphed: 'Book defensible as medical work.'[23]

On his return to London Bradlaugh betrayed uncharacteristic disquiet. He had seen the plates when Watts first received them. Finding himself in a hurry he had said, 'Yes, it's indictable, I'll speak to you again about it',[24] and had forgotten. Now, faced with the summons to Cook, Bradlaugh gave his opinion that it was Watts's duty as the publisher to present himself at Bristol. Like Watts, Bradlaugh was ignorant of the changes Cook had made.

In Bristol, on learning the true state of affairs, Watts hastily withdrew from Cook's defence and declared that whatever the outcome of the case he would no longer sell the *Fruits of Philosophy*. When, at the quarter sessions on 19 December, Cook was convicted of circulating indecent material and sentenced to two years in prison, Watts telegraphed to London for the book to be removed from sale. Nevertheless he was arrested on 8 January, and brought before the City magistrate at the Guildhall. At the hearing the prosecution announced that Watts had agreed to withdraw the book and destroy the plates. Bradlaugh, who was present, displayed astonishment, claiming that he thought Watts would defend the book. The hearing was adjourned to the Central Criminal Court.

'The case is looking rather serious but we must face it', Bradlaugh said to Mrs Watts after the Guildhall hearing. 'I would the prosecution had been against any other book.'[25] He did not like its style and he thought the medical advice was out of date. (It was.) Yet how could any of them repudiate a work that had been sold for years by their respected colleagues James Watson and Austin Holyoake? Bradlaugh was also very concerned at the reaction of the free-thought party. He knew from experience how sensitive it was to the subject of neo-Malthusianism. In 1861 the row had been colossal over his undiscriminating praise of *The Elements of Social Science*. How was he to ask the party for help in Watts's case?

It was not like Bradlaugh to betray unease; Kate Watts seized upon it as a means of rescuing her husband. If she had known how strongly Bradlaugh felt about the book she would have opposed it more firmly herself. Strong-minded, attractive, with a flair for acting, Kate was accustomed to getting her way. They could not afford for Charles to go to prison. He had just spent a large sum of money on new offices, machinery, and type for the *National Reformer*. Moved by her appeal, Bradlaugh cast about for a solution. He might try for a grand jury hearing on a technicality; if it returned 'no bill' the issue would be lost without discussion.[26] His prowess in litigation rested upon such devices, which he loved to spring upon his unprepared opponents. Cheered by this he went to tell Annie what he proposed to do. He was taken aback by her response.

She insisted that the book was defensible, that it must be defended, and that the free-thought party would respond to an appeal. None of the arguments Bradlaugh put to her as to the risks and cost, and the hardship to the Watts family, moved her at all. She was carried away by the idea of promoting a noble cause—the Knowlton book could deliver the working class from the scourge of poverty inflicted by too many children.[27] Bradlaugh argued with her for some hours. But the woman who had fashioned herself in his image had vanished, to be replaced by the headstrong creature so troublesome to Frank Besant. Watts's plea of guilty to an offence Annie held to be disgraceful was a violation of her integrity; her fiery sense of honour was aroused.

She told Bradlaugh she would never allow her work to appear over the imprint of a man who pleaded guilty to publishing obscene literature. Bradlaugh had rarely encountered a stronger will; he bent before its force.

That evening Annie invited Mr and Mrs Watts to Oatlands. Bradlaugh was there, accompanied by his daughters, but Annie occupied the centre of the stage. To Watts, who said that imprisonment would kill him, she replied that in pleading guilty he would ruin himself. She would withdraw her books from him—'the thing is too disgraceful'. She produced a draft of a circular addressed to the free-thought movement asking for money and support for a plea of 'Not Guilty'. She had dashed it off after Bradlaugh's visit earlier in the day. She offered it to her guests with an assurance that she and Bradlaugh would look after Kate and the business if Watts had to go to prison.[28]

If Bradlaugh had been driven back by Annie in this incandescent mood what chance had Watts? In any case he was anxious to do his best for secularism without prompting from her. His dealings with all comers were as genial as those of Annie's favourites, the miners; he had the same bluff manners but not as large a share of shrewdness. He trusted Bradlaugh absolutely and deferred to the latter's more potent way with words.[29] Bradlaugh told Watts it had taken him a long time to come round to Mrs Besant's way of thinking but now he was convinced. There was little use in saying more. Watts acquiesced in the printing of the circular; he and Kate went home to contemplate the alarming prospect of a glorious martyrdom.

Annie went to Plymouth to fulfil a lecturing engagement. Upon arrival she was handed a telegram from Bradlaugh informing her that the Wattses' resolve had not endured; Charles was to plead guilty after all. Ignoring this Annie stirred up her audience to a pitch of enthusiasm for the *Fruits of Philosophy*; as a medical work it was defensible, she said; as literature it was no more indecent than the works of Fielding, Sterne, or Smollett. She extracted £8 from them for the defence.

That Watts had changed his mind was due to Kate, his wife. After the Oatlands meeting she had gone home and read the book. It struck her as, in Bradlaugh's words, 'a nasty thing'.[30] She spoke to Charles. Next morning a note was sent to Bradlaugh asking for a change of plan. Face to face at Turner Street

Bradlaugh's anger quickly turned vindictive. Unless Watts adhered to the course laid down in Annie's circular, all business between them must cease.

Half-frantic though he was, Watts would not renege on his wife; he forfeited his livelihood instead. He had no redress in law; no contract existed either for printing the *National Reformer*, or for his position as sub-editor (it was bestowed on Annie Besant). He had simply trusted Bradlaugh. All the money spent on refurbishment was lost and Bradlaugh pursued him for the return of stock. When, as legal adviser, George Henry Lewis gave his opinion that the circular should not be sent out, and money certainly not solicited unless the case were to be defended, Annie grew furious in her turn. Forced to return £8 to secularists in Plymouth she disguised her embarrassment with a caustic note. The £8 'was given to help Mr Watts fight, not to assist him in running away'.[31]

Less than a week after the Oatlands meeting a partnership agreement was drawn up between Mrs Annie Besant, journalist, and Charles Bradlaugh. Dated 20 January 1877 it set up the Freethought Publishing Company to issue their separate works by joint assent, the profits to be equally divided. Bradlaugh was credited with £1,000, representing the copyright of the *National Reformer*; in the event of his death this was to pass, not to his natural heirs, his daughters, but to Annie Besant.[32]

CHAPTER 10

THE KNOWLTON TRIAL

'WE intend to publish nothing we do not think we can morally defend. All that we publish we shall defend', Annie Besant and Charles Bradlaugh declared in the *National Reformer* of 4 March 1877.

They had decided to issue a new version of the *Fruits of Philosophy* under the imprint of their Freethought Publishing Company. As a result of Charles Watts pleading guilty at the Old Bailey in February, Knowlton's pamphlet stood convicted of an offence against the law. CB and AB saw that verdict as a threat to the freedom of publication. To safeguard that, and to test the right to circulate neo-Malthusian views, they decided to invite prosecution under the Obscene Publications Act.[1]

While Annie exulted in the prospect of battle in a noble cause, Bradlaugh was far from sanguine. His legal knowledge caused him to suspect the hazardous nature of their enterprise. Acquittal would turn upon the content of the pamphlet; he could not easily convince himself that it was wholly innocent. Annie objected to the style,[2] but was confident a jury must find in their favour as long as their motive in publishing was acknowledged to be pure. She could not imagine that anyone could question hers.

The 1857 Act left the definition of obscenity extremely vague. In 1868 a case, *Regina* v. *Hicklin*, helped to clarify the matter. Hicklin, a fervent Protestant, defended his publication of *The Confessional Unmasked* as a lawful attempt to expose alleged errors of the Church of Rome. The judge, Sir Alexander Cockburn, thought the test was whether the tendency of the material was to deprave and corrupt those whose minds were open to such criminal influences and into whose hands a publication of that sort might fall. Concerning Hicklin's plea of justification, Cockburn said, 'The question . . . presents itself in this simple form. May you commit an offence against the law in order that thereby you may effect some ulterior object which you may have in view which may be an honest, and even a laudable one? My answer is, emphatically, No.'³

With this precedent in mind the partners did their best to edit Knowlton. His tendentious subtitle, *The Private Companion of Young Married Couples*, was replaced by *An Essay on the Population Question*. A preface was added over the names of Bradlaugh and Besant, relating the history of sale in England since 1832 by James Watson, George and Austin Holyoake, and Charles Watts. 'For the last forty years the book has been identified with Freethought,' the preface declared, 'advertised by leading free-thinkers . . . and sold in the headquarters of Freethought literature.'⁴

In England up to that time very few works containing specific advice on 'preventive checks' to conception had been published.⁵ Their authors, Francis Place, Richard Carlile, Robert Dale Owen, and Charles Knowlton, were all free-thinkers, but that was by the way. Paradoxically, the impulse to disseminate advice which the great majority of persons found offensive came from a philosophy which was concerned with achieving the greatest good for the greatest number—utilitarianism. It was as a disciple of its leader Jeremy Bentham that Francis Place distributed leaflets among the poor, in the 1820s, containing advice about limiting their families.

Utilitarians had a ready-made laboratory for testing ideas in Scotland, at the New Lanark cotton spinning mill run by Robert Owen, in which Bentham and others held shares. While publicity was given to educational reform in New Lanark's schools, neo-Malthusian experiments were naturally taboo. Nevertheless there is some evidence that Place and Owen at least considered introducing such knowledge into New Lanark, whose burgeoning population Owen was anxious to restrain, and that Owen collected information as to means in France, where the practice was well known.⁶

In 1825 Place's leaflet was reprinted by his radical journalist friend Richard Carlile. The following year Carlile brought out his *Every Woman's Book; or, What is Love?* It gained a reputation for coarseness. A copy found its way to New Harmony in America, where Robert Owen had gone to establish one of his experimental villages. His eldest son Robert Dale Owen was editor of the village newspaper, on whose press a prospectus for Carlile's work was printed, thereby causing both Owens to be accused of wanting to destroy female chastity.[7] The younger man defended himself with *Moral Physiology; or, A Brief and Plain Treatise on the Population Question*, published in 1831. It was a thoughtful analysis of the economic, moral, and social problems of human fertility, and it called for men and women to be free to gratify their natural instincts without fear of consequences.[8]

The *Fruits of Philosophy* followed a few months later, Knowlton taking his economic and social arguments from his predecessor without acknowledgement. His interest in that aspect of the question was perfunctory, and he lacked the sympathetic attitude conveyed in *Moral Physiology*. His language betrayed a certain callousness which jarred on those engaging in general discussion of the subject as it does, to an extent, even to the modern ear.

The decision to republish with the new preface split the free-thought party. Perceived as the driving force, Mrs Annie Besant was assailed by those persons who objected to being associated with a subject they so thoroughly disliked. They dismissed her courage as uncomprehending foolishness and the high purpose that sustained her as a selfish desire for notoriety.[9]

Even staunch supporters like Moncure Conway and his wife remonstrated when Annie, copying Bradlaugh, announced she would conduct her own defence. Deeply embarrassed by accusations that he was sheltering behind a lady, Bradlaugh pleaded with her to change her mind; it was too dangerous a course, he argued: whether she faltered or appeared too bold, she risked alienating the court. Moreover, such behaviour in pursuit of such an object was particularly unbecoming in a mother of a young daughter. Bradlaugh and many others warned Annie that Frank Besant would hardly fail to use any weapon she put into his hands.

'Dear Mrs Besant, if you intend to publish the Knowlton work it means ruin to you as a lady', George Holyoake wrote; 'at that I am concerned.'[10] He went on to express his long-standing distaste for *Fruits of Philosophy*, which he had taken over jointly with his brother Austin, when they bought James Watson's business in 1853. Holyoake assured Annie that he had never allowed his name to appear on the pamphlet, which he had taken care not to publicize, providing copies only on demand: he considered it no part of free-thought literature. 'I bore "the flag" into prison in worse days than these', Holyoake reminded Annie, 'but I can't think of resting the defense of liberty of publication upon a quack like Knowlton.'

'Thank you for your nice little note',[11] Annie replied, 'I think it probable that we shall follow the line of Mr Bradlaugh in the *National Reformer*, January 21. Objection to style of pamphlet as unduly coarse but maintenance of right of discussion of sexual problems, i.e. revise carefully, publish matter, but refine style.'

This airy spirit marked all her utterances before the trial. 'I am inclined to think that she hardly realized at all the gravity of her situation', Hypatia Bradlaugh wrote; 'any true sense of the possibilities involved was perhaps somewhat obscured by the atmosphere of excitement and admiration in which she was living.'[12] There was truth in that assessment, though it did not sufficiently recognize Annie's code of conduct, which demanded conspicuous gallantry in the face of peril.

In March her vivid imagination was alerted by the arrival when Bradlaugh was away of stocks of the new edition of *Fruits*. In order to forestall a police raid which she convinced herself was imminent, Annie enlisted Alice and Hypatia in concealing several hundred of the books. Some were buried in the garden, others hidden in the conservatory, in flower boxes topped with faded geraniums; still others went behind the lavatory cistern. Hypatia condemned the move as proof of how childish Annie was. When Bradlaugh heard what she had done he was extremely angry; her action threatened to compromise the reputation for openness and honesty by which he set great store.[13]

On Friday 23 March he and she together delivered a series of copies: to the Guildhall for the magistrates, to the Police Department, and to the City Solicitor who was expected to prosecute. On Saturday afternoon, when they had announced they would sell the pamphlet, a crowd filled the narrow space outside the Freethought Publishing Company in Stonecutter Street, next to Farringdon Street. Inside Alice and Hypatia were kept busy wrapping copies while their elders took the money. Five hundred copies went in twenty minutes. Bradlaugh noted detectives among the purchasers; however no arrests were made.

The following Thursday the authorities were re-notified and friends invited to Stonecutter Street to witness the arrest. When nothing happened Bradlaugh and Annie Besant took a cab to the police station to enquire about the delay. They were told the papers would not be ready until the following week. They also learned that Bradlaugh's old antagonists the Christian Evidence Society had asked the Government to prosecute. Pleased that their challenge was to be met at the highest level they began to prepare for trial. A defence fund was established whose subscribers were listed in the *National Reformer*; George Holyoake refused to contribute. Sureties for bail were offered by, among others, members of the Dialectical Society; in that there was a particular significance.

Annie had been a member for only three years but Bradlaugh's association went back to the founding of the Society in 1868, when he had been involved in a public outcry as a result of remarks made at one of the first meetings by the Vice-President, Lord Amberley. The Society followed the principle of absolute liberty of speech; its prospectus—described as 'somewhat hectoring' by the Revd C. M. Davies—directed its members to examine all evidence before believing anything. The Society was unusual in that women were admitted to its unbridled deliberations. In July 1868, with Amberley in the chair, the subject had been how married persons might limit their families.[14]

Bradlaugh's contribution was that working people were beginning to debate the matter. When he spoke Amberley was devastatingly frank. He objected to celibacy, he said; he wished he could hear the proposals of the medical men in the room as to the best means of limiting families. American ladies, he knew, were in the habit of 'keeping back' their families, but the means they employed seemed to him to be dangerous. He should like to hear a discussion on whether some innocuous method might not be discovered.[15] Made public, these remarks infuriated the medical profession, most of whom regarded 'checks' as quackery. Amberley, who had hoped to follow his father Lord John Russell into politics, was forced to retire into private life.

Bradlaugh needed no reminder of the disastrous outcome of the Amberley affair. The damaging effect of another neo-Malthusian scandal upon the electors of Northampton had figured in the argument he had had with Annie. The knowledge that their support of Knowlton might distress loyal party workers increased the burden Bradlaugh shouldered before the trial.

'Upon Mr Bradlaugh lay the whole responsibility of the defence', Hypatia wrote; 'his was the mind that planned it and he had to conduct the fight not merely for himself but for the woman beside him.'[16] Since Annie was unfamiliar with the procedure of the courts Bradlaugh had to tell her not only the things it was desirable she should say, but those which were better left unsaid.

Arrest by warrant was not at all like responding to a summons. When, on 7 April, they were taken into custody at the police station Annie's jauntiness evaporated. Her account of the proceedings, though still defiant, registered distaste and apprehension. She was alarmed when the door of the 'funny iron barred place' in which they had to go snapped shut on them; she stood uneasily while the police examined and measured her and removed her watch, keys, and purse. But she tried to pull away when the housekeeper took her into a neighbouring cell to be undressed and searched. 'It is extremely unpleasant to be handled', Annie told the readers of the *National Reformer*; 'the woman was as civil as she could be . . . but it strikes me this is an unnecessary indignity to which to subject an unconvicted person.'[17] Worse was to come.

They were taken under escort to the Guildhall where, for $2\frac{1}{2}$ hours, they were held in adjoining cells, dimly visible to each other through the bars while the daily procession of drunks, thieves, and prostitutes came up before the magistrate.

The record of the Bradlaugh/Besant trial contains bitter objections by both defendants to the indictment. To them and their contemporaries the words it used were shaming—to hear them read aloud was punishment before a verdict. They were accused of corrupting the morals of youth, of inciting them and others to 'indecent, obscene, unnatural and immoral practices', of bringing them to a state of 'wickedness, lewdness, and debauchery' by printing and publishing a certain 'indecent, lewd, filthy, bawdy, and obscene book'.[18] When the magistrate at the second hearing offered to dismiss Annie from the case the Conways begged her to agree because 'we foresaw that evil tongues would be busy with her reputation'.[19] She refused. The trial was fixed for early in May at the Central Criminal Court.

Bradlaugh, who had no desire to find himself, let alone Annie, in the dock at the Old Bailey, hastily applied for the case to be removed to a superior court. The Lord Chief Justice was now Sir Alexander Cockburn, the same who had given the definition of obscenity in *Regina* v. *Hicklin*. After reading *Fruits of Philosophy*, he and another judge decided that the purpose of the book could not be condemned outright; its object appeared to be the legitimate one of promoting knowledge and not, under cover of science, of circulating 'Impurity'. Whether it was obscene or not was a suitable question for a special jury to decide. A writ was granted for a hearing in the Court of Queen's Bench in June.[20] The time intervening permitted so many extra copies of the pamphlet to be sold the Freethought Publishing Company gained £2,000. His enemies said this was Bradlaugh's only reason for obtaining a delay.

Annie's special task before the trial was to compare the text of Knowlton with medical works in order to demonstrate that equivalent material had been freely published. She also combed back numbers of the leading journals, and pamphlets by the dozen, in the hope of finding opinions that supported the principal argument of the defence—that family limitation was a proper subject for general discussion.

Bradlaugh tried to enlist the help of witnesses. Many of those he asked refused. George Holyoake, for example, evaded all attempts to bring him into court to testify to the long period when *Fruits* had been openly on sale. Confronted by a process server with a subpoena Professor Henry Fawcett, who was blind, put his hands behind his back to avoid it. The defence wanted him as one who had published support for the Revd Malthus's views. Fawcett was a distinguished economist and political theorist and MP for Hackney. His wife, Millicent Garrett Fawcett, was a leading suffragist. The professor declared he would send her out of the country rather than permit her to appear as a witness in the Bradlaugh/Besant case.[21]

Fawcett's extraordinary vehemence may have been due to fear that he would be cross-examined about the late John Stuart Mill, whose close friend he had been. Annie was to quote extensively at her trial from Mill's *Principles of Political Economy*. It discussed Malthus's doctrine and the growing need to limit population; no one had dreamed of prosecuting it. Mill had stated that there must be restraint within marriage; he had not touched upon the question of 'preventive checks'. But he was suspected of approving them. Immediately after his death in 1873, an article in *The Times* shocked his friends by accusing Mill in his youth of having been caught distributing the leaflets published by Francis Place.[22] Whatever the cause of Fawcett's reluctance, he entertained 'very strong opinions as to the objectionable character of the work you have published', Mrs Fawcett told Bradlaugh. 'If we were called as witnesses we should effectively damage your case.'[23]

Mrs Fawcett and many of her supporters in the campaign for women's suffrage regarded the use of mechanical devices to interfere with nature not as liberating women, but as further subjecting them to men's desires.[24] 'The world is not so virtuous that we can afford to remove any of the barriers which society has set around it',[25] the *Liverpool Mail* wrote, commenting on the trial. The *Daily Telegraph* remarked that there was no difference between the publishers of 'this vile work' and people who offered poisoned food for sale. The *Evening Standard* called it poison, too, and the *Englishman* denounced the *Fruits of Philosophy* as 'the most wicked work ever written'.[26] Yet the demand for the book showed no sign of tailing off. Hundreds of appreciative letters poured in to Stonecutter Street. Most were from poor men and women; a significant number, Annie noticed, came from clergymen's wives.

Though individuals and many newspapers expressed their sense of decency outraged, it seems there was a current, as yet submerged, of private interest unprejudiced by cant. And among those who detested every word of Knowlton were many who defended its publication in the interest of liberty.

So many members of the public occupied the court on 18 June that lawyers who had counted on observing an interesting case were crowded out. Because the Royal Courts of Justice in the Strand were still under construction, the trial took place in rooms that opened off the ancient, vast, and echoing space of Westminster Hall. To be tried in these romantic surroundings, and by the Lord Chief Justice of England, made Annie very proud; the splendour seemed to her to dispose of the imputation that she and Bradlaugh would knowingly or willingly publish an obscene book.[27]

The prosecution was conducted by the Solicitor-General, Sir Hardinge Giffard,[28] but the Government kept well clear of the affair. In spite of persistent enquiry the defendants never did discover by whom they were accused. The best guess was a member of the City establishment; an alderman perhaps, someone who, Annie suggested, might have been prompted by malice.[29]

Bradlaugh moved at once to have the indictment quashed because it did not set out the passage deemed to be obscene. When the Solicitor-General indicated that virtually the whole of the work was in question, Bradlaugh argued that it ought to be given word for word. The Lord Chief Justice reserved the point so the trial began.[30]

Sir Hardinge Giffard confined himself to reading out chapter 3 'Of Promoting and Checking Conception'. His tone was one of outraged contempt; this was a 'dirty, filthy book' he told the jury. The question he wanted them to consider was

> whether a book of this sort, published to everybody would not suggest to the unmarried as well as to the married, and any persons into whose hands this book might get—the boy of 17 and the girl of the same age—that they might gratify their passions without the mischief and the inconvenience and the destruction of character which would be involved if they gratified them and conception followed.

This caused a stir in court; it raised the spectre of free love, the charge brought against Robert Dale Owen, the accusation that those who wrote and published advice on contraception really intended that not only men but women should abandon chastity, to the utter ruin of civilized society. In choosing this line of attack the prosecution went against the spirit in which the writ for the Queen's Bench trial had been granted. But Bradlaugh and Besant were already suspected of wanting to subvert chastity through their much publicized association with *The Elements of Social Science*. According to some commentators it was to confound the heinous threat contained in George Drysdale's book that a lesser offender, *Fruits of Philosophy*, had been brought to trial.[31]

When she spoke in her own defence Annie Besant was quick to refute the Solicitor-General's allegation. It was a calumny upon Englishwomen to suggest they kept chaste only by fear of maternity, she declared; women who entertained such an idea—sex outside marriage—were already depraved and not to be corrupted by this book. The *Fruits of Philosophy* contained useful information that it was desirable should be made known. In so saying, Annie was the first woman publicly to advocate birth control.

She complained of 'coarse imputations' in the indictment, especially that of bad intent. Here she defied the ruling given by Sir Alexander Cockburn in *Regina* v. *Hicklin*. If the object of publishing Knowlton's book was good it was a lawful and not an immoral act, she told him to his face. Unless there was bad intent the indictment failed. Turning to the jury she warned them that if they found her guilty she would not feel herself guilty and would do again what she had done.

Annie then dealt with the 1857 Act which she deplored as vague in its definition of obscenity. What was obscene? Was *Tristram Shandy* obscene? She threatened to read it to the court. The Act was not meant to include medical books, she was sure. If Knowlton's work was to be convicted, publishers of medical works would find themselves out of business.

The real significance of the case, Annie declared, was that if her defence succeeded, a right would be established that did not yet exist, the right to public discussion of family limitation. She proceeded to an exposition of the subject which lasted for the rest of that day and the following day and which was very fully reported in the newspapers.[32]

Over-population was one of the most important social questions of the day, she told the court; unless checks were supplied by science they were left to checks of war, famine, disease, and— a recent source of scandal—baby farming. She wanted birth-restricting checks not death-producing checks. She quoted from the 1867 *Report of the Committee on the Employment of Young People in Agriculture* which spoke of the poor herded together 'like swine'. She cited a coroner who had estimated thousands of women guilty of child murder by overlaying their infants. She took issue with Charles Darwin. His theory of the survival of the fittest might do for the animal kingdom but it did not work with humans.

Darwin had opposed neo-Malthusianism, holding that the most gifted people ought not to be hindered from raising large numbers of offspring.[33] But, Annie Besant argued, it would not be the most able but the least able who would increase, as they were the least considerate, and careless of the consequences. Here the Lord Chief Justice intervened, remarking it was a point Mr Darwin might consider. The defence had hoped to bring the author of *Origin of Species* into court but he had declined, courteously, pleading weakness arising from poor health.[34]

The court then heard that Millicent Garrett Fawcett had declared that pauperism never would be cured until the rapid and continuous growth of population could be checked. Of course Mrs Fawcett had not breathed a word about preventive checks, Mrs Besant assured her hearers.

No one had upon whose word it was desirable to dwell. Instead it had to be argued that *Fruits of Philosophy* was no different from any medical book; from those, for instance, given as prizes to students of both sexes and tender years at the government-supported School of Science and Art in South Kensington.

I hold so thoroughly that the Government is right in putting that information in the hands of boys and girls ... [Annie declared] that I say deliberately to you as the mother of a daughter whom I love that I believe it will tend to her happiness in her future as well as to her health, that she shall not have made to her the kind of mystery about sexual functions that every man and woman must know sooner or later.

These words, so obviously inspired by her own unfortunate experience, would eventually be turned against her.

As the judge remarked, she conducted her defence with great ability and tact, earning the respect as well as the sympathy of the court. When to speak of sexual matters without hyperbole was the height of impropriety; when impropriety was equated with immorality; Annie's use of plain unsentimental language was heroic. Her presentation of the case challenged one of the most formidable assumptions upholding Victorian society—that knowledge was too dangerous a thing for women to possess.

Bradlaugh's argument was essentially the same as Annie's. By the third day, when he stood up, the court was tiring. After several hours of him reading from Carpenter's *Animal Physiology* and Acton's *Functions and Disorders of the Reproductive Organs*, the Lord Chief Justice interrupted. He did not think Mr Bradlaugh quite grappled with the case; it did not advance the defence to multiply instances of medical advice: 'if it was a medical work no-one could say these details were not necessary, but the whole case against you is that it is not a medical work.'

Bradlaugh's most telling moment came when he spoke simply of the poverty he had known as a boy, and had seen around him, and of his passionate desire to alleviate the suffering it caused.

After him came Alice Vickery, who had trained as a midwife in Paris. She gave evidence of the practice of suckling babies for up to two years by mothers who feared another pregnancy. The only doctor who appeared for the defence was Charles Drysdale, Senior Physician at the Metropolitan Free Hospital (brother of the author of *The Elements of Social Science*). He told the court he had observed how, in a family, the first and second children were likely to be well, the third rickety. He applauded the French for not bringing up children to starve; they used preventive checks.[35] His attitude contrasted with that of the majority of his professional colleagues. The *Lancet* had denounced 'beastly contrivances', 'filthy expedients for the prevention of conception'. Those who used them were liable to all sorts of afflictions ranging from nervous prostration and mental decay to 'galloping' cancer and intense cardiac palpitations. As for the French, their indulgence in 'sexual fraudulency' marked them as a nation of greedy selfish persons who put comfort before all else.[36]

Summing up in her own defence Annie Besant delivered an earnest plea to the jury not to shame her with an adverse verdict. 'Will you convict me of indecency, will you send me to prison to herd with poor degraded creatures, contact with whom would be agony to me? Think what such a verdict would do to me ... a verdict of guilty—you cannot give it. If you believe in truth, in justice ... you cannot give it.'

Addressing the jury the Lord Chief Justice remarked that there never was a more ill-advised and mischievous prosecution. Knowlton's *Fruits of Philosophy* had been published for forty years and had never been in general circulation. Now it was in high demand. Turning to the Solicitor-General, Cockburn sharply rejected the prosecution's suggestion that the book was a sham, that its real intention was to facilitate intercourse outside marriage. 'I think a more unjust accusation never was made', he declared, at which those in court broke into applause.

Cockburn told the jury they must decide whether the book contained details inconsistent with morality. 'Are they ... things that might be resorted to in order to remedy ... evils, or are they such that we had better bear the evils that we know of rather than encourage means which have demoralizing results?' If the jury thought the book outraged public decency, no impression that the defendants were activated by a sense of duty ought to prevent them from finding them guilty.[37]

The jury were out for $1\frac{1}{2}$ hours, during which they could be heard arguing loudly. On their return the foreman announced:

'We are unanimously of the opinion that the book in question is calculated to deprave public morals but at the same time we entirely exonerate the defendants from any corrupt motive in publishing it.' In that case, the Lord Chief Justice said, he had to pronounce the defendants guilty. It came out afterwards that six of the jurymen had not intended to assent to a verdict of guilty. 'Among the other six were some who in discussion admitted to religious as distinct from moral motives' in reaching their conclusion. Sentence was postponed until 28 June.

On the Sunday intervening CB and AB appeared to tremendous acclaim at a packed meeting at the Hall of Science. The *Fruits of Philosophy* was openly on sale. Both speakers told their audience, among whom were many young women, that it would continue to be sold. Mrs Besant was in euphoric mood. She declared that a new trial was a virtual certainty as the verdict was inconsistent; the jury had not used the word 'guilty'. The courteous manner of the judge, his helpful intervention while she was speaking, caused Annie to conclude that he was on her side. She told her audience that one of the most highly trained intellects in the country, that of the Lord Chief Justice, had openly endorsed their teaching. She would not flinch from continuing the work.

When the defendants appeared for sentencing the Solicitor-General produced affidavits sworn by some who had attended the Hall of Science meeting. Cockburn was seriously annoyed. By putting the book on sale after it had been condemned Bradlaugh and Besant were guilty of an aggravated offence; their names were on the preface. While the jury's exoneration of their motive would have been reflected in a lighter sentence, now he would have to be far more strict. And he delivered a severe rebuke to Mrs Besant for claiming that his summing up supported her. He had merely held the balance in the case, Cockburn insisted.

When Annie rashly tried to deny that she or Bradlaugh had said they would continue to sell the book, Bradlaugh intervened to confess they had indeed done so.[38] The Lord Chief Justice grew more impatient by the minute. Annie's request for a new trial was summarily dismissed, as was Bradlaugh's argument, put forward at the beginning, that the indictment must be quashed as it did not set out the words deemed to be obscene. That was a matter for the Court of Error, Cockburn remarked, and proceeded to sentence them. They were to serve six months in prison, pay a fine of £200 each, and enter into personal recognizances of £500 for two years. Bradlaugh, who had gone very white, cautiously advanced a plea that sentence be suspended until after the Court of Error had pronounced. 'Certainly not', the Lord Chief Justice snapped.

Disaster thus fell on both defendants but more especially on Annie Besant. Social ruin was the least of it. Almost certainly prison would destroy her morale, which depended on restless intellectual activity and the physical freedom to go where she pleased. Darkness, dirt, noise, smells, rough company, disapproval, and restraint would impose suffering out of all proportion to the offence.

The prisoners had left the dock and the wardress's hand was on Annie's arm when the judge relented. On consideration, Cockburn told them, he would stay execution until the Court of Error had examined Bradlaugh's argument about the indictment, if they would promise not to sell the book in the meantime. They promised. Cockburn released them on their own recognizance of £100. But as a married woman Annie could not enter into her own recognizance. To save her going to prison after all the court turned a blind eye when Bradlaugh paid £100 for both of them.[39]

CHAPTER 11

LIFE-ENHANCING CHECKS

As soon as the threat of imprisonment receded, Annie's spirits soared. As a result of the trial she had discovered a role more suited to her nature. Her mind was more constructive than sceptical, Moncure Conway noted.[1] While she had eagerly endorsed Bradlaugh's opposition to forms of Christian belief, his doctrine, which turned upon a failure to find evidence that God existed, could not content her for very long: her inclination to worship, her yearning for a point of reference against which to measure herself and the world, was bound to be frustrated.[2]

Now, in championing the cause of family limitation, she was led to assert the possibility of human improvement by means of a power to which all might defer. She no longer called that power God, but knowledge; as before she vowed to be its servant.

She became the very active Secretary of a revived Malthusian League, which had briefly flourished under Bradlaugh in the 1860s; Charles Drysdale was its new President. The League sought the abolition of all penalties on the public discussion of the population question, and urged that the uncertain state of the law might be changed so that such discussion could never again be deemed a misdemeanour.[3] Two hundred and twenty members enrolled in the first week.

Publicity surrounding the Knowlton trial was responsible for that, as it was for the brisk sale of the book in the summer. While Bradlaugh skirmished in the courts against local worthies who tried (unsuccessfully) to have it suppressed before the verdict of the Court of Error, Annie sat down to write a more suitable treatise on family limitation. Her *Law of Population: Its Consequences, and Its Bearing upon Human Conduct and Morals* was published by instalments in the *National Reformer* beginning on 7 October 1877. Reprinted as a pamphlet which cost 6*d*. it sold 40,000 copies in the first three years.[4]

The argument, which repeated evidence Annie had put forward at the trial, was charged with messianic fervour; she had convinced herself that mankind could be saved and—in time—perfected. The *Law of Population* cited India as an example of what might occur in the future. It was a favourite subject of the editors of the *Reformer*; Bradlaugh and Besant taught free-thinkers to deplore the mischief wrought in the subcontinent, on the one hand, by the servants of imperialism, on the other, by Christian missionaries.

India now gave Annie a splendid opportunity for anti-Tory propaganda. The magnificent durbar held to celebrate Queen Victoria's assumption of the title of Empress of India at the beginning of the year had coincided with one of the worst famines of the century; from the Deccan to Cape Cormorin 500,000 people died.[5] Under the British Raj death-producing checks to population like disease and war had been reduced, Besant stated. But this partial application of scientific knowledge had destroyed the balance of nature, leading to a vast increase in population, but not in food. How could the British sit down with folded hands and contemplate a famine that killed tens of thousands? Yet these people were starved to death according to a natural law—the 'irrefragable' law of population—and the same consequences would occur sooner or later in other countries; even in such progressive countries as England.

The only humane and rational course was to match science with science by adopting birth-restricting checks, Besant declared, and proceeded to an examination of existing methods. This 'medical portion' of her treatise was not printed in the *National Reformer*. None of the methods she described was foolproof, Besant warned her readers—just as well, for her description of the so-called 'safe period' was dangerously vague. She condemned the practice of syringing after intercourse—the method approved by doctors administering the Contagious Diseases Acts.[6] Prostitutes so treated were not meant to be mothers, Besant wrote; 'they are to be kept fit for use by Her Majesty's soldiers'—she condemned the Acts as infamous, and the method obnoxious. The check she preferred was one that Richard Carlile had described, as long ago as 1827, in his *Every Women's Book*; a fine sponge dipped in water. Besant was careful to distinguish the methods which prevented life from abortion, which destroyed it; that was a criminal offence.

Her strongest denunciation was reserved for the practice of celibacy, either by remaining unmarried, or by delaying marriage—the only remedy the Revd Malthus advocated. Celibacy was unnatural, Besant declared: 'to be in harmony with nature men and women should be husbands and wives, fathers and mothers.' Until nature evolved a neuter sex celibacy would be a mark of imperfection. 'Very clearly has nature marked celibacy with disapproval, the average life of the unmarried is shorter than that of the married; the unmarried have a less vigorous physique, are more withered, more rapidly aged, more fanciful.' Early marriage was best both physically and morally, Besant declared: 'it guards purity, softens the affections, trains the heart and preserves physical health. It teaches thought for others, gentleness and self control.'[7]

How could Annie write such words in the light of her experience with the Revd Frank Besant? Either she did not connect the ideas set out in the *Law of Population* with her own life, or her attitude to marriage had altered since her flight from Sibsey. Perhaps there was truth in both these alternatives. Regarding the first, there were many occasions on which Annie Besant publicly contradicted herself, voiced opinions demonstrably at odds with her personal experience—so many occasions that they damaged her credibility in public life. That was the penalty she paid for her dramatic genius. Those who heard her felt betrayed when they discovered that whatever it was she, as orator, had made them believe was not a conviction that necessarily endured with her.

On the second point, Annie's kind words for marriage may have been inspired by the domestic harmony she and Charles Bradlaugh were enjoying together at the time. In February 1877 Bradlaugh, who had inherited some money from an admirer, moved to Circus Road, St John's Wood, just round the corner from Oatlands. Though vastly more salubrious than his lodging in the East End, the accommodation was almost as eccentric. He and the girls had the top floor of the house, where his books filled the only room of any size, and a huge dark room in the basement. The middle of the house was occupied by a firm of music publishers. Meals were taken in the basement where the ramshackle chairs from Turner Street stood forlornly in the middle of the floor. Hypatia presided here over a little maid, a pale shoot growing in the dark, 'much given to fainting'. More often than not Bradlaugh dined at Oatlands, where Annie kept an excellent cook and parlour maid.[8]

Each morning, after he had attended to business brought to his door as the most famous poor man's lawyer in the country, Bradlaugh took his papers over to Annie's book-lined study, where he sat working hour after hour. She was equally busy with her own work. They spent the day together until, about ten o'clock, he would take himself off again to his own lodgings.[9]

'In their common labours, in the risks and responsibilities jointly undertaken', Hypatia Bradlaugh wrote in her published memoir of her father, 'their friendship grew and strengthened.'[10] But the closeness of their relationship did not please Alice or Hypatia, and it infuriated all those who deplored the influence Mrs Besant had gained over their leader since his illness. Privately Hypatia wrote, 'She took possession of my father and by her airs of proprietorship and superiority she aroused the bitterest jealousy among those who had known and worked with him for many years.' Hypatia complained that it never occurred to Annie to consider the feelings of other people: 'she was the most tactless person I ever knew.'[11]

Hypatia's anger was particularly aroused by a letter Annie wrote to an Edinburgh acquaintance, G. A. Stewart, asserting that Alice and Hypatia were favourable to the 'intimacy' which existed between their father and herself. Stewart asked the girls if this was so. An outraged Hypatia declared that it was absolutely contrary to the truth. Conscious of having offended them, one day when Bradlaugh was out of the room, Annie knelt down beside the girls to say that, while she knew it was not true, she hoped it would one day be true.[12]

There is no direct evidence of the nature of the friendship; as far as is known all letters were destroyed. For Bradlaugh to defend Annie against all comers, as he did for many years, was to allow insinuation to flourish unchecked. He was indifferent to whispering campaigns, as when he refused all explanation of his treatment of his wife.[13] But the Knowlton trial had made him and Annie Besant two of the most famous people in the country; their lives were at the mercy of the public's curiosity. Most people who did not know them concluded that they were lovers. One who did know both of them, the journalist W. T. Stead, believed their relationship was 'pure'. Nevertheless, he told his friend Olga Novikoff, 'Except Mrs Besant no lady is associated with Mr Bradlaugh, and the association is not savoury.'[14]

Moncure Conway was one of the friends who, after Bradlaugh's death, publicly denied that he and Annie had been lovers, an impression for which, privately, he held Annie responsible. Bradlaugh was a man of nice ideas of honour, he wrote, in all matters relating to sex and marriage not merely chaste, but exceptionally conservative in opinion. 'he was cruelly slandered in relation to a refined and eminent lady . . . I happened to know these two leaders of Freethought had definitely sacrificed their happiness for the sake of their example, and the honour of their cause.'[15]

Annie was as touchingly eager to bestow love as she was to receive it, and quite unguarded in her expression of it. But people who knew her very well believed her to be chaste; they suspected that her marriage to Frank Besant destroyed all further sexual inclination.[16] There can be no doubt that she was surprised and hurt by the antagonism of the Bradlaugh girls, which continued even after Susannah Bradlaugh's unexpected death at the time of the Knowlton trial. Their attitude prevented Annie from becoming what she yearned to be—a member of a family.[17]

Her fondest memories of Bradlaugh were of companionship, of the occasions when he and she left work for a time to go for long walks in Richmond Park or Kew Gardens, where they had tea in a 'funny little room with watercress *ad libitum*'. Annie's favourite excursion was to the Lee Valley, east of London. Bradlaugh was a great fisherman and knew every eddy of the river, and taught Annie all he knew, except how to conquer her distaste at the sight of a fish wriggling on her hook. 'In those days he would talk of all his hopes of the future', she wrote, 'of his duty to the thousands who looked to him for guidance, of the time when he would sit in Parliament as Member for Northampton.'[18] As a result of the Knowlton affair that time was nearer than they imagined.

When Annie arrived to lecture in Northampton on 4 August she was received with cheers. The courage she had displayed during the ordeal of the Queen's Bench trial had elevated her status to that of heroine in many people's eyes—one young woman insisted on kissing the hem of her skirt.[19] At Northampton the subject of her lecture was political: 'A free Press, how it was won, and how it should be maintained.' The hall was crowded.

The audience, which sat shoulder to shoulder and knees to the spine of the man in front, were no longer confined to secularists and the more embattled of their Christian opponents; they were ordinary working people—miners, cutlers, weavers, spinners, shoemakers, Annie noted, thereby describing the pattern of her journeys through industrial England.[20] Bradlaugh had often boasted that he commanded a constituency, formed by his supporters, larger and more politically aware than that of many MPs.[21] Now, as a result of the Knowlton trial, that extra-parliamentary constituency materialized, giving its allegiance, for the moment, as much to Annie Besant as to Charles Bradlaugh. Annie quickly divined the message her audience came to hear: she no longer concentrated on attacking Christianity but talked to them of peace, anti-imperialism, and social justice. These were aspects of the theme which governed her thought and action from this time on—the necessity for men to live as brothers. Wherever she strayed she was constant to this fixed idea, and, though she turned away from Bradlaugh's teaching, she honoured him for having awakened her to the inspiration of her life's work.

In the spring of 1878 the country was again in ferment over the risk of war posed by England's support for Turkey as a shield against Russian territorial ambition. At Huddersfield, in March, Annie was cheered to the echo when she demanded a general election on the issue of the Prime Minister's juggling with the Eastern Question.[22] She and Bradlaugh were acclaimed by their new working-class supporters when they appeared at peace demonstrations called to oppose the jingo militants. Looking back on that exciting time she wrote: 'the Freethought party may well take the credit for being first in the field against the Tory policy and for having successfully begun the work carried on by Mr Gladstone in his Midlothian campaign.'[23]

The free-thought party to which she referred—with unaccustomed tact—was no longer the same she had joined four years before. The strain imposed by the Knowlton affair had brought about a split.[24] While Bradlaugh and Besant were preparing for the trial, the publisher Edward Truelove was arrested for selling Robert Dale Owen's *Moral Physiology*. His accusers, the Society for the Suppression of Vice, did not rest until Truelove, who was 66, had been sentenced to four months in gaol, which he had to serve. Bradlaugh and Besant were criticized for bringing disaster to an old man, and for not doing enough to help him. At the annual conference of the National Secular Society just before the Queen's Bench trial, certain members tried to force Annie Besant's resignation as a Vice-President for exercising undue influence in the decision to defend Knowlton's book. When the motion failed a number of prominent secularists, among them Mr and Mrs Charles Watts, George Holyoake, and Harriet Law, departed to form a separate body, the British Secular Union. However, more than 600 new members flocked to Bradlaugh and Besant in the course of the following year.

Their prestige within the National Secular Society and in the world at large was greatly enhanced by the decision of the Court of Error which sat in January 1878 to consider Bradlaugh's objection to the indictment in the Knowlton case. The three judges decided that, since it had not set out the words complained of as obscene, it had not been formed in accordance with the rules. The judgement was set aside, and Bradlaugh and Besant went free.[25] They kept Knowlton's book on sale until it became clear that the authorities had no stomach for another battle—or so they thought—then it was quietly withdrawn to make way for the *Law of Population*. Preliminary moves were made to indict the latter but no warrant was ever issued. The two partners in the Freethought Publishing Company rejoiced in victory. Even as they did so the Church of England prepared to punish Annie Besant by appealing to the law.

With advice and encouragement from the bishop of Lincoln,[26] Frank Besant briefed counsel to prepare a petition for custody of his daughter Mabel to be restored to him. This would state that by lecturing and publishing Mrs Besant had propagated atheism, and had associated herself with an infidel named Charles Bradlaugh, 'whereby the truth of the Christian religion is impeached'.[27]

CHAPTER 12

BESANT v. SOCIETY

FRANK BESANT and his advisers obtained firm evidence on which to base a plea for the return of his daughter when Mabel was sent to a day school in St John's Wood. They discovered that the Headmistress had been instructed to excuse her from all religious teaching.[1] In March 1878 Mabel fell ill with scarlet fever contracted at the school.[2] When Annie's anxiety was at its height she was served with notice of her husband's petition. 'I am anxious and have expressed to the said Annie Besant that Mabel Emily Besant should be kept free from all association with Charles Bradlaugh. She has not acceded to my request',[3] the petition stated, and claimed that it would be detrimental to Mabel's morals and happiness to leave her in her mother's care. Pending a decision, the 7-year-old child was made a ward of court.

The case appeared in the lists as *Besant* v. *Wood*, the defendant being Annie's trustee for the Deed of Separation, her brother Henry Wood. He was already eager to divest himself of responsibility for his sister's affairs and, faced with the prospect of Chancery proceedings, which were notorious for tedium and inordinate expense, he declined to appear. In his absence Annie, as a married woman with no separate legal identity, had to have someone to act as 'next friend': the choice fell naturally on Charles Bradlaugh. Given the terms of the petition, he and Annie regarded the new action as another opportunity for a public contest between free-thought and the Christian religion.

Annie opened the campaign with a letter chiding *The Times*[4] for giving notice of the case as one in which the vicar of Sibsey alleged his wife was not a fit person to have charge of Mabel. It was not herself but her views that were in question, Annie declared; the judge must decide if her atheism and neo-Malthusianism must deprive her of her daughter. The judge was to be Sir George Jessel, Master of the Rolls. *Besant* v. *Wood* qualified for a person of his eminence because it was the first dispute to arise under the Custody of Infants Act 1873. Jessel remarked that it would surely throw up some 'nice points of law'.[5]

His legal knowledge was the admiration of his colleagues; his mind was formidable, his memory amazing: Jessel did not have to look things up, and he rarely wrote them down. In the course of his career, now approaching its end, he wound himself up to such a pitch he could not abide interruptions or delay. The time he won from legal matters was devoted to the management of London University, his beloved Alma Mater. The fact that Jessel was a Fellow of University College in Gower Street was to have a damaging effect upon Annie's career later on. He was the first Jew to be a Law Officer, a Privy Counsellor, and a judge.[6]

Sir George was famous for his brusqueness to counsel and witness alike. But when Annie appeared in his court on 18 May he was downright rude. 'Appear in person? A lady appear in person? Never heard of such a thing. Does the lady really appear in person?' When Annie said she did, Jessel warned her that she would not be heard at greater length than the case required, or allowed to go into irrelevant matter, 'as persons who argue their case usually do'.[7]

Jessel's hostility at this early stage arose from his desire to hear the case in chambers, which was usual when marital relations and the welfare of children was involved.[8] The defence had objected on the grounds that as a married woman—feme covert—the mother could not be a party to any compromise which might be offered there. On that point Jessel was unsure; moreover he suspected that Bradlaugh and Besant were after maximum publicity.

The case was indeed attracting a great deal of attention from those who were campaigning for a relaxation in the very restrictive laws affecting married women. Among those in court was a leader of the women's movement in the United States, Colonel Thomas Wentworth Higginson.[9] He was a friend of Moncure and Ellen Conway, who supported Annie by their presence at this ordeal as they had before.

The case before the Master of the Rolls turned upon the Deed of Separation signed in October 1873.[10] Frank Besant had to establish that he did not know of his wife's atheistic views at the time he committed his daughter to her care, or his petition failed.

Annie had to make the court recognize her legal standing. She argued that the Deed made her the child's guardian appointed by the father in full knowledge that she was not a Christian.

Frank's backers had enabled him to brief expensive counsel: Mr Ince, QC, and a colleague, Mr Bardswell. Together they painted a picture of Mabel's mother completely under the influence of Bradlaugh, the 'bellowing blasphemer'.[11] During their association she had written shocking books like *The Freethinker's Text Book*, and the *Law of Population*. She and Bradlaugh had published a pamphlet which stood condemned as indecent and obscene. During the trial Mrs Besant had flaunted her intention of teaching young children physiological facts contained in that pamphlet—here Ince, QC, threw back in Annie's face her words about her daughter's happiness depending on a knowledge of the facts of life. A girl brought up by such a mother must be shunned by respectable women, Ince concluded.

'Most impertinent and unworthy of a gentleman', was Annie's scalding comment on insinuations about her friendship with Charles Bradlaugh. He had never even talked of religion in Mabel's presence she told the Judge. She denied that she had ever intended to teach young children the facts contained in Knowlton's *Fruits of Philosophy*. However she admitted that she had given instructions for Mabel to be excused from religious teaching at her school, and that she had withheld the Bible from her wanting her to be old enough to understand it before she read it for herself. Mabel was a happy child, free to do much as she pleased. For instance, 'I left her to sing hymns all over the house, only telling her the words were foolish', Annie said. She had taken her daughter occasionally to churches, Roman Catholic and others, 'so that she might see there were such religions in the world; I tried this to counter balance the effect of her being turned entirely in the direction of one when she went to Sibsey.'

In spite of his foreboding Jessel was impressed; 'I must say you have conducted your case very ably and temperately', he conceded.

For the defence, Dr Charles Drysdale said that Mabel, a sensitive and nervous child, needed the continued care of her mother.[12] For the petitioner, Dr Philip Heasly said there was no reason to suppose that her removal from the custody of one person to another would in any way endanger her health.[13]

Those in court watched impatience gaining on the Master of the Rolls. When the time came for him to pronounce judgement he popped like a cork at the moment of release. This was as unpleasant and painful a case as any that had come before him, he began, the more so for being argued in public. The question was, if the Deed of Separation between the Revd Besant and Mrs Besant was valid, ought he—Jessel—to enforce it?

According to the Custody of Infants Act 1873 he was not to enforce it if he so much as doubted it was for the child's benefit. In that case the father's rights were the same as they had been before the passage of the Act—the father had a legal right to the custody of the child, and a legal duty to control and direct the child in its religious and secular education.

There was no advantage to Mabel physically in living with her mother, Jessel declared. In going to her father she would have the companionship of her brother Digby. 'As a man of the world, and speaking as a father, I am satisfied that solitary children are not so likely to make good men and women as children brought up in the society of brothers and sisters.' Jessel was glad to say that Mrs Besant had been kind and affectionate to the child. But she had taken it upon herself not merely to ignore religion, not merely to believe in no religion, but to publish that belief. As a man of the world Jessel had to consider what effect upon a woman's position this conduct would have. As a man of the world he must conclude that it must quite cut her off from the great majority of her sex. 'Not only does Mrs Besant entertain those opinions which are reprobated by the great mass of mankind (whether rightly or wrongly I have no business to say, though I, of course, think rightly), but she carries these speculative opinions into practice as regards the education of the child.' To deprive Mabel of religious education was not only reprehensible but detestable, the Master of the Rolls declared; on that ground alone the child ought not to remain another day in the care of the lady.

But, Jessel was sorry to say, there was another ground which he would gladly avoid if he could—the publication of an obscene book, the *Fruits of Philosophy*. Having had his attention drawn to some pages he was sorry to say he could entertain no doubt of its being obscene: 'although the conviction was set aside on a technicality, no Judge, so far as I am aware, has doubted the propriety of that conviction.' Mrs Besant's character was to be assessed, not only by the publication of the book, but by the conviction following from that publication. '*One cannot expect modest women to associate with her*'. Mabel Emily Besant must forthwith be given up to her father.[14]

The sensation created by Jessel's words was as great as if he had sentenced Annie to a public flogging. Whatever animus provoked his outburst—it may have been revenge for the verdict of the Court of Error—his remarks gave judicial sanction to the view that Annie was an outcast from society. Moncure Conway described, with feeling, how on hearing the Master's admonition, all eyes turned to his wife Ellen, who was sitting next to Annie.[15]

Throughout her life Annie was steadfast in the belief that the world was the object of a contest between good and evil, light and dark. Jessel was evil; Jessel had the sneer of Mephistopheles. 'The old brutal Jewish spirit regarding women as the mere slaves of men breaks out in the coarse language that disgraces him rather than the woman at which it is aimed', she told the readers of the *National Reformer*. To Hebrew bigotry he added the time-serving morality of a 'man of the world', 'sceptical to all sincerity, and contemptuous of all devotion to an unpopular cause'.[16]

Bradlaugh, who feared that Frank Besant's advisers would not lose a moment in claiming Mabel, sent his daughters from the court to carry the child away from home. They took her to Willesden Junction Station to meet Annie who, as soon as the hearing was over, had to set out for Manchester, where she had a lecturing engagement. 'Thus the poor mother was able to take her farewell from her child in peace instead of having her torn at a moment's notice from her arms',[17] Hypatia wrote, mistakenly, according to Annie's *Autobiography*. That describes how, when the messenger from Sibsey eventually arrived, Mabel had to be forcibly carried away, 'shrieking and struggling . . . near frantic with fear and passionate resistance'.[18]

In Mabel's absence the house was cold and empty; loneliness, which Annie feared above all else, invaded it. She was briefly sustained by the prospect of rejoining battle. On 26 June she asked leave to appeal against Jessel's decision without a 'next friend', Bradlaugh having deliberately refused to act again in that humiliating capacity.[19] This was granted on condition that Annie paid £20 into court as a security. Then came July, the month for Digby's annual visit to his mother; he did not appear. Upon enquiry the vicar's advisers let it be known that a mother who was not fit to have care of her daughter could hardly expect to be entrusted with the son.[20] Later Frank Besant explained that he feared an upsurge of publicity if Digby was delivered to his mother.[21]

Shock and humiliation drove Annie frantic. Her solicitor George Lewis realized something urgent must be done. On 22 July he wrote to the vicar of Sibsey to say that Mrs Besant had come to the determination to make any sacrifice for her children, from whom she would not live separate. 'She has therefore now decided to go home and live with you.' What day would he be ready to receive her? When no answer was forthcoming, a second letter was dispatched saying that, if Mrs Besant did not hear, she would be compelled to come to Sibsey.[22] The reason for this move, Lewis explained later, was to frighten Frank Besant into giving up the children since Annie, as a married woman, could not sue. But if her husband had refused to receive her at Sibsey, the way would have been open for Lewis to start proceedings for a judicial separation on the grounds that her husband would not live with her.[23] Frank was determined to thwart any such attempt by Annie to obtain her freedom. On 5 August, therefore, he sought and obtained a temporary injunction restraining his wife from molesting or annoying him.

This was served on Annie as she lay in bed racked with the pain of rheumatic fever and, at times, delirious. The illness marked a similar collapse to that she had suffered after Mabel's near death with bronchitis. According to her *Autobiography* the fever was extremely grave and lasted for many weeks; during it Bradlaugh sat beside her, writing steadily, feeding her at intervals with the ice and milk she would take from no one else. Hypatia was somewhat aggrieved by this account; she and Alice had done their best to nurse Annie through this illness, which she felt was not all *that* severe.[24] By the end of August Annie was well enough to go with her and Alice to convalesce in North Wales. A room had been booked at a hotel in Betws-y-coed, but when the proprietors realized who Annie was, they refused to admit her. She and the Bradlaugh girls went on to Capel Curig where they found more hospitable lodgings. Annie devoured *Les Misérables*, and rode up Snowdon on a pony escorted by a group of young men, members of an Oxford reading party who were staying in the village.[25]

After a debilitating illness, and with crucial legal actions pending, any other person might have avoided extra work. On her return from Wales Annie forgot her troubles in research for a series of articles to be published in the *National Reformer* towards the end of the year. 'England, India and Afghanistan; A Plea for the Weak against the Strong' was inspired by the further adventures of the Tory Government which, in order to counter the perennial Russian menace, had got itself involved in a ferocious Afghan war.[26]

Annie had met administrators of the Raj at Cheltenham, and in Thomas Scott's salon; she had attended Moncure Conway's seminars on Indian religion at South Place Chapel; she had often presided at meetings when Bradlaugh belaboured missionaries for their interfering zeal; she was, therefore, perfectly well aware of the appalling problems that beset rulers and ruled in India barely twenty years after the Mutiny had been bloodily suppressed. Nevertheless 'England, India and Afghanistan' contained a proposal for reform which was several generations before its time. The means by which, in 1878, Annie Besant hoped India would achieve self-government were virtually the same as those she commended to the multitude that witnessed her inauguration as President of the Indian National Congress in 1917.[27]

In the course of a lucid survey of the British connection with India, Annie deplored the destruction of ancient, indigenous forms of government, particularly the system of land tenure which had allowed an effective form of local democracy. Her analysis conspicuously favoured Hindu tradition, religion, and law above those of all other races. Hindus were the aristocracy of the East, Annie wrote, 'learned, acute, subtle, dignified, courteous'—was she describing someone she had met? People like these must be capable of governing themselves better than the British could. 'I would let the supreme power gradually pass, not into the hands of the princes', she wrote, 'but into the hands of the Indian people so that a mighty self governing nation should slowly arise.' She dismissed the 'Russian menace' as a shibboleth. An India oppressed by English tyranny might well look to Russia for freedom, but with India free, 'the Cossack must seek other hunting grounds'.

Once the article was in print—it coincided with news of British disasters in Afghanistan—Annie's attention reverted to the legal business interrupted by her illness. She faced two separate actions. One was in the Court of Appeal to overturn the Master's decision which had taken Mabel away from her, the other, in Chancery, was to prevent a permanent injunction being granted to her husband. The Court of Appeal sat on 24 March 1879. Annie, who again appeared in person, claimed that she was Mabel's legal guardian. In support of the argument that her unbelief was no ground for depriving her of her child she mounted a defence of free-thought which Bradlaugh praised as 'unequalled in its boldness'. In particular Annie confronted the example of the poet Shelley, which might have been turned against her. His children were removed from him, not on the grounds of his infidelity, but because of his immoral way of life—an argument that tacitly defied the court to prove her immoral too.

As for depriving Mabel of religious instruction, Annie reminded the judges that, under the 1870 Education Act, such training was not compulsory in schools. The standard of education Mabel would receive greatly concerned her mother. If the child stayed with her father it would be third rate, that was all a country parson could afford. In the country she would be thrown together with people of a lower rank, farmers and their sons and daughters, 'perfectly good respectable people', Annie conceded, but 'not her equals by birth'. And if one was to consider so foolish a point as matrimony for a child of 8, Annie went on, when she lived at Sibsey there were five daughters of a clergyman unmarried there, and six in the next village.

Her long, laboriously constructed argument was to no avail. With infinite courtesy Lords Justice James, Bramwell, and Bagally damned her for her infidel opinions. Her conduct in propagating them, though done with good intent, would be regarded with disgust by Englishmen and women. 'The Court cannot allow its ward to be brought up in opposition to the view of mankind generally as to what is moral, what is decent, what is womanly, or proper merely because her mother differs from these views ... the child might even grow up to write such things herself!' The appeal was dismissed, with costs, to the extent of the £20 Annie had already paid.[28]

When the Chancery proceedings arrived before the Master of the Rolls, the pleadings were about as odd as any he had seen,

Jessel remarked: neither party wanted to live with the other.[29] 'The object of the lady—by no means an improper object—is to be free from having anything to do with her husband, and to acquire certain rights which only women who are separated can have.' With an eye to dispatch, and a desire to be helpful, Jessel suggested that a judicial separation would put an end to the litigation. He told Annie he could grant her one, 'though not today'. She was delighted. The Master directed George Lewis to arrange the preliminaries with Ince, QC and Bardswell. On behalf of their client Frank Besant they refused even to discuss it. Jessel, who had assumed both parties to be in agreement, was nonplussed. Now, to obtain a judicial separation which was contested, Annie would have to prove she went in fear of her husband, and produce evidence of his cruelty. The hearing was adjourned so she could do so.

Among the affidavits she obtained was one from the sympathetic doctor who had attended her at Cheltenham. Her own evidence, the gist of which was published in the *National Reformer*, included details of Frank shaking and pushing her while she was pregnant, and of his hostile behaviour and language.[30] 'The lady has brooded over what she supposes to be her wrongs, which are purely imaginary',[31] Ince declared, and refused to call his client. Though Frank Besant's failure to refute the charges was tantamount to an admission, all such incidents were held to have been condoned by the signing of the Deed in 1873. Jessel was forced to decide that it raised an absolute bar to the granting of a judicial separation. So did the letters threatening Annie's return to Sibsey; they demonstrated that she could not have been in fear of her husband. Jessel reproved both parties for trying to manipulate the law, and ordered the injunction restraining Annie from molesting or annoying her husband to be made permanent.[32]

So Frank and Annie Besant were bound together in law until death might part them. And, although the Court made reasonable provision for Annie to see Digby and Mabel, she found the supervision of their meetings by persons appointed by their father so irksome that she soon forsook the opportunity altogether in the hope (which was fulfilled) that they would seek her out when they were free to do so.

For nearly ten years the children's lives were divided between boarding school and holidays at Sibsey where they went in awe of their father. 'I suppose no form of training can induce more autocratic tendencies in a man than . . . scholarship, schoolmastership, and control of a country parish', Digby Besant wrote. 'The Vicar did not make friends easily; he was too austere and his temper was too sharp . . . he did not attempt to hide the fact that intellectually he was head and shoulders above nearly all the people with whom he came in touch.' Frank kept to his study, while the children had the garden, and friends in the village, 'so far as the Vicar would permit'.[33] For warmth they went to the kitchen, to the housekeeper Mrs Baines. Otherwise they were on their own. 'These were Calvinistic days', Digby wrote, 'and to me, a delicate, nervous and imaginative child the fear of hell fire was ever present. Many a night have I leapt trembling out of my bed at Sibsey lest I should find the horizon in flames and the Day of Judgement arrived.'[34] Frank was fearful too. Crates of examination papers filled the house to overflowing; the vicar marked them to a rigorous timetable in the despairing hope of earning an annuity large enough to support him if illness, or any other reason, deprived him of his living. His anger if the papers were disturbed by dusting regularly reduced Mrs Baines to tears.

Yet he kept the friends he made, his son admitted. And in his extremity his neighbours rallied to his support. When Annie threatened to carry the case over custody of Mabel to the House of Lords an appeal was launched for contributions to the cost of the Revd Besant's struggle to secure the education of his children 'in the pure morality and faith of Jesus Christ.'[35] The bishop of Lincoln headed the list with £10. Altogether twenty-eight clergy, nineteen Justices of the Peace, and four Members of Parliament from the area supported Frank Besant. Bradlaugh pilloried them in the *National Reformer* as persons who had taken Mrs Besant's little daughter away from her. 'There is no one on that list who will be forgotten by me', Mabel's mother told the newspaper-reading public.[36]

CHAPTER 13

TUTORIALS

FRUSTRATION at her defeat now betrayed Annie into behaviour that was often ill considered and occasionally extreme. She allowed her distress at losing Mabel to be paraded in the interest of propaganda for the secularist cause. Portraits of the little girl in boots, short skirts, ringlets, and a simper were printed by the dozen with the caption 'Mabel Emily Besant, deprived of her mother, May 23 1878'.[1] The *National Reformer* advertised a package of books—*The Freethinker's Text Book*, *The Gospel of Atheism*, *The Fruits of Christianity*, and the *Law of Population*—as, 'works by Mrs Besant used in justification of taking away Mrs Besant's daughter'.[2]

The campaign was prolonged, and public sympathy, which had been aroused in some quarters by Jessel's brutal words, turned to pity, a suspect emotion at the best of times, and detestable to one as proud as Annie. Too much under strain, she could no longer cope with adverse situations that once she would have mastered with a phrase. When members of a deputation from the National Sunday League, whose aim was to liberalize Sabbath activities, waited on her to tell her their new President, Lord Thurlow, so objected to her neo-Malthusian views that he refused to take office while she remained a Vice-President, her poise abandoned her: she declined to discuss the matter and drove them from her house.[3] She was equally fierce in her advice to readers of the *Reformer*; she told them she had made up her mind not to join any society which regarded free-thinkers with dislike, or even with toleration. 'Those who seek our aid must treat us with the respect they demand for themselves.'[4]

Annie had already learned that positive action was her best relief from animus. Through intellectual effort she had thrown off the bonds imposed on her by the Church of England. Now the same ability would rescue her from the worst effect of her defeat in the courts. She resolved to overcome the system whose servants had unjustly deprived her of her daughter; the first step would be to gain a degree in law.[5] It was only a year since London University had amended its Charter to admit women students; so far no female had graduated, let alone been admitted to the Bar.[6] Annie was never intimidated by lack of precedent and, fortunately, the time was ripe. Other women, equally well endowed with capacity and the will to act, saw their purpose founder on lack of opportunity.

The boldness of her aim caused her critics to assume that the extravagant praise heaped on her for her performance in the courts had turned her head. But her attempt to qualify in law attracted serious encouragement. Bradlaugh, for one, was delighted at her interest in a subject dear to his heart; his famous thoroughness would keep her up to the mark.[7] As a founder member of the Dialectical Society, no doubt his influence was instrumental in the Society's invitation to her to join a committee examining the need for changes in the law.[8] Nevertheless it was a valuable endorsement.

The undertaking differed from Annie's previous adventures in that it had to be conducted according to someone else's rules: discipline, not drama, would be required. Before proceeding to a degree course at London University students had to matriculate by passing examinations in five subjects, including mathematics and a science. Annie had never submitted herself to public examination. Careful tuition would be needed; hard work too. 'Mrs Besant, thinking that it may add to her usefulness to the cause, intends to try to take advantage of the opportunity for women obtaining degrees in the London University,' Bradlaugh announced in the *National Reformer*. 'The necessary studies in preparation for the severe examinations will occupy so much of her time that for many months to come she will be able to lecture only on Saturday and Sunday. Miss Hypatia Bradlaugh is studying with Mrs Besant for the same object.'[9]

That Annie and Hypatia embarked on the enterprise at all was largely due to encouragement from Joseph Hiam Levy, who taught political economy at the City of London College, where they started classes. Levy, who was an enthusiastic advocate of women's emancipation,[10] had been attracted to Bradlaugh and Besant by their defence of the Knowlton pamphlet. Subsequently, as Dialecticus, he often deputized for Bradlaugh in writing articles on politics and economics for the *National Reformer*. It was Levy who arranged for Annie and Hypatia to be tutored in science by Edward Bibbins Aveling,[11] a young man Annie had already met in her capacity as sub-editor of the paper. In January 1879 he had begun to contribute articles signed 'E.D., (D.sc. Lond.)'.[12] The disguise was necessary to safeguard his career. Though Dr Aveling (like Sir George Jessel) was a Fellow of University College London, which had been founded on non-sectarian lines, he taught science in an Anglican establishment. King's College, in the same University, was hardly more liberal in matters of religion than it had been in the days when Walter and Frank Besant were undergraduates there.[13] The risk that Aveling ran of losing his job by writing for a free-thought paper won him the immediate sympathy of Annie Besant.

They had other things in common. Both had taken refuge in free-thought after a close acquaintance with orthodox religion; Annie as a parson's wife, Aveling the son of a Congregationalist Minister. His family, like the fondest part of hers, was Irish. And he had been at Harrow during the years her mother kept the boarding house for boys—when Annie was safely shut away at Fern Hill. At school, and afterwards at University College, Edward won prizes and medals in every scientific subject, and seemed destined for a brilliant career in medicine. But his chief love was the theatre. He wrote plays, acted in them, and worked up a reputation as a critic; he had even briefly managed a group of strolling players. This histrionic flair enhanced his effectiveness when lack of money to pursue his medical studies forced him to become a teacher.[14]

Even after their relationship had collapsed in acrimony,[15] Annie was generous in her assessment of his ability. 'Clear and accurate in his knowledge, with a singular gift for lucid explanation, enthusiastic in his love of science, and taking vivid pleasure in imparting his knowledge to others, he was an ideal teacher.'[16] She derived her pleasure from widely different aspects of his teaching. In his tutorials he provided facts, for which she always hungered; and a masterly exposition of the ruling theory—he was a convinced Darwinist. At the same time he allowed himself romantic speculation.[17] Aveling posited a connection between medieval alchemy and astrology, and modern chemistry; he traced the origin of psychology, the science of the mind, to ancient myth and legend—stories which were familiar to Annie from her study of the Fathers of the Church. He followed Comte in suggesting that all branches of science were related in a manner yet to be discovered. Annie was captivated by what she termed his 'luminous and beautiful' scientific theories.[18]

Shortly after he began to teach them Hypatia was thrown from Kathleen, Annie's mare. The accident occurred on the same day that the Court of Appeal decided in favour of Frank Besant; it was a measure of Annie's distress that she ordered Kathleen, though not injured, to be shot.[19] Hypatia, who was concussed, was advised not to read for several months, so Annie and Edward were left to study tête-à-tête. When Hypatia rejoined them she found Annie no longer interested in law: 'It was science which attracted her in the person of Dr Aveling.'[20] He was nearly 28; Annie 31; Charles Bradlaugh 46.

Now Aveling was not a handsome man. He was undersized and had the eyes of a basilisk, and it was said of him that he would have been interesting in a Zoological museum as a reptile but impossible as a man. Short of actual deformation he had every aesthetic disadvantage except a voice like a euphonium of astounding resonance and beauty.[21]

Thus George Bernard Shaw who found Aveling's power of attracting women as surprising as his own.

Perhaps the attraction was due to Aveling's manner, which was dramatic: he offered admiration, sympathy, affection in large measure; 'he was as gushing as a boy',[22] one of his few male friends remarked. He indulged in fits of poetic dreaming. He was well turned out, with a hint of the dandy about his clothes, as befitted one who liked the best of everything. Among the members of Bradlaugh's circle only Hypatia did not take to him at once. She suspected that he sought out her and her sister Alice in order to 'flatter and pump' for influence with their father.[23] She communicated her animosity to the dog Lion, which, on the first occasion Aveling appeared in Annie's drawing-room, kept up a low growl from his place at Bradlaugh's feet. Annie, on the other hand, greeted her guest with effusion. Aveling was given his own key to Oatlands.[24]

Among the Bradlaugh papers is a letter by Aveling of birthday greetings addressed to 'My dear friend', dated 26 September 1879—Annie was born on 1 October.

I have already gained so much from you, and have been so greatly strengthened by your strength, and made so much more desirous to be all that is good by the sight of your goodness, but for the sake of your future, as well as for the sake of the world's, I hope your life may last for many years to come, and that throughout such of those years as form also a part of my life, I may be able to sign myself as I do now, Your affectionate Friend, Edward Bibbins Aveling.[25]

The tone of respectful adoration is maintained in Aveling's other letters to Annie Besant that have survived.

In July 1879 Aveling made a public declaration of his commitment to secularism in the *National Reformer*; henceforth he wrote under his own name. Annie's favour, joined to his eagerness to serve, and his obvious ability as a lecturer, led to a rise within the movement as spectacular as hers had been. Soon he came third in the list of speakers after Bradlaugh and Besant. Every December at a soirée and ball CB and AB addressed the rank and file; in 1879 the newest member of the triumvirate also delivered a homily.[26] By that time he was referred to—affectionately no doubt—as 'Edward D.sc'. A few weeks later he was called upon to perform the ceremony of 'naming' a child which, for earnest secularists, replaced the sacrament of baptism.[27] In May 1880 he joined Annie as a Vice-President of the National Secular Society.

To her Aveling's speech was music, his language exquisitely chosen, his charm artistic, his scholarship deep, and his knowledge wide. 'Our friends will not wonder that we, who know him, rejoice that our mistress, Liberty, has won this new knight.'[28] But did she—or Bradlaugh—really know him? '[Aveling] would have gone to the stake for ... Atheism but he had absolutely no conscience in his private life. He seduced every woman he met and borrowed from every man', Shaw remarked.[29] He was an odd mixture of fine qualities and bad, the schoolmaster Henry Salt declared: 'his duplicities were the result less of a calculated dishonesty than of a nature in which there was an excess of the emotional and artistic ... with an almost complete lack of the moral.'[30]

Aveling's public espousal of free-thought seriously damaged his teaching career, and left him more than usually short of money. In a bid to remedy this he proposed that classes should be held at the Hall of Science, in the smaller of the two meeting rooms, with the object of preparing students for the examinations set by the government-supported Institute at South Kensington. For practical experiments he offered the use of his rooms at 13 Newman Street, to which he referred as his laboratory. He had already done some private teaching there but the necessary equipment was lacking. Bradlaugh and Besant, who were anxious for the idea to succeed, agreed to remedy the deficiency out of the funds of the Freethought Publishing Company; Bradlaugh also personally guaranteed the rent.[31]

The next hurdle was peculiarly difficult. A clergyman had to be found to serve on the management committee of the new course in order to qualify it for recognition from South Kensington and to obtain the grant which, with students' fees, would provide a salary for Aveling and the other teachers. Fortunately for everyone the Revd Stuart Headlam agreed to serve. He was a most unusual priest whom Annie was to come to know extremely well. Constantly at odds with his superiors over his unorthodox behaviour, Headlam could not look for preferment in the Church: he was, and would remain, a curate. Unlike Voysey, whom he somewhat resembled in his recklessness, Headlam was serene in his acceptance of the fundamental beliefs of his religion. His time and energy were devoted to practising Christian ideas in places where they did not often penetrate: the theatre, for example—about which he was as passionate as Aveling. His first curacy, in the early 1870s, was at St Matthew's, Bethnal Green. The better to answer his congregation's awkward questions about secularism he began to frequent the nearby Hall of Science, where he and Bradlaugh enjoyed their differences. Headlam's uncompromising integrity drove him to appear as a witness for the defence at the Knowlton trial.[32]

He was a dedicated supporter of Christian Socialism, whose founder, the Revd F. D. Maurice, had briefly offered hope to Annie in the ordeal of her marriage. One of its tenets was education for the underprivileged—including women. This was another link to Edward Aveling, whose first teaching post had been at the pioneering North London Collegiate School for Girls.[33]

Annie was a credit to J. H. Levy, Aveling, and Headlam: she gained matriculation in the shortest possible time. By July 1879 she was free to study for a degree and chose science, not law.

At the same time she undertook to help in teaching at the Hall of Science. Beginning in the autumn, the classes attracted a growing number of young men and women, not all of whom were secularists. Those that were received special encouragement from Dr Aveling. 'It is necessary that Freethinkers should think accurately', he admonished them; 'in the struggle after ... carefulness of reason and ... carefulness of speech, no better helper exists than the study of science.'[34]

Annie helped Aveling with experiments and studied animal physiology under him. In 1880 she gained a First Class in botany and animal physiology and taught elementary physiology to a class of thirty. Not to be outdone Hypatia Bradlaugh gained a First in mathematics, Alice the same in botany. Hypatia taught elementary chemistry, Alice, French. By 1881 the results were easily good enough to qualify for the precious grant from South Kensington.

It was secured against vociferous opposition in Parliament, for by that time politicians had joined churchmen in their desire to destroy Bradlaugh and Besant. The two leaders of free-thought were now perceived as threatening the political stability of the nation; on the one hand they encouraged measures of reform which some did not hesitate to call communist;[35] on the other persons desirous of overthrowing British rule in Ireland.

These two issues came together at a huge meeting in St James's Hall in Regent Street on 10 February 1880 to discuss land reform. Bradlaugh presided; behind him on the platform were Annie Besant, Stuart Headlam, Edward Aveling, and Joseph Arch, Secretary of the National Agricultural Labourers' Union, besides representatives from trade unions and politically active working men's clubs. The subject they spent all day and part of the evening discussing was one which all radicals considered fundamental to the advancement of social justice. Misuse of land deprived agricultural workers of a decent living and condemned industrial workers to harsh conditions. Why should they live cramped in narrow streets when land lay empty on the edge of towns because its owners did not care to build? A more immediate question was what would happen to food prices and agricultural wages as a result of the recent appearance in the markets of cheap grain from the American prairies?

But the reason why Arch, who was a Methodist preacher, and others who held no brief for free-thought agreed to attend a conference under the auspices of the notorious National Secular Society was that a dreadful emergency had fallen upon the countryside. 1879 was the kind of year that fortunately occurs only once or twice a century, when it rained relentlessly and the sun was not seen from one month to another. What little corn was harvested turned black in the stook and thousands of animals, particularly sheep, died from standing on perpetually sodden ground.[36]

For weeks before the conference Bradlaugh and Besant hammered away at the need for land reform. Annie applied her customary vigour—'Away with parks and ornamental lakes', she cried; land must be compulsorily cultivated or it would be confiscated by the state. At Bradlaugh's insistence reasonable compensation would be paid. His programme, which she publicized, called for an end to laws of primogeniture and entail whose effect was to create vast estates on which the land lay idle. The Game Laws were to be repealed, and the system of rents and taxes overhauled. Like Arch, who was promoting his own scheme for reform, Annie wanted to see a race of English peasant proprietors.[37]

When the conference assembled Bradlaugh found himself faced with a Pandora's box of opinions and demands. He was apparently taken unawares by a motion put forward by supporters of the London Trades Council calling for the outright nationalization of the land. With the help of a speech by Annie urging reasonable compensation by the state, the motion was defeated. Nevertheless some newspapers raised a cry of communism. Their readers were even more alarmed by reports that sharing the platform with Bradlaugh and his other guests was a leading Irish nationalist, Michael Davitt.

He was there to enlist support for the Irish Land League which he had set on foot the previous autumn, by which the ideas of the political economist Henry George concerning the equitable distribution and use of land were to be applied to direct action in relieving the oppression of Irish peasants by their British landlords. Although the supreme council of the Irish Republican Brotherhood, to which Davitt belonged, rejected the plan, preferring to conspire in secret, the majority of Fenians supported it, as, at that time, did Charles Stuart Parnell, leader of the Irish Members of Parliament at Westminster.

Before his release on ticket of leave in December 1877 Davitt had served over seven years in Dartmoor for his part in the Fenian conspiracy of ten years before. Annie was thrilled to appear in public with a colleague of the unfortunate men Lawyer Roberts had tried to save from the gallows. But Davitt was not the first Fenian leader she had encountered since that trial. In 1878, at the height of the crisis over the Eastern Question, she had let it be known that her views were based on information from a traveller.[38] This was the French socialist and soldier of fortune Gustave Cluseret who, after fighting for the North in the American Civil War, had thrown in his lot with those Fenians who had taken refuge in the United States.

Cluseret became known to Annie as a result of his contacts with Charles Bradlaugh. In 1867 he and another of the Fenian leaders, Captain Kelly, asked Bradlaugh to assess the likely response of the English working class to the independent Irish state they proposed to establish by force of arms.[39] Years later Bradlaugh privately admitted to helping Cluseret and Kelly draft the proclamation of this new state. In so doing, his friend and biographer John Mackinnon Robertson remarked, 'he recognized the right of Irishmen to use force against the force of England'.[40] This commitment, which ran counter to Bradlaugh's detestation of violence, made a significant impression on Annie Besant. In the great struggle of the 1880s Bradlaugh, with Annie's enthusiastic help, was to try to do more for the Irish cause than almost any other Englishman, always excepting Gladstone.

It was no small part of Bradlaugh's initial attraction for Annie Besant that, in 1867, he had made a strenuous and unpopular attempt to secure a reprieve for the Manchester martyrs, and followed it up with a pamphlet pleading for conciliation of the Irish. Annie quoted it at length in her *Autobiography*.

Oh before it be too late, before more blood stain the pages of our present history, before we ... arouse bitter animosities, let us try to do justice to our sister nation. Abolish ... the land laws which in their iniquitous operation have ruined her peasantry. Sweep away the leech like Church which has sucked her vitality ... Turn her barracks into flax mills, encourage a spirit of independence in her citizens, restore to her people the protection of the law so that they may speak without fear of arrest, and beg them to plainly and boldly state their grievances.[41]

In 1879 it was Davitt's desire to bring Irish grievances out into the open through the Land League that prompted Bradlaugh to invite him to the London conference.

Davitt's long years of suffering in the nationalist cause, his good looks, great height, and eloquence gave him as strong a presence on the platform as Iconoclast.[42] At the evening session of the conference the delegates were roused to acclaim when Davitt denounced rent as a criminal tax upon the industry of the people, and landlordism as robbery. But they were shocked into silence when he called the Commons an assembly of land sharks, and the Lords a senate of idle aristocrats. His language reminded his audience of the last occasion, in France in 1870, when citizens had destroyed a system they had been invited to despise. In moving support for the Irish in their struggle against what he termed feudalism, Stuart Headlam, who followed Davitt, voiced his regret at hearing Parliament vilified.[43]

The conference voted to set up a Land League with Bradlaugh as its President; his demands became its programme. Annie Besant was a Vice-President, as were Edward Aveling, Stuart Headlam, and Joseph Arch—like Bradlaugh he hoped to enter Parliament at the next election. One of the most assiduous in taking the League demands to working men's clubs up and down the country was Herbert Burrows. He attended the conference as a delegate of a radical association in Tower Hamlets, perhaps the grimmest part of the East End. A clerk in the Inland Revenue, his tallness made him seem grotesquely thin—an emaciation which already conveyed an air of martyrdom. Burrows, whose father had been a Chartist, turned to socialism in 1870, the year of the Paris Commune. Since then his every spare moment had been spent in the quest for, and encouragement of, an insurgent spirit among English working men.[44] Some years after this he would be closely concerned in the devastating change to Annie's life. But if she noticed him at all at this particular time she soon forgot him. On 8 March, faced with growing unrest in Ireland, the Prime Minister, Lord Beaconsfield, decided to ask for a new mandate to employ drastic measures of repression.

The general election which took place at the end of March revealed that public support had deserted the Tories. It was Bradlaugh's fourth attempt to win Northampton in twelve years; once more he stood as a radical hoping to attract Liberal support. Annie, Aveling, and the Bradlaugh girls drove about the town to show his colours to the crowd. But only Aveling went to support the candidate while the votes were counted; ladies were barred from the climax at the Town Hall.

Annie, with Alice and Hypatia, had to wait at the George Hotel, as she had waited in 1874. This time

we knelt by the window listening—listening to the hoarse murmur of the crowd . . . And now silence sank and we knew the moment had come, and held our breath, and then—a roar, a wild roar of joy and exultation, cheer after cheer, ringing, throbbing, pealing, and then the mighty surge of the crowd bringing him back, their member at last, waving hats, handkerchiefs, a very madness of tumultuous delight and the shrill strains of 'Bradlaugh for Northampton' . . . And he, very grave, somewhat shaken by the outpour of love and exultation, very silent, feeling the weight of responsibility more than the gladness of victory.[45]

CHAPTER 14

'LET THE PEOPLE SPEAK!'

QUALMS about responsibility hardly worried Annie; the power bestowed by her gift as orator dispensed with doubt when she was on the platform. To use her own words: those rows of eager upturned faces told her she was 'ruler of the crowd, master of myself'.[1] There would be occasions in the next few years when public order would depend on whether she could maintain that balance.

Whenever a new Parliament assembled, its members were required to swear an oath of allegiance to the Crown. Although Bradlaugh was by no means the only atheist returned, he was the only one who gave advance notice of his intention to affirm.[2] Though affirmation was coming into use in the courts, no one had yet had recourse to it in Parliament. When Bradlaugh presented himself in the House, for the first time, the Speaker, Henry Brand, declined to sanction this procedure; instead he referred the decision as to whether affirmation was acceptable or not to a Select Committee. Its decision was that he could not affirm. Bradlaugh therefore declared that he would take the oath in the form prescribed; it contained the words, 'So help me God.' It was objected that, as one who did not believe in God, Bradlaugh could not swear. When the Speaker referred that objection to another Select Committee, its members found themselves trapped in an argument that slowly inflamed the nation. The more the issue was debated, the more distraught everyone became.[3] For his part, Bradlaugh pointed out that, as the elected representative of Northampton, he had a moral obligation to be sworn so that he might take his seat. 'Any form that I went through, any oath that I took I should regard as binding on my conscience to the fullest degree', he promised, and wrote in the same sense to *The Times*.[4]

The letter was a mistake; it upset secularists who could not stomach Bradlaugh's acquiescence in the oath, and it gave point to his enemies' assertion that he and Besant would do anything for publicity. To many people the fact that 'brawling, swaggering Bradlaugh',[5] author of the *Impeachment of the House of Brunswick*, dared to flaunt his conscience before the Crown in Parliament was intolerable. 'As a mere speculative freethinker, Mr Bradlaugh might possibly have passed muster', *The Times* remarked, 'but as a social reformer, with republican opinions and a very aggressive mode of displaying them he is thought to be deserving of Parliamentary ostracism.'[6] As co-publisher of the *Fruits of Philosophy*, and author of the widely selling *Law of Population*, Annie Besant's name was synonymous with that of Bradlaugh. Indeed it made the more inviting target for those who hesitated to attack a properly elected Member, as Bradlaugh unquestionably was.

Besant was pilloried; in the Commons Chamber, as the author of cheap and pernicious literature aimed at persuading the masses to adopt neo-Malthusianism;[7] in the smoking room, as Bradlaugh's mistress; at the various by-elections which punctuated his dogged fight to take his seat, as a red revolutionary and the chief exponent of the unspeakably shocking ideas contained in the *Elements of Social Science*. The dispute dragged on for six years and was a source of frustration and bad temper among those who were genuinely concerned to find a way round it; a device to serve the ends of conspicuously ambitious politicians; a rod in Bradlaugh's hand whenever the temptation to behave as a demagogue got the better of him.

There was a succession of crises in which Annie Besant played a part. On the first of these occasions, in June 1880, Bradlaugh was summoned to the Speaker's Table to hear Brand deliver the decision of the House against his taking the oath. Bradlaugh 'respectfully' refused to withdraw, contending that the order to do so was against the law, since he was a democratically elected Member. In spite of another vote specifically directing his departure, he refused to move. The Serjeant-at-Arms had to take him into custody while the House pondered what on earth to do next. He was lodged in a room high up in the tower of Big Ben.

On hearing this, Besant declared that Bradlaugh was a prisoner, the most awkward prisoner the House of Commons had ever had;

in his person it was imprisoning the law. The situation appeared to her to be a conflict between constitutionalism and the people's will. A special issue of the *National Reformer* was already set up in type with an account of the debate about Bradlaugh's right to take the oath. Annie simply added a few words. 'As this issue goes to press I go to Westminister to receive from [Bradlaugh] his directions as to the conduct of the struggle with the nation into which the House of Commons has recklessly plunged.'[8] She found her colleague surrounded by letters and telegrams of support; many were from foreigners astounded that the Mother of Parliaments should abuse a man who had done so much to support freedom fighters in Italy, in France, in Spain, in Poland. 'I wish you good luck in the name of Jesus Christ, the Emancipator, whom so many of your opponents blaspheme', Stuart Headlam telegraphed, to the huge annoyance of his bishop.[9]

Annie and his daughters dined with the 'prisoner', then Annie went home to compose a leaflet, *Law Breakers and Law Makers*, whose rousing style was worthy of a Mirabeau.

Let the people speak. Gladstone and Bright [who had voted for Bradlaugh to be allowed to swear] are for Liberty, and the help denied them from within the House must come to them from without. No time must be lost. While we remain idle a representative of the people is illegally held in prison. Northampton is insulted and in this great constituency every constituency is threatened ... Tory squires and lordlings have defied the people and measured their strength against the masses. Let the masses speak.[10]

The leaflet was barely on the streets before Bradlaugh was released as a result of Members' own distaste for a situation which had caught them unprepared. Even so, they found the tide of popular indignation lapping all too closely at their feet. Four thousand people crammed into Westminster Hall to cheer their hero as he returned to the Lobby, and a few days later, the largest crowd ever seen gathered for a protest meeting in Trafalgar Square. The Prime Minister intervened, hoping to restore control so that his followers could turn their attention to the very pressing problems—especially over Ireland—that faced his new government. Gladstone moved an amendment whose effect would be to permit Bradlaugh, and others, to affirm. A judicious rumour, that if the Prime Minister's motion failed, the Government would resign and hold a general election dominated by the issue of religious freedom, frightened a sufficient number of his party into voting for it. So, on 2 July, Annie looked down from the Ladies' Gallery to watch Bradlaugh affirm his allegiance to the Crown and, in a totally silent House, walk proudly to a seat.[11]

That, unfortunately, was far from being the end of the affair. As the law stood, an indemnity of £500 might be levied on MPs each time they voted without having taken the oath. As soon as Bradlaugh took part in a division a writ was served on him on behalf of a private citizen, claiming £500. In the long-drawn-out case that followed Bradlaugh claimed that by affirming he had disposed of the matter: counsel for the plaintiff—none other than the prosecutor in the Knowlton case, Sir Hardinge Giffard—argued that, in this context, affirmation was not legal. Until the case came into court Bradlaugh conducted himself as an ordinary Member of Parliament, voting over 200 times. Besides his constant attendance at the House, and a huge correspondence, he was obliged to keep up his lectures to provide himself with income. Health and private life were sacrificed to this enormous effort. For the first time since before the Knowlton trial, Annie found herself excluded from the most important part of Bradlaugh's daily occupation, and, in so far as she had no vote, reduced to the ranks where once she had commanded.

Her morale was so obviously low that Bradlaugh, hoping to restore it, sent her as his substitute to the first international gathering of free-thinkers: in Brussels at the end of August. She was much admired on the platform; useful in committee, and everywhere in demand for her ability to provide instant translations from French into German or English. Predictably, the conference voted to set up an international free-thought association; just as predictably Mrs Besant collected a Vice-Presidency of the new body. She also captivated the most considerable figure at the conference—Dr Friedrich Büchner, a leader of the German materialist movement.[12]

If Bradlaugh personified free-thought at the moment Annie Besant embraced it, the 56-year-old Büchner embodied the doctrine of monism, to which, since she first caught a hint of it in Tyndall's Belfast address, Besant had given great attention in her search for a satisfactory system of belief. Aveling's lectures, particularly those on physiology, were a means to her of understanding the premiss that every natural and spiritual force was derived from one source only—matter. This was the central thesis of Büchner's many works. The friendship between them, which began at Brussels, was maintained over the years through correspondence. Besant was granted permission to translate his works, beginning with the comparatively simple *Mind in Animals* and proceeding to the enlarged edition of his weightiest publication, *Force and Matter*.

That autumn, and for the first time since she began to recount her experiences, a note of weariness crept in to Besant's column in the *National Reformer*. In November she had an operation, possibly in an attempt to relieve the constant heavy colds from which she suffered. 'Writing in bed is tiring', she explained in a severely shortened Daybreak, 'and my head not yet of the strongest.' Bradlaugh and others chided her for overworking.[13]

She was fit enough to come to Bradlaugh's aid when, in March 1881, he lost the case over the legality of his affirmation. When his appeal was dismissed, he was obliged to vacate his seat. The ensuing by-election at Northampton was fought with great bitterness. One of the most effective anti-Bradlaugh pamphlets was by a Henry Varley, who dragged in Besant's name the better to smear her colleague.[14] Varley dwelt with relish on the brutal words of the indictment in the Knowlton case—'lewd, filthy, obscene'—and on the words of the 'English Marseillaise' Besant had composed to be sung at secularist gatherings. The last verse rhymed alarmingly;

> We are sworn to put tyranny down
> We strike at the Throne and the Crown
>   To arms Republicans
>   Strike now for Liberty.

In spite of all this, Northampton narrowly re-elected Bradlaugh. Since the case he lost had disposed of the option to affirm, he again announced his intention of taking the oath. At the end of a debate fraught with bad temper, the Commons voted once again to refuse to allow him to do so. Once again Bradlaugh resisted Speaker Brand's call for his withdrawal from the Chamber. This time a confrontation was avoided by the House adjourning. As things stood, a harassed Gladstone remarked, Bradlaugh might come down to the House every day for a game of romps with the Serjeant-at-Arms. He thankfully accepted a proposal by Bradlaugh's fellow Member for Northampton, Henry Labouchère, that a new Affirmation Bill be introduced as soon as possible to enable Bradlaugh to take his seat without offending the moral sensibility of the nation. In the meantime Bradlaugh agreed to refrain from visiting his presence on the House, a respite Gladstone welcomed. His government was preoccupied by the necessity of introducing legislation to ease the conditions of tenants in Ireland. Over the past year these conditions had provoked outbreaks of violence, forcing the introduction of repressive measures. The need to repeal these measures, to alleviate Ireland's suffering, to deal justly with her poverty-stricken people, now engaged the full force of Annie Besant's oratory. The long-drawn-out crisis was to delay the appearance of a Parliamentary Oaths Bill, as Bradlaugh's hope of deliverance was called.

When the session became advanced without the Bill appearing, he began to have recourse to threats. Denied an interview with the Prime Minister, and having had no answer to a letter, on 29 June he warned Gladstone that he felt impelled to act in order to vindicate the clear right of his constituents to be represented. On 14 July he addressed a formal warning to the Speaker, the Clerk, the Serjeant-at-Arms, the Doorkeeper, messengers, and the police inspector on duty, that on 3 August he would present himself at the Table to take the oath; if force was applied he would resist.[15]

In the short time intervening before the appointed day, Besant flung herself into helping to promote meetings, demonstrations, and petitions in support of Bradlaugh. On 2 August she took her place on a platform between the lions flanking Nelson's Column in Trafalgar Square. Cheap summer excursion trains from all parts of the country had brought crowds into London; some 15,000 people were present in the Square to hear Bradlaugh ask for a vote in favour of his taking the parliamentary oath; scarcely a dissenting arm was raised. But the crowd was so large that even Bradlaugh had difficulty making himself heard. Suddenly the mass of eager uplifted faces dissolved before Annie's eyes as several thousand excited people surged away down Whitehall towards Parliament Square in the mistaken impression that Bradlaugh was to go to the House that night. Police struggled to keep them out of Palace Yard where the Prime Minister, emerging on his way home to Downing Street, received an ovation he neither expected nor desired.[16]

It is clear from Besant's account of events that Bradlaugh already had misgivings about the whirlwind he had raised. On 3 August when, accompanied by Edward Aveling, he left Circus Road in a hansom adorned with purple, green, and white ribbons, his parting words to her were: 'The people know you better than anyone save myself: whatever happens, mind, whatever happens, let them do no violence. I trust to you to keep them quiet.'[17] Annie, and those two stalwarts Alice and Hypatia Bradlaugh, went down to the House in another cab bearing a petition, expecting to gain admission to the Lobby. Crowds lined Whitehall and Parliament Square but, as a result of consultations between the Speaker, the Home Secretary, and the Commissioner of Police, this time they were under strict control. Besant was stopped and sent to join a group of several hundred people with petitions who were allowed one by one through a gate into Westminster Hall. There Besant made for the steps leading to the Lobby, a forward position she was allowed to retain only after a brisk argument with four policemen.

Meanwhile Bradlaugh mingled with other Members waiting in the Lobby for the Speaker's procession to go by. T. P. O'Connor was struck by his manner, which was smiling, debonair; with his shiny new silk hat, large black tie, half French, half clerical, tied across his expansive white shirt front, he looked like a bridegroom. As soon as prayers were over Bradlaugh advanced to the door of the Chamber where the Deputy Serjeant-at-Arms told him he had orders not to admit him. 'Those orders are not legal', Bradlaugh replied, and made to enter. Four messengers seized him by the arms and the collar of his elegant black frock coat; his top hat and cane went flying as he grappled with them. The vaulted ceiling of the Lobby rang with shouts—'It is not legal' from Bradlaugh and 'kick him out' from an over-excited Alderman Fowler, MP (who was probably the moving spirit behind the Knowlton prosecution). O'Connor was appalled at the change in Bradlaugh, whose face was

haggard, fierce, terrible—almost the death mask of one who had just been guillotined, so deadly was the pallor, so fierce the look at once of anger, pain, defiance. Bradlaugh, the great giant of a man, the representative of a constituency, was being dragged with torn coat and shirt, down the steps of the House by a band of attendants and policemen, resisting, shouting, defiant, apparently astounded as much as others at the sudden transformation which had changed him from the erect and smiling and respectable figure into this hunted and outraged outcast.[18]

In Westminster Hall, Annie, who knew nothing of this drama, was alarmed by the increasing restlessness of the group below her in the body of the Hall, angry at being denied entrance to the Lobby.

Suddenly with the impulse that will sway a crowd into a single action [she wrote] there was a roar—'petition, petition, justice, justice', and they surged up the step, charging at the policeman who held the door. Flashed into my mind my chief's charge, his words, 'I trust you to keep them quiet', and as the police sprang forward to meet the crowd I threw myself between them with all the advantage of the position on the top of the step that I had chosen.

There, as George Lansbury, MP, recalled years afterwards, she held the crowd in check through sheer force of personality until their anger subsided, and a messenger summoned her to Bradlaugh, who was by then in Palace Yard.[19]

The Times reported that he had told the inspector in charge he would come again with a force at his back that would compel them to admit him. Asked how many, he indicated there would be thousands. These were dangerous words, spoken in the heat of the moment, and Bradlaugh immediately sought to retrieve them. He meant he could bring such a force if he would, he told *The Times*.[20] The shock of the manhandling had a grave effect on him; a day or so later he suffered a dangerous attack of erysipelas in one arm. His exhausted appearance alarmed those who saw him at the Hall of Science meeting on the Sunday following his ejection from the House. His voice was too feeble for him to say more than a few words so Besant took command. She was in no mood to copy Bradlaugh's self-imposed restraint.

If she had known what was happening in the Lobby, she told her audience, she would have let all their supporters show Mr Speaker Brand something of the same sort his mob of police had shown to a Member of Parliament. She advised them to make good use of their time (before Bradlaugh's next attempt to take his seat). 'A little drilling would not be bad because otherwise the police have a great advantage over you', Besant declared, and urged them to join Volunteer regiments for the drilling and the training; Volunteer regiments were perfectly legal and the more that joined them the better it would be for their cause in the future. Mr Bradlaugh was going to the House of Commons again; he was not going to let the officials know, but he would let a large number of his own supporters know. This was unashamedly provocative:[21] talk of drilling echoed a famous speech in Ireland by the nationalist leader John Dillon, which was condemned in England as seditious. Besant's speech was made the object of a parliamentary question—whether the Government had taken measures to prevent the riot it seemed to threaten? Gladstone replied that he had no reason to suppose the means of protecting the House were inadequate.[22]

It was to take four interminable years and three more election campaigns before Bradlaugh was admitted to Parliament in 1886. They were years disfigured by suffocating legal wrangles, constant polemic in the newspapers, and from the pulpit, and a discreditable draining dispute in Parliament where the 'Bradlaugh mess'[23] played havoc with party loyalties. When in 1883 the Commons voted on the Affirmation Bill which Gladstone hoped would put an end to the affair, a cartoon was displayed in the 'No' lobby. Members trooping in to vote were confronted by Gladstone and Bradlaugh embracing atheism, whose face was recognizably that of Annie Besant.[24] The Bill was lost by a narrow majority, so Bradlaugh remained excluded.

Meanwhile, like Bradlaugh, Besant charged head on at all those obstacles society placed in her way to defend itself from her unorthodox behaviour. Opposition came even from people on whose understanding she might have expected to rely. 'I have never been able to get hold of a copy of "Fruits of Philosophy"', her sometime science tutor T. H. Huxley wrote, 'but if it violates the principles which I am told it does with respect to the safeguards on sexual intercourse among unmarried people, and if Mrs Besant has made herself responsible for that doctrine ... I have no objection to her exclusion.'[25]

Huxley's remark (which he must have qualified if he had bothered to discover that Besant condemned sex outside marriage) was prompted by the uproar among his fellow members of University College London, following her application in the spring of 1883, with Alice Bradlaugh, to be admitted to a class in botany. The College feared publicity as a result of admitting ladies whose names were too well known, and whose purpose, they suspected, was rather to serve the cause of Bradlaughism than to acquire scientific knowledge. As soon as he received their application forms the College Secretary, Talfourd Ely, returned them with a request that the applicants call to see the Lady Superintendent, Miss Rosa Morison. Although she had only just taken up her post as a result of the recent decision to open certain classes to women, Miss Morison proved equal to the confidence placed in her. She told Annie to her face that she could not admit her 'because there is a prejudice against you'. No further explanation was conceded in spite of the storm Annie and her supporters raised. 'Though I gave Mrs Besant no reasons', Miss Morison wrote to Talfourd Ely, 'I may mention in confidence to you that I was guided in my course of action by a judicial authority respecting a trial in which Mrs Besant had been engaged'—a guarded reference to Sir George Jessel's remark that, as a result of the technical conviction at the Knowlton trial for circulating obscene literature, no modest woman would associate with Annie Besant.[26]

These words, spoken in a judicial capacity by a late Fellow of the College (Jessel had just died), carried great weight. The College Council was unanimous in its decision to endorse Miss Morison's action and equally unanimous in declining to give the reason. Its silence exposed it to attack for allegedly departing from the principle of impartiality in religious matters which the founders had laid down in 1827. For it was supposed that while Mrs Besant might have been excluded on social, as well as religious, grounds, Alice Bradlaugh's only possible offence was to be her father's daughter.

As a Fellow himself Edward Aveling circulated a petition calling for an Extraordinary General Meeting to consider the Council's action and its silence. In a furious letter to the *Daily News* Aveling accused it of preferring the success of its classes to justice for individuals. His efforts were to no avail. Extracts from the judgement of the late Master of the Rolls were read at a closed meeting of the Senate, the supreme governing body of the College, and its approval of the Council's action communicated to the Extraordinary General Meeting.[27] There, although an attempt was made to distinguish between Miss Bradlaugh and Mrs Besant in order to admit the former, the vote was overwhelmingly against both ladies. Although he had signed the requisition for that meeting, T. H. Huxley abstained from voting at it. While he wanted to be assured that the Council had no intention of interfering with religious and philosophical freedom (he was himself an unbeliever) he felt it was proper that it should concern itself with the character of those who were to be admitted to the College. He had 'a strong feeling that the freedom of thought should be carefully distinguished from laxity in morals. Freethinking does not mean free love.'[28]

The fear that Besant would 'start a propaganda' among the students and drive them away with her obnoxious opinions was the reason why University College shut her out.[29] That fear closed many other doors to her. Besant accepted such rebuffs if she sympathized with their perpetrators' cause. Thus she returned a gentle answer to a correspondent who demanded to know why she did not join the campaign against the Contagious Diseases Acts: the people who conducted it would not welcome her, she replied. But she publicly derided Mr Sowerby, Secretary of the Botanic Gardens in Regent's Park, for refusing her a ticket; he was nervous for his daughters, who sometimes walked there, she told her readers.[30] In a private letter to the Director of Kew Gardens, Sir Joseph Hooker, she was more polite. Mr Sowerby had refused her because he objected to the opinions attributed to her in theology and politics. 'As these questions do not affect the study of Botany and I am exceedingly anxious to make my study practical, I venture to ask you to allow me to avail myself of the great advantages of the gardens under your control.'[31] Entry was granted, with the request that the visits be confined to the early morning.

Several attempts were made in the House of Commons by a particular opponent of Besant, Sir Henry Tyler, MP, to move the withdrawal of recognition from the Hall of Science school. His contention was that atheists, who publicly proclaimed scientific truth to be at odds with the story in the Bible, ought not to be allowed anywhere near classrooms. The situation appeared so precarious that Besant renounced the salary provided by South Kensington. Tyler's constant sniping bored the House to the point where no less a person than A. J. Mundella, Minister responsible for education, was moved to endorse both school and teachers.

But Tyler did succeed in obtaining Aveling's dismissal from a teaching post at the London Hospital, and brought about the final rift between the Revd Stuart Headlam and his bishop. The bishop complained that no laws of toleration or Christian charity could justify a clergyman placing the teaching of the young in the hands of avowed atheists. The students at the Hall of Science school were not children, Headlam replied, and the information they received about God's wonderful world must make them more careful in their theological and natural studies than most. In any case, as Chairman of the school he did not expect the bishop to defend him. 'Defence, My Lord is about the last thing which I . . . and any priest who goes a little out of the ordinary way in trying to fetch the wanderers home to the flock, have learnt to expect from their bishops.' Headlam had to resign his curacy and was never preferred again. As one whose aim was the same as hers, Besant paid him a glowing tribute: 'deeper than the wrangling of our shibboleths is the firm underlying rock of unity and brotherhood between all who recognize duty, love and work as the three great foundation stones of all noble human life.'[32]

Duty, love, work; these were vital threads in the pattern of Annie's life as it was formed in the years after 1875, when she embraced free-thought. At the beginning of the 1880s that pattern appeared likely to continue as long as she was able to sustain the effort it required. Such an effort might have been expected to assuage the demands her talent imposed on her, and the suffering it brought in its wake gratify her desire for self-sacrifice. With the support of such affectionately admiring comrades as Charles Bradlaugh and Edward Aveling, surely Annie was at peace with herself? She was not. A crucial thread was missing. Annie did not mention faith in her praise of Stuart Headlam—naturally enough, since she had rejected the Christianity that inspired him. Though it was not yet apparent to her colleagues, she was no longer secure in her unbelief, the mainspring of her action. She had grown susceptible to socialist ideas which appealed to the Christian Headlam as much as they did to the free-thinker Aveling. But Charles Bradlaugh regarded socialism with disapproval—detestation even. To him it was an alien system of thought that promised violence. Socialism was to undermine the trust Bradlaugh placed in Annie Besant.

CHAPTER 15

SOCIAL REFORM, NOT SOCIALISM

STRIDENT in proclaiming her latest belief, yet shallow in her understanding; so volatile that she changed direction as often as she attached herself to another masculine associate—this was the impression of Annie Besant her critics liked to foster.[1] Among the evidence they cited was her 'conversion' to socialism under the influence of Edward Aveling, with whom it was supposed she had an affair, a conclusion that was given credence by Annie's unconventional behaviour in allowing him the free run of her house.[2] This did not do justice to either her intelligence or her sincerity, and it misinterpreted the nature of her relationship with Aveling, which, on the evidence of his letters, was not a sexual liaison.

There were good reasons why Annie should have been attracted to socialism when it began to make headway in the early 1880s. Like others who had striven to obtain the election of a Liberal government, she was swiftly disillusioned by its acts on coming into office. 'From 1874 to 1880 the conservatives had ruled, going from bad to worse. As leader of the anti-Tory discontent Gladstone seemed to gather all the forces of progress under his leadership', an acquaintance of hers, Ambrose Barker, Secretary of an East End radical club, recalled. When the general election resulted in the greatest number of Liberals since the Reform Bill of 1862 arriving in the House, great things were expected of them. But 'almost their first action was to introduce a Coercion Bill in Ireland . . . with that, and the neglect of social reforms which they had been led to expect, the working classes were grievously disappointed', Barker wrote; 'organisations came into being advocating social change'.[3]

One of these organizations was the Democratic Federation, founded in 1881 by Henry Hyndman specifically to oppose 'the monstrous tyranny of Gladstone and his Whigs in Ireland and their equally abominable policy in Egypt'. Among the radicals he invited to a preliminary meeting were Professor Beesly, Helen Taylor, Joseph Cowen, MP and Herbert Burrows, who, unlike the others, was wholly committed to the ideas of Karl Marx. Hyndman had become a Marxist in the early 1870s through reading *Das Kapital*.[4] A wealthy man of immense self-esteem and some irascibility, Hyndman preached the revolution of the proletariat without abandoning top hat or silver-mounted cane. He approved of Annie Besant; she had cleared her mind of those theological and conventional prejudices that stunted the intelligence of so many able women, he remarked. But he dismissed Bradlaugh as little more than a bully, and secularism as 'the fanaticism of negation'.[5]

From the moment they appeared, Bradlaugh saw organizations like the Democratic Federation as a challenge to the position he had built up over the years. His rapport with the working class was a source of tremendous pride. 'He was closer to the people than anyone', Thomas Burt, the miners' leader, judged; 'he loved them.'[6] Bradlaugh met the socialist challenge as he had the Christian; he opened the Halls of Science all over the country to their speakers, and the columns of the *National Reformer* to their theorists, secure in the belief that he and his colleagues like Besant and J. H. Levy could demolish them. Socialists, therefore, became well known to Annie Besant long before she dreamed of joining them.

Bradlaugh's strategy turned out to be ill conceived. The publicity afforded by the many channels of the National Secular Society served to make German philosophic socialism better known. The mistake was due to Bradlaugh's obsessive hatred of Germans as people, and of socialism. He judged both in the light of what he took to be inevitable at no distant date—either a huge European war, or a terrible European revolution. Socialism was an alien creed—a violent creed—imported into England by German refugees, people who were very dangerous because they were so earnest. Their aim was to destabilize the working class—his class—to neutralize the kind of individual effort that had brought him success; to suppress the freedom of expression which he had fought to extend—all in the name of a collectivist ideal that seemed to him wholly sinister.[7] Bradlaugh believed that, just as the spread of collectivist ideas in Ireland had engendered violence, forcing the British Government to reintroduce coercion, so the result of socialism would be similarly violent. 'You are driving poor people into danger', he told Marx's followers; '. . . you are giving excuses for coercion, you are trying to lead my people wrong, and therefore I bar your way.'[8]

Annie Besant, who agreed with Bradlaugh, was appalled to discover that Aveling thought otherwise. 'Karl Marx' influence is becoming a dangerous one over some with whom I work, and I would be grateful for any information throwing light on his literary and political honesty', she wrote, in November 1883. Her letter was to Sedley Taylor, a radical cleric, a Cambridge don (another protégé of Thomas Scott) who had attacked Marx in *The Times* for allegedly distorting quotations for his own polemical ends. Besant hoped Taylor would supply her with material she could use to divert Edward Aveling from his course.[9] Her efforts were not successful. By early 1884 Aveling had revealed himself as an enthusiastic partisan of Marx; letters to the press, lectures, and long articles in praise of his new belief poured from his pen.

Besant herself now resorted to the public prints ostensibly to give Aveling a chance to recant. When the first number of *Justice*, the official organ of the Democratic Federation, arrived at the office of the *National Reformer* in January 1884, she used it as the occasion for a swingeing attack on the 'school of Marx'. 'The citizen who, in England, with free press and free platform for political and social agitation, with vote that he can use to send to Parliament the man with whom he agrees, either counsels resort to violence, or permits such counsel to be given in his presence without rebuke ... is the worst enemy of social progress and of social unity', Daybreak proclaimed.[10] Aveling had been among those present at a party to launch *Justice*, when there had been talk of the coming revolution (understandably, since Hyndman was obsessed with a belief that it was imminent). Besant took issue with the editor of *Justice* for omitting any mention of the protest Dr Aveling must have registered against such talk: not to protest was to be guilty of nothing less than murder. Aveling promptly riposted; the meeting had been correctly reported; there was no question of murder; he hoped the revolution, when it came, would be one of thought.[11] He sent Besant a long defence of the speech by William Morris, who had also been present, and of the scientific side of Dr Marx's socialism. The *National Reformer* did not have space to print either.

On 17 April, Charles Bradlaugh and Henry Hyndman met in St James's Hall in the presence of an invited audience of socialists and free-thinkers to debate the question, 'Will Socialism Benefit the English People?' Professor Beesly was in the chair, Besant in the audience—though she would have preferred a seat on the platform, and said so. Hyndman, who summarized the Marxist case as presented in his book *England for All*,[12] was resigned to taking second place to his formidably experienced adversary. Certainly Bradlaugh was the louder of the two. 'We are for reform', he told the audience, 'we will cure gradually. If we try to cure immediately we will poison or destroy. Class war is murder; class war is fratricide; class war is suicide.'[13] This peroration was considered to have won the day. Bradlaugh did not understand that socialism, through collectivism, would eventually arrive at the higher individualism, Hyndman complained. He was consoled by George Bernard Shaw: 'our man has played at longer bowls than you know', Shaw told a triumphant Bradlaughite.[14]

Socialism was becoming as fashionable as asceticism had once been, Daybreak jeered.[15] In May a course of lectures on the subject of the moment was held at South Place. Herbert Burrows thought it a good venue: 'many of the people who go there are ready for more light', he told Hyndman.[16] When he spoke, Burrows followed his leader's line—that the poverty of the workers was due to lack of control over the means of production, distribution, and exchange. Invited to comment afterwards, Besant revealed that her thoughts were in something of a muddle. Instead of demolishing her opponent she suggested that, without profit, exchange was impossible, for profit was really the wage of the distributor. 'Everyone ought to work but everybody ought not to produce; non-productive work is essential to the well being of the community', she ventured.[17] It was at this meeting that Bernard Shaw proclaimed himself to be a loafer, a remark which provoked Besant to attack him in Daybreak. She had often heard Mr Shaw speak; she had never understood why he was so 'shrewish and crooked' in discussion. Now she knew. Self-discontent was bound to turn into mere bad temper, she warned him.[18]

It was only a few months since Shaw had embarked on a career as a public speaker. He had been attracted to socialism in 1882 after hearing Henry George speak on land reform, and he became passionately committed to it on reading a French translation of *Das Kapital* (there was as yet no English version).[19] Aged 27—nine years younger than Annie—Shaw was not yet in control: of his voice, of his talent, of his miserably underfed and elongated person. But, as soon as they came to close quarters, his charm proved irresistible to Annie. Two weeks after her nannyish rebuke she retracted it in terms that verged upon the arch. Mr Shaw was not a loafer, she told readers of Daybreak. Or, if he was, he was a poor and hard-working loafer; she had discovered that he was a struggling writer who devoted all his spare time to helping others without pay by preaching a faith in which he devoutly believed.[20] This was a rare recantation: Annie was obviously intrigued by the strength of Shaw's commitment.

On 4 May she delivered a direct personal attack on Edward Aveling. *Justice* reported him as saying that he had been converted to socialism only after five years' deep study. 'I am sure that Dr Aveling never said anything so untrue', she exclaimed. 'It is less than five years since Dr Aveling joined the Freethought Party... As his friends closed their doors on him I opened mine, and save for the time when he was with his pupils, and night time, he made my house his home. All his work was carried on with me.' With Bradlaugh and herself he had constantly discussed politics and Aveling had been 'quite at one' with them, though his political knowledge was very small. 'In fact he never touched Socialism at all... until in 1882, he took to reading at the British Museum and unfortunately fell into the company of some of the Bohemian socialists, male and female, who flourish there.'[21]

The popular impression of Bohemians as disorganized, untidy people, which Annie here invoked, was in sharp contrast to her own appearance and way of life. Her poise reflected nearly ten years in the public eye. When not appearing at the Hall of Science she dressed with quietly impressive taste as befitted the chatelaine of a sixteen-roomed mansion in its own grounds. She moved to 19 Avenue Road, the principal thoroughfare in St John's Wood, in June 1883. Her object in taking on so large a house seems to have been to unite the two establishments of herself and Charles Bradlaugh under one roof. But he proved impossible to dislodge from Circus Road. At Avenue Road there was a housekeeper, Mrs Kenwood, who lived in with her family, and a companion for Annie, Elizabeth Cracknell, besides the usual complement of birds, dogs, and ponies.[22] Edward Aveling attended daily, except on the occasions when he tutored students at his rooms in Newman Street. One of these, to whom Aveling devoted an increasing amount of time, was Eleanor, youngest daughter of Karl Marx.[23]

With her dark and dishevelled hair and picturesque clothes the 26-year-old Eleanor undoubtedly qualified as a 'Bohemian socialist'. 'Lives alone, is much connected with the Bradlaugh set, evidently peculiar views on love, etc., and I should think, has somewhat "natural" relations with men. Should fear the chances were against her remaining long within the pale of "respectable" society', was Beatrice Webb's tart verdict on Miss Marx whom she encountered in the tea room of the British Museum.[24]

The public animosity between Besant and Eleanor Marx was inevitably put down to sexual rivalry. While Annie was undoubtedly upset and jealous at Aveling's defection from his role of devoted squire, the situation was far more complex than a dispute between two women over the same man. It was not in Besant's, or Bradlaugh's, interest to reveal that for at least two years, possibly longer, they had known that Aveling was a scamp. This was the origin of the bitterness they displayed in the political contest which, by 1884, appeared to be increasingly favourable to socialism.

Shaw remarked that up to 1883 foreigners who observed the conditions of industrial England were appalled by the apathy of the more prosperous inhabitants, and the resignation with which the poor accepted their lot.[25] In August that year the *Pall Mall Gazette* acquired a new editor: W. T. Stead, whose genius for publicity and social concern was to make him a power in the land for nearly thirty years. His first coup was to discover a penny pamphlet, by an anonymous author, with the inspired title *The Bitter Cry of Outcast London*. It gave an obviously authentic, detailed, and horrifying picture of conditions in the East End.[26] Week by week the *Pall Mall Gazette* drove its message home. 'South Kensington awakened to the fact that there were some two or three million people in the brick and mortar wilderness beyond the Bank of England, many of them in woeful distress',

Hyndman wrote. 'It became quite the proper thing to do to go down East.'[27] Soon, from dozens of hastily established 'missions' and 'settlements' charitably minded persons went forth into the streets of miserable houses whose light was stolen by the vast wall of the docks, or by railway embankments where trains rattled past at the level of their chimney pots.

'Bradlaugh and Mrs Besant are furious at the new Socialist "rage" in London which threatens to cut short their "wittles"', Friedrich Engels wrote to Marx's middle daughter in February 1884, describing a visit from her sister Eleanor, accompanied by Aveling. 'Bradlaugh turns out the most mysterious innuendoes about Mohr's [Karl Marx] having preached assassination and arson and having been in secret league with Continental governments—but nothing tangible. I want to get him to come out a bit more before I unmask my batteries.'[28] Engels was referring to the fact that Bradlaugh had once accused Marx of spying for Bismarck,[29] and more recently had attacked Eleanor Marx for her failure in a newspaper article sufficiently to condemn violence as a means of achieving reform.[30]

On a personal level, it appears that Annie tried to warn Eleanor about Aveling's unreliability, and that Bradlaugh threatened him with dire consequences if he persisted in pursuing her.[31] She had not yet recovered from the shock of Marx's sudden death the year before, and her immaturity was, in any case, a source of anxiety to her friends. In December 1883 Eleanor denounced Annie Besant for allegedly circulating a scurrilous rumour about Edward Aveling.[32] The evidence is incomplete, but the probability is that this rumour was to the effect that Aveling had concealed the fact that he was married. Among the Bradlaugh Papers is a statement signed by Aveling that reads like a confession extracted under duress. It begins, 'Written, Thursday, January 26, 1882, 6.45 a.m. I married Isabel Campbell Frank eight or nine years ago. Both of us were very young...' The marriage largely occurred as a result of pressure by Aveling's employers—heads of women's schools. Relations broke down over religion. A High Church woman, Isabel filled the house with clergy whom Aveling the free-thinker could not tolerate, and with whom he alleged she committed adultery. When he left her he promised to continue to support her, but lost touch. 'This matter has not troubled me as perhaps it ought to have done', he wrote, and pleaded that other questions might be put in writing and could be answered in writing, 'as that makes it a little less painful'.[33]

Notes made by Bradlaugh reveal the extent of the other damage Aveling did. He was already in debt when he joined the staff of the *National Reformer* in 1879. Since he had no credit, and impressed by his ability as a scientific teacher, Bradlaugh and Besant advanced the cost of apparatus, took 13 Newman Street in his name, and eventually found £120 in rent when he refused to pay. It was alleged he stole two cheques destined for a builder doing repairs to his rooms. Though Bradlaugh tried to call a halt in 1882, Aveling went on borrowing behind his back. Once he had to extract 'Edward Dsc' from Holloway Gaol where he had been committed by a swift-footed creditor. As partners in the Freethought Publishing Company, Bradlaugh and Besant agreed to pay Aveling's creditors 20s. in the pound—which they eventually managed by instalments. This at a time when Bradlaugh was facing heavy costs of legal action arising from his attempt to take his seat in Parliament. By 1883 the partners were seriously alarmed at the effect upon the Company as guarantor of Aveling's debts. They were also very concerned for the good name of the Hall of Science school, at risk from Aveling's habit of borrowing from his pupils, and from his alleged seduction of several of them. And they were desperate not to hand a propaganda victory to their political opponents at the height of the furore over the Affirmation Bill.[34] When Hypatia asked to be allowed to end her lessons with the—to her—odious Dr Aveling, Bradlaugh urged her to carry on. 'I desire especially to avoid even the semblance of disagreement [with Aveling] for it would be open to the enemy to say that I quarrelled with Holyoake, with Foote, with Charles Watts, and now with Aveling', Bradlaugh wrote.[35]

He and Besant endured in silence even after Aveling joined the opposition, becoming a member of the executive of the Democratic Federation in 1884 (which now added Social to its name). Nothing was said, for example, to prevent the annual conference of the National Secular Society re-electing Aveling a Vice-President. Annie continued to print friendly remarks about him until the end of July.[36] On 29 July Aveling wrote a curiously formal letter to Bradlaugh from the Nelson Arms at Wirkworth in Derbyshire where he had gone with Eleanor Marx.

Dear Sir, As you are President of a Society of which I am a Vice President I feel that I ought to tell you of the important step I have just taken. I have decided to live as man and wife with Miss Marx. You know that we cannot live legally. But I hope we shall be able to do so morally... You will let me know if this step... will make any difference to my official position.[37]

Bradlaugh's reply was to tell Aveling to refer to the Secretary of the NSS. But he added these ominous words: 'In such course you must remember that other questions of conduct would probably arise.'[38] Almost at once, and with the full approval of Annie Besant, Bradlaugh sent a circular to all branches of the National Secular Society with notice of a motion: 'that in consequence of repeated and increasing complaints of borrowings of money by Dr Edward Bibbins Aveling—some of these under circumstances involving very grave blame—and of other serious irregularities in connexion with moneys received by him, Dr E. B. Aveling ceases to be a Vice President and a member of this Society.'[39]

Edward's dunning letters addressed to Annie were all distressingly alike. The earliest surviving is dated 28 April 1880; it contained a complete list of all his 'difficulties'. Those marked 'thanks to you [Annie]' were partially settled. Others were pressing, some urgently. Was there any possibility of clearing at least those marked XXX and ridding him of this nightmare? 'You have helped me so much lately. For my own expenses and so forth this week £10 will suffice. I have no words to thank you and am faithfully your Edward Aveling.'[40] When Aveling protested at the injustice of Bradlaugh revealing how far he was in debt, he was told he could either earn silence by his immediate resignation, or defend himself by appealing to the branches of the NSS.[41]

Although secularists were asked to keep the matter to themselves, Herbert Burrows got wind of it, probably through Annie Besant. Burrows had been in the Black Country all summer, dashing about from one industrial centre to another, forming lodges among the miners, glass workers, and iron workers that would be affiliated to the Social Democratic Federation. Overexcitement, fatigue, and congestion of the lungs probably were responsible for the feverish tone of the letter Burrows addressed to the Secretary of the SDF, H. H. Champion, on 15 September. Referring to the circular Bradlaugh had sent the NSS branches, Burrows described the charges made against Edward Aveling as having serious implications for the SDF. As a member of the executive Aveling must publicly refute the charges or resign at once to avoid damaging the reputation of the SDF.[42] Since Chartist times political movements had been rendered useless through corruption, Burrows argued: 'He who would free others must himself be free. He who would purify others must himself be pure. He who would raise others must show that he himself does not voluntarily fail.'[43]

This diatribe was not taken too seriously by the other members of the SDF executive. They settled for a statement in *Justice* by Aveling admitting that he was in debt and hoping to be able to repay at some future date.[44] Bradlaugh, meanwhile, had fired off another circular to the members of the NSS with a report of a meeting he had with Aveling on 28 August. The latter had come to this unaccompanied, though he had been urged to bring a witness, had heard the charges read, and had admitted them. Since he offered to resign, Bradlaugh proposed to make no further statement.[45]

On 22 October Aveling wrote to Annie Besant; the terms he used have to be read in the light of her knowledge of his relationship with Eleanor Marx. Aveling did not act the contrite former lover, nor was he defiant. The words he used are consistent with his attitude to Annie when first they met and he offered her his respectful adoration. 'I know that [Bradlaugh] wants to give me the opportunity of retrieving the past. You will not frustrate me I believe... How I miss and ache for you! When you come back [Annie was in Scotland] may I come sometimes and just get what I call the rest of your presence? I may be able to bear up then Help me Help me.'[46] Later he wrote: 'Am I ever to see you again? If only I could be with you quietly in Scotland for an hour or a day—at rest. When do you return? Are you well? I *am* so tired.'[47]

Though they were never published, Bradlaugh's charges against Aveling were already suspected by others; many people shunned him. Some followers of Marx refused to grant him the preeminence in the movement Engels tried to claim for him as Eleanor Marx's 'husband'. Some even stayed away from Engels's house for fear of meeting Aveling.[48]

In Scotland, in October, Annie presided at a farewell party for a young journalist and free-thinker, John Mackinnon Robertson, given by the Edinburgh Secular Society. Robertson was to replace Aveling on the *National Reformer* to which he had already contributed.[49] Since 1878 he had been a leader writer on the *Edinburgh Evening News*. Though young for such a role—he was the same age as Shaw—his prickly integrity and delight in learning had already earned esteem. Robertson, who came from the Isle of Arran, had left school at 13. Like Bradlaugh he was self-educated. On one important matter he and his new editor were at odds. Having examined the case for socialism, Robertson considered it proven in theory, if not yet in practice. A bachelor, he was cherished by his friends; in her speech Annie apologized for stealing him. On arrival at St Pancras Station Robertson was carried straight to 19 Avenue Road, where, for the next three years, he was to be a lodger.[50]

CHAPTER 16

BRADLAUGH ESTRANGED

'MRS BESANT is feeling it necessary to turn Socialist but does not want anybody to tell her so', *Justice* remarked after hearing Annie lecture in June 1884.[1] For most of that year she tried to reconcile her instinct to approve ideas put forward by the Democratic Federation with loyalty to Bradlaugh. The struggle was not as prolonged as the ordeal she had endured at Sibsey before leaving the Church—this time there was no need to abjure her faith in free-thought—but once again, in the beginning, she tried to compromise. 'Social Reform not Socialism', was the title of the lecture that alerted *Justice* to her change of attitude. In it Annie argued for the adoption of a programme of reform but, as *Justice* did not fail to point out, could not bring herself to accept that responsibility for it must be carried by the state. Instead she proposed that it might be implemented by members of a Social Reform League who would be elected to Parliament—Bradlaugh's programme in effect.[2]

In September, one of the readers of Daybreak, W. P. Ball, claimed to detect a whiff of socialism in an item about Board school children, whose standard of achievement was reported to be very low.[3] Annie called for common sense: if children were half-starved, as these apparently were, they must be fed before they could be expected to attend to lessons. (Her fierce concern for children's welfare had its origin in her ever present sense of loss over Digby and Mabel.) By December Daybreak was demanding why, if people did not feel 'pauperized' by paying for police, drainage, street lighting, and street cleaning on the rates, they should be so very much opposed to any call for school meals on the rates.[4]

As self-appointed guardian of her ideological virtue, W. P. Ball could not let this pass, and challenged her to debate 'Creeping Socialism' in the New Year. When they met he charged her with being a socialist at heart, and accused her of only remaining in the liberal ranks because of the 'powerful and incisive' mind of her co-editor on the *National Reformer*.[5] In fact Bradlaugh's influence had begun to wane the previous April as a result of the famous debate with Henry Hyndman. Annie rejoiced at his victory on the night, but afterwards, away from the magic of 'Mr Bradlaugh's commanding eloquence and personal magnetism',[6] she found that Hyndman's denunciation of the capitalist system remained in her mind and on her conscience. But so long as Edward Aveling continued his behaviour she could hardly endorse his well-publicized enthusiasm for the new belief.

His banishment removed that bar in August, and shortly after that, Robertson arrived under her roof. He was preparing a series of articles examining the points of similarity between radical and socialist ideas. Annie grasped the opportunity of abandoning her co-editor's hostile stance, and began to preach *rapprochement*, 'treating the Radical as an unevolved Socialist rather than as the anti-Socialist', she explained.[7]

To those who could read between the lines she had already given notice of an important difference of attitude between Bradlaugh and herself in her *Autobiographical Sketches*, which began appearing by instalments in 1884, in the magazine *Our Corner*, whose sole proprietor and editor she was. Bradlaugh was as dismissive of the socialist dreams of the past as of its contemporary apologists, even when the dreamers were good Englishmen like the followers of Robert Owen and their successors, supporters of the Chartist and Co-operative movements. In the *Sketches* Annie went out of her way to praise William Prowting Roberts, whose account of life on an experimental farm as Treasurer of the Chartist Co-operative Land Scheme had formed part of the political education of the young Annie Wood, as had his imprisonment for sedition in 1839.[8] It was the Chartist Roberts who first taught Annie not to bow before authority when it was represented by unjust laws; it was that same spirit which in the summer of 1884 led her publicly to support Michael Davitt long after Bradlaugh had distanced himself from all the Irish nationalist leaders.[9]

Shortly after he arrived at Avenue Road, Robertson was taken for a ramble on Wimbledon Common, when he and Annie discussed *Our Corner*. She was anxious to enlist more writers, offering dinner at Avenue Road as an inducement. Robertson found himself pursuing old acquaintances. 'We, that is my hostess and I, want the magazine to be as good as possible', he told an Edinburgh friend, Patrick Geddes.[10] *Our Corner*'s aim was to combine a mission to instruct with wide popular appeal; its systematic approach (somewhat reminiscent of Mrs Beeton's *Book of Household Management*) was designed to make reading easy. There was a Corner for each subject—Politics, Young Folks, Art, and, whenever Annie had anything particular to say, for the Publisher. These were linked by poems, extracts from novels, and puzzles, the last under the regular heading 'Nuts for Sharp Little Teeth to Crack'.

The magazine 'had the singular habit of paying for its contributions', Bernard Shaw wrote, 'and was, I am afraid, to some extent a device of Mrs Annie Besant for relieving necessitous young propagandists without wounding their pride by open almsgiving'.[11] Six months after she had first become aware of his penury, Annie offered Shaw serialization in *Our Corner* of one of his five unpublished novels. She and John Robertson disagreed about the choice; Annie favoured the second, *The Irrational Knot*. As its title implies this was a cautionary tale about marriage; it would be a timely illustration of the theme Annie had not neglected to address—that the religious institution of marriage ought to be replaced by a contract the parties might abrogate if they so desired. Robertson preferred Shaw's third novel, *Love among the Artists*, and told Annie that *The Irrational Knot* was immoral (thus agreeing with Macmillan's reader who had described it as a novel 'of the most disagreeable kind . . . odd, perverse, and crude . . . with too much adultery . . . in it).[12] Shaw sprang to its defence. '*The Irrational Knot* is very long, highly moral and deeply interesting', he admonished Robertson. At the same time, in answer to Annie's request to print the paper on socialism he was due to give to the Dialectical Society the following week, Shaw explained that he had not had the time to prepare one; 'I . . . shall orate extemporaneously.'[13]

The engagement, on 21 January 1885, was important to him because of the 'very prestigious'[14] nature of the Society and its practice at the end of each lecture of inviting cross-examination of the speaker from the floor. Writing a long time afterwards Shaw said it was rumoured that Mrs Besant, as a leading Bradlaughite, would come down to destroy him. However, when he finished she kept her seat, leaving another member to oppose. When he sat down she rose and 'to the amazement of the meeting' utterly demolished *him*. 'At the end she asked me to nominate her for election to the Fabian Society [Shaw had just become a member of its executive] and asked me to dine with her.'[15] The dinner, at which John Robertson and Elizabeth Cracknell were also present, took place ten days later.[16] But it was nearly five months before Annie's nomination was submitted to the Fabian.[17] Shaw's story, though dramatic, made Annie appear more impulsive even than she was, and thoughtlessly disloyal to Bradlaugh, when in fact she agonized over the effect of her conversion on him.

I dreaded to make the plunge of publicly allying myself with the advocates of Socialism because of the attitude of bitter hostility they had adopted towards Mr Bradlaugh [she wrote]. On his strong tenacious nature, nurtured on self reliant individualism the arguments of the younger generation made no impression. He could not change his methods because a new tendency was rising towards the surface, and he did not see how different was the Socialism of the day to the Socialist dreams of the past ... My affection, my gratitude all warred against the idea of working with those who wronged Bradlaugh so bitterly. But the cry of starving children was ever in my ears, the sobs of women poisoned in lead works, exhausted in nail works, driven to prostitution by starvation; made old and haggard by ceaseless work. I saw their misery was the result of an evil system, was inseparable from private ownership of the instruments of wealth production; that while the worker was himself but an instrument, selling his labour under the law of supply and demand, he must remain helpless in the grip of the employing classes, and that trade combinations could only mean increased warfare—necessary indeed for the time as weapons of defence—but meaning war, not brotherly cooperation of all for the good of all.[18]

At the beginning of April Annie wrote to Ellen Conway from St Leonards-on-Sea: 'I have been very ill. I was obliged to leave town, for my strength obstinately refused to return, and it became monotonous to lie on the sofa day after day.' Her state was not improved by the prospect of losing her companion Lizzie Cracknell, who had become engaged to Captain Bingham.[19] It was almost a family affair; from Burma, where he was an administrator, the captain sent notes on tropical flora and fauna to *Our Corner*. When Lizzie had to go home to organize her trousseau, Annie thought of Shaw; in June he took over as art critic of the magazine.

Bradlaugh suspected that Annie's month-long illness was due to overwork and exposure to cold, rain, and fog on the nights she travelled to teach at the Hall of Science, unrecompensed, he pointed out, except for satisfaction in her work. 'No other woman in the [free-thought] movement has ever faced the heavy combination of literary, teaching and lecture work undertaken by Mrs Besant the past few years.'[20] Some writers have transferred this illness to the previous summer and made it a consequence of Aveling's 'desertion'.[21] If nothing else the date must rule that out. But Annie was vulnerable to mental stress which brought on physical collapse, as in the case of Mabel. This time her illness may have been aggravated by concern for Bradlaugh's reaction to her apostasy.

It was barely a year since he had paid a touching tribute to her. Under the terms of their partnership in the Freethought Publishing Company, the survivor would inherit the stock and copyrights. But the Company's affairs were in such a parlous state that, in drafting his will, Bradlaugh feared its liabilities would exceed its assets. 'I have nothing to leave the good woman who has stood side by side with me and born calumny and slander for my sake save my tenderest love', he wrote and, in the event of his death, urged secularists to rally round and repay the devotion she had shown.[22] In her *Autobiography* Annie claimed that her defection made no difference to their private friendship. But at the same time she was forced to admit that it caused Bradlaugh to lose faith in her judgement, when he ceased to consult her about his own affairs. Hypatia dated the beginning of their 'estrangement'—her word—to the mid-1880s because there was no opening for Annie in his parliamentary work: 'she was never content with a subsidiary position.'[23]

Bradlaugh's heroic fight to take his seat lasted for six years, during which his financial resources drained away in legal costs, and his health was damaged fatally. In January 1886 Mr Speaker Peel did what many people, including Annie, wished his predecessor had been brave enough to do—he administered the oath as a matter of course to every Member who came before him, including Bradlaugh. This simple remedy broke the fever that had raged for so long. As sitting Member for Northampton Bradlaugh, who voted as an extreme radical supporter of the Liberal Party, was accepted by the House. A few months later

Annie proudly told Moncure Conway that he was making his way very rapidly and steadily, 'and in the gossip of the lobbies is already spoken of as a future Minister'.[24]

As for her, it was not enough to profess socialist ideas; she had to 'live the life'. This was no less a challenge than the one she had faced on embracing free-thought. 'In "polite society" a Socialist is a wild beast to be hunted down, with soldiers, if he lives under Bismarck', she wrote, 'with sneers, abuse and petty persecutions if he lives under Victoria.'[25] This time there was no equivalent of Charles Bradlaugh to stretch out his hand to her; she had to propose herself to one of several organizations. At the beginning of the year the Social Democratic Federation had split when William Morris, Edward and Eleanor Aveling, with others, went off to found the Socialist League. Hyndman was left with a body mainly composed of working men, with a few middle-class revolutionaries like himself and the ever faithful Herbert Burrows. Shaw refrained from joining the SDF, 'not because of snobbery, but because I wanted to work with men of my own mental training'.[26]

Even if Annie had been able to contemplate marching under Hyndman's banner, or subordinating herself to the Avelings, she would not have joined the SDF or the Socialist League. The Fabian Society attracted her because of its origin in the Fellowship of the New Life. This was a group that had come together in 1883 under the influence of Thomas Davidson, a self-professed evangelist in the style of Annie's old friend and patron Thomas Scott. Where Scott directed his efforts towards freeing religion from its sectarian chains, Davidson focused his on the regeneration of society, a subject that fascinated Annie.[27]

Members of the Fellowship were all high-minded people with a desire to serve humanity, but some were less spiritually inclined than others. In 1884 the pragmatists separated from the saints in order to form their own society. The question was whether a leaping salmon ought to be admitted to their tranquil pond? In sharp contrast to the famous Mrs Besant, the forty or so Fabians were young, predominantly male, impecunious, and quite unknown. Shaw, who liked unusual people, was a good choice to sponsor Annie, but even he seems to have anticipated opposition, and moved to head it off. If anyone was so steeped in bigotry as to object to Mrs Besant's nomination—by himself and Sidney

Webb—he, Shaw, would leave the socialist party and run for election as a conservative, he wrote to a fellow member of the executive on 4 June.[28] This jocular threat produced the desired effect; on 19 June the minutes of the Society recorded the admission of Mrs Annie Besant.[29]

In under a year she was at the forefront of the propaganda effort. When William Morris enquired about her lectures she sent him a list with titles that included 'The Unemployed'; 'Why I am a Socialist'; 'How Poverty may be Destroyed'. Annie told Morris she was preparing a pamphlet specially for circulation among the workers whom she had facilities for reaching. 'In the country I am finding considerable response to lectures of the type of the one of which I enclose a report. ['Why I am a Socialist']. But I find the much divided state of London Socialists a considerable hindrance when I urge practical work.'[30]

In 'Why I am a Socialist' Annie declared that it was because she was a believer in evolution—the lesson Aveling had taught her. Evolution, which had 'sorted out' natural history, was destined to solve questions of psychology, morals, and sociology; it would make intelligible and consequent all that was now supernatural and inconsequent. She was a socialist, Annie continued, because the present system threatened a revolution; not one conducted by men of genius and directed by men of experience and knowledge as was the French Revolution, but 'a mad outburst' of misery, starvation, and recklessness that would end in desolation. She was a socialist, she concluded, because the poverty of the workers was inevitable under the present method of production and distribution of wealth.

A longer essay, 'Modern Socialism', appeared in *Our Corner*. It began with a historical survey of socialist communities, including that at New Harmony, Indiana, founded by Robert Owen (whose granddaughter Rosamond Dale Owen was a member of the Fellowship of New Life). Then it dealt with justification.

Great changes are long in preparing and every thought that meets ultimately with wide acceptance is lying inarticulate in many minds ere it is syllabled out by some articulate one, and stands forth a Spoken Word. The *Zeitgeist* has its mouth in those of its children who have brain to understand, voice to proclaim, courage to stand alone. Some new Truth then peals out... melodious to the ears attuned to the deep grand harmonies of Nature but terrible to those accustomed only to the subdued lispings of artificial trifles and the

1. Annie Besant in the 1870s

2. Frank Besant in Sibsey Churchyard with the Churchwarden and Parish Clerk, 1906

3. Annie Besant in her lecturing dress, 1885

4. Charles Bradlaugh, in his fifties

5. The Match Girls' Strike Committee, Herbert Burrows and Annie Besant standing centre

7. Annie Besant, 1897, in her Indian robes

6. Helena Petrovna Blavatsky

9. Charles Webster Leadbeater

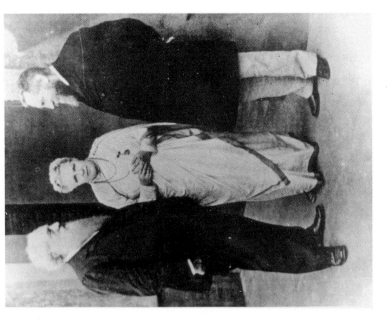

8. Annie Besant with Colonel Olcott (*left*) and Charles Leadbeater, 1890s

10. Krishnamurti (*right*) and Nityananda, 1910

11. Annie Besant and Krishnamurti arriving at Charing Cross Station, 1911

12. Dr Besant in Trafalgar Square, c.1925

murmurs which float around the hangings of courtly halls. Forever the new Truth condemns some hoary Lie . . . and thus the race makes progress and humanity climbs ever upwards towards the perfect life.[31]

This painfully laboured peroration drew unkind comments: 'St Athanasius in petticoats' was one.[32] And while the apocalyptic language and millennial vision expressed by Annie must have struck a chord among New Lifers, they made *Justice* nervous. 'Joan of Arc of the proletariat', it called her, 'chastening herself with prayer and fasting for her great mission.'[33] The comparison, though here intended to be flippant, was made in all seriousness later on. Annie was one of those rare persons who desire the mystic's inward-looking life at the same time as they pursue worldly success.

It was, for instance, second nature for her to apply herself immediately to gaining influence among her new colleagues. Like members of the National Secular Society, Fabians were very sociable: there were tea parties, country walks, outings to the theatre, soirées and conversaziones. The house and grounds at Avenue Road proved eminently suited to summer gatherings. One of these was a 'Russophile party'—Shaw's description[34]—that Annie gave for exiles from the tsarist regime. Among the guests, who included Bradlaugh, were two most prominent opponents of the regime, Nicolai Tchaykowsky and the nihilist known as Sergius Stepniak. The result was the formation of a Society of the Friends of Russia, which was to be a clearing house of accurate information about conditions in the country.[35]

As summer waned Annie was drawn into a bitter struggle to defend the right of free speech, which had come under threat from the authorities' determination to keep hold of a rapidly worsening economic situation. Unemployment was high and rising; the many poor went in fear of winter. There was unrest, oppression, and misery in Ireland. When radicals and socialists addressed these issues the tension rose. Their remedies—especially those prescribed at street corners by socialists—were very distasteful to the middle class, who began to fear for their property, if not yet for their lives. One of the petty persecutions complained of by Annie was the increasing tendency of the police to move people on.[36] In September she and Bernard Shaw were Fabian delegates to a Vigilance Committee composed of representatives from the working men's clubs, the National Secular

Society, the SDF, the Socialist League, and other radical organizations to discuss how this harassment might be overcome.[37]

Perhaps the committee's most effective measure was to have its members standing by for a summons—by telegram if necessary—to bail out speakers who had been arrested. 'Oh the weary, sickening waiting...for "my prisoner", the sordid vice, the revolting details of human depravity to which my unwilling eyes and ears were witness',[38] Annie wrote, though she gave no hint of weakness at the time. She would march into a police court, make her way to the witness box, and make the magistrate listen to her by sheer force of style and character, Shaw remarked.[39]

'We have had a miserably wet and gloomy autumn and the social and political environment is as unsettled and stormy as the weather',[40] Annie told Ellen Conway in December 1885. The months that followed were even worse. 'A feeling of uneasiness heightened by the all pervading black fog culminated in absolute panic in the afternoon', the *Annual Register* wrote, in chronicling the events of 8 February; 'the rumour became current that a mob of 50,000 men was marching from Deptford and Greenwich towards the West End wrecking and looting on their way. Throughout the night a repetition of false alarms kept up the excitement.'[41] The 'mob' was a deputation of the unemployed which Henry Hyndman conducted through Piccadilly and St James's. When they were jeered by gentlemen at the windows of their clubs, they responded by throwing stones at the windows. The police were considered to have lost control. The Government's answer to these 'West End Riots', 'Socialist Riots', was to try to relieve the situation that had provoked them by extending outdoor relief, and to declare its firm intention of prosecuting socialists who preached sedition.[42]

It was in the aftermath of these disturbances that the executive committee of the Fabian Society met on 17 February at 63 Fleet Street, the offices of the Freethought Publishing Company. The members present were Bernard Shaw, Sidney Webb, Mrs Charlotte Wilson, and Edward Pease; according to the minutes Mrs Besant attended by invitation. It was resolved that she be permitted to use the official Fabian design for a heading to a report of Fabian and other socialist affairs in *Our Corner* (this appeared monthly from then on). And that she draw up a scheme for organization in the provinces on the lines she had proposed: 'viz;

those adopted by the National Seccularist [*sic*] Society, with modifications'. It was also resolved that the next meeting of the whole Society be at Mrs Besant's.[43] When this meeting took place Annie gave notice that she would move for two members to be added to the executive and the present members re-elected. On 19 March two new members were duly added; one was Frank Podmore, the other Annie Besant.[44] Her advance to the executive had been as spectacularly rapid as her upward progress through the ranks of the NSS to become a Vice-President—an office she retained.

Annie knew precisely what she wanted—to use her new position of influence to further the long expected transformation of English political parties which, she told Moncure Conway shortly before the Fabian election, was just beginning: 'the moderate Liberals are drifting Tory-wards, and the antagonism between them and the Radicals is proving more marked... the balance of party strength leaves the Commons at the mercy of Parnell, and the situation will be too intolerable to last long.'[45] Annie foresaw an alliance of socialists and radicals whose aim would be representation in Parliament. Bradlaugh thought she was tilting at windmills, but gave her valuable assistance behind the scenes—it was his legal expertise for example that sustained her confident appearances in search of bail before the magistrates.

She was the moving spirit behind the conference of delegates from fifty-three socialist and radical organizations held at the South Place Institute on three days in June 1886. They discussed the aims and methods of democracy.[46] It was she who on 17 September, at a joint meeting of Fabians, the SDF, and the Socialist League at Anderton's Hotel, proposed the resolution (seconded by Hubert Bland) 'That it is advisable that Socialists should organize themselves as a political party for the purpose of transferring into the hands of the whole working community full control over the soil and the means of production and distribution of wealth'. The motion was carried by 47 votes to 19; William Morris being one of those most energetically dissenting. (The meeting was so noisy the hotel manager invited them not to return.[47])

This favourable vote resulted in the formation of the Fabian Parliamentary League (whose membership was optional). It was 'composed of Socialists who believe that Socialism may be most quickly and most securely reached by utilizing the political power already possessed by the people'.[48] Annie was also the founder of the Charing Cross Parliament, which met on Fridays and was taken very seriously by those Fabians who attended it. Details of procedure were scrupulously followed, and much care went into preparing speeches.[49] It was an attempt to make Parliament more real to those who, one day, might enter it. If Annie had succeeded in her broad aim the history of the Labour movement might have begun some years before it did, and she have been one of the first women members of any party. But she encountered obstacles; nothing so positive at first as open opposition, more a passive resistance.

Bernard Shaw wrote at length about her connection with the Fabians,[50] and Edward Pease, the Secretary, noticed her time as a member in his History of the Society.[51] Neither did justice to her political initiatives. Pease was cool; she made her reputation in other fields, he wrote; 'she was just a little of an outsider'.[52] Elsewhere he remarked that at the time few outside the working class regarded her with respect (Fabians of course were not working class).[53] Shaw supported Pease in suggesting that Annie found herself excluded from the inner circle. The Society was so successfully directed by the little group of men already in possession when she joined that she must have found, 'as other women later found', that as far as what might be called indoor work, 'she was wasting her time as fifth wheel to the coach'. Shaw paid tribute to Annie's courage and energy in becoming a sort of expeditionary force; 'always to the front when there was trouble and danger, carrying away audiences... founding branches... dashing into the great strikes and free speech agitations'.[54] But he, too, contrived to present her as an unharmonious, even slightly ridiculous figure.

She was heroic in her methods, as in her power, courage, and oratorical genius, he wrote. Fabianism was not heroic; it was in reaction to the heroism by which socialism suffered so much in the Commune... Fabians wanted to make socialism as possible as liberalism or conservatism for the pottering suburban voter who desired to go to church because his neighbours did, and to live always on the side of the police.[55] (Elsewhere he contended that Fabians were just as revolutionary as the SDF in the beginning, a proposition that Pease rejected.) But it is arguable that Annie's work, especially in connection with the Fabian Parliamentary League, was done precisely with the average voter in mind. 'We circulated questions to be put to all candidates for parliamentary or other offices, stirred up interest in local elections, educated men and women into an understanding of the causes of their poverty, won recruits for the army of propagandists from the younger of the educated middle class', she wrote.[56]

It was left to Edward Carpenter, a socialist who had his own millennial vision, to make a more detached and generous assessment of her contribution. She became the exponent in succession of large and important blocks of modern thought, he said; 'She helped to batter down the ruins and remains of the stupefied old Anglican Church; she gave the general mind a wholesome shock on the Malthusian question; she dotted out clearly the main lines of the socialist movement.'[57]

All this at a time when she was nervous about her health and, apparently, under mental stress. 'She brooded', according to Hypatia who, in 1886, came to live at Avenue Road with her husband, Arthur Bonner, printer to the Freethought Publishing Company. Annie had erysipelas in the face, and colds that lasted for weeks at a time. By the end of the year she was convinced that she had heart disease and got into the habit of carrying a pot of Brand's Essence of Meat about with her in case she felt faint. After her lectures she showed Hypatia how cold and numb her hands were.[58] It was hardly surprising that Annie betrayed signs of exhaustion; she was constantly on the move. Either she lectured or she went to other people's lectures and asked informed questions afterwards. She still had engagements to fulfil for the NSS, which sometimes took her out of London, so that she had to excuse herself from important socialist events. She edited the monthly *Our Corner*, and co-edited the weekly *National Reformer*. She taught classes at the Hall of Science; she churned out pamphlets. She even asked Ellen Conway to look out for an American paper that might take a 'London Letter': 'I want a little more writing than I have.'[59] She kept up her studies for a degree in science until she had failed chemistry three times; then she abandoned the idea, convinced that the examiners were prejudiced against her.[60]

From her home at Avenue Road she went down to the office in Fleet Street; from there to the Hall of Science on the other side of the City. Up to Wildwood Farm on Hampstead Heath to discuss anarchy with Charlotte Wilson; back to Adam Street for the Dialectical Society,[61] and a long way west to Kelmscott House in Hammersmith to hear William Morris, and stay to supper afterwards. Shaw was forever escorting her to the station, putting her into hansom cabs, walking her to her gate. When they travelled together he paid her fare and, along with all his other financial transactions, noted the cost; it was sometimes more than he spent on supper.[62] Annie began by calling him Mr Shaw. By mid-1887 she was writing to him as 'My dear Bernard'.[63] He called her Mrs Besant even in the absolute privacy of his almost indecipherable shorthand diary.[64]

Annie wrote virtually nothing about herself and Shaw; he a good deal. His account of her varied according to when he set it down. Unfortunately, in extreme old age, he grew waspish and inclined to score over his contemporaries, none of whom was alive to see fair play. So he called Annie dull, lacking in taste, and without sex appeal.[65] But for the most part of his long life he remembered her with affection and respect, though not neglecting flaws in her character: the two that counted most were lack of humour and overweening pride. 'Comedy was not her clue to life. She had a healthy sense of fun but no truth came to her first as a joke. Injustice, waste and the defeat of noble aspiration did not revolt her by way of irony or paradox, they stirred her to direct and powerful indignation and to active resistance.'[66] Shaw teased Annie by pleading poverty and, when she tried to remedy it, rebuffed her. This exercise was supposed to cure her pride. For instance, he said he needed an umbrella; when she gave him one he pronounced it in poor taste. Furious, she threw it over a hedge in Regent's Park; Shaw presented her with a drawing of a field sprouting small umbrellas.[67] 'Although I succeeded in making her laugh at me I never succeeded in making her laugh at herself or check her inveterate largesse.'[68]

Annie often found his behaviour beyond a joke; she resented being made to look foolish, as when he had tricked her into calling him a loafer. He was a man with 'a perfect genius for "aggravating" the enthusiastically earnest',[69] she wrote—a neat description of herself. Yet she put herself out to amuse him, practising assiduously for the duets they played on Mondays at Avenue Road. When Shaw stopped coming, Annie stopped practising,

Hypatia, who found these musical evenings tedious, thankfully recalled.[70]

Shaw's Irishness was part of a personality that fascinated Annie.[71] They came from a similar background—Protestant Ascendancy fallen on hard times. Shaw was curiously reluctant to acknowledge Annie as a compatriot; perhaps he thought she made too much of it herself.[72] Whatever he was, he was never dull, as Bradlaugh could be dull. Like Bradlaugh, Annie was everything on the platform, nothing off, Shaw remarked, in unkind old age.[73] On the other hand she found even his appearance in Dr Jaeger's woollen clothing system compelling;[74] according to another of his admirers he was 'a strange and wonderful looking man, tall and thin as a whipping post with a massive head...one side of his face Christ like although the other was Mephistophelian'.[75]

Each appears to have mistaken the other's feeling, not surprisingly, for each was the central character in his or her production and paid scant attention to the supporting cast. To take Shaw's version first. He became alarmed when, in 1887, he thought a serious 'intrigue' was threatened, and blamed himself for leading Annie on. According to his diary his decision to draw back strained relations with her for a time, and, on 23 December, she returned his letters.[76] Fifty years later Shaw said that Annie drew up a contract setting out terms on which they should live together as man and wife. He rejected them—'Good God this is worse than all the vows of all the Churches on earth. I had rather be legally married to you ten times over.' In recording this, Hesketh Pearson, Shaw's biographer, commented: 'the parting was something worse for the lady than a *mauvaise quart d'heure*. Her hair turned grey; she even thought of suicide.'[77]

This is a good example of Shavian self-regard. Annie's distress was due to other causes, not to her thwarted desire for a common law marriage to him. Annie knew that Shaw had a jealous and demanding mistress; they discussed his liaison with Jenny Patterson, a 44-year-old widow, in their correspondence; the letters returned by Annie were found by Mrs Patterson. From the passionate complaint she addressed to Shaw after reading them, it is clear that Annie referred to her as Shaw's mistress with contempt and would not have dreamed of taking her place—which is what Shaw's version of the contract implied.[78] Shaw, who frequently complained of Jenny's possessiveness—she followed him about and made scenes—was very anxious to avoid tying himself down. He was philandering in all directions—with Eleanor Marx Aveling, Mrs Hubert Bland, May Morris. 'As soon as I could afford to dress presentably I became accustomed to women falling in love with me. I did not pursue women, I was pursued by them',[79] he remarked complacently.

'Let us be honest', Annie wrote, 'I have not worked with any man in close intimacy who has not fallen in love with me, but I have managed to steer through and . . . keep my friend.'[80] Though her manner may have invited gallantry, she did not want a lover but a co-worker. She, who could not stand being alone, had not been without the constant support of a male colleague, whether it was Bradlaugh, Aveling, or John Robertson, since the beginning of her public career. Robertson especially responded sensitively to the romantic atmosphere she created to console herself for loss of respectability. In Annie's company of parfit knights, Robertson played the part of devoted squire; he went everywhere with her and was always on call to assist in her manifold activities while pursuing studies of his own. In March 1887 he was overworking, and consequently bilious, he told Patrick Geddes. 'One of these days I'll take flight abroad and settle down somewhere (Athens or Germany), to produce one of my magnum opus... At thirty it is time to be building as you know.'[81]

Robertson's manner was distinguished by an irritable bluntness that was allowed to influence his literary work. Whenever he and Shaw met they 'sparred'.[82] Shaw, who disliked getting as good as he gave, was irked by Robertson's reviews—especially the one which labelled his novel *Cashel Byron's Profession* a brilliant failure. Shaw flattered himself that it was Annie's infatuation with him that drove the Scot abroad.[83] By the autumn of 1887 Robertson was in Darmstadt living with Dr Büchner, with the intention of learning German to equip himself for the scholarly History of Christianity he contemplated.[84] In his absence Annie fixed on Shaw to take his place and offered him an agreement—which may have been in the form of a knightly vow—to live at Avenue Road and join in her work, which she increasingly regarded as inspired.

She was not greatly upset when he refused. Letters exist contradicting his claim to have been the cause of her contemplating suicide; they do not even mention him. They describe her profound distress at having brought down Bradlaugh's fury on her as a result of her conduct during the affair known as Bloody Sunday and how she turned for comfort, not to Shaw, but to the messianic figure of William Thomas Stead.[85]

CHAPTER 17

THE NEW MODEL LEADER

IN 1887 Annie's vigorous propaganda for the socialist cause was continually assessed for the damage it was thought to be doing to Bradlaugh's parliamentary career, while each time he denounced her new associates, the newspapers hailed a breach between them. In October the rumours were confirmed when Annie resigned as co-editor of the *National Reformer*. Under the heading 'Personal', which had signalled so many crises in their joint career, she declared that she had been brought to it by complaints of a divided editorial policy between herself and Bradlaugh on one issue only—socialism. As socialism became more and more a question of practical politics, differences of theory tended to produce differences in conduct, and that gave rise to inconvenience, she conceded. She would continue to contribute articles to the paper and, as joint proprietor, would naturally bear her share of responsibility for everything it printed.[1]

While Bradlaugh expressed regret at the ending of their partnership, he was undoubtedly relieved. The *Reformer*'s circulation was declining, as was the membership of the National Secular Society; something had to be done. He pinned his hopes on an autumn series of provincial lecture tours. There was another reason for her departure that she dared not name to Bradlaugh, Annie wrote (after his death). 'I saw the swift turning of public opinion, the gradual approach to him among Liberals who had hitherto held aloof, and I knew that they looked upon me as a clog and a burden and that were I less prominently with him his way would be the easier to tread. So I slipped more and more into the background.'[2]

A graceful valediction but untrue. Of all the names that figured in the sensational events of that autumn none was more prominent than that of Bradlaugh's supposedly self-effacing friend. In 1887 the course of events that Annie chronicled in *Our Corner* was rushing to a climax that she felt could not be long delayed. To her the plight of the poor and unemployed was simply unendurable. In Ireland, where the Government had introduced a new Coercion Bill, she thought the country 'in reasonable distance of civil war'.[3] As for comrades in America and on the Continent, whose fortunes she followed in her monthly 'Fabian and Other Socialist Notes', they were often at risk of their lives. Irrepressibly militant herself, Annie was drawn to fellow activists in the fight against oppression. Herbert Burrows now became her constant companion; Michael Davitt sent front-line dispatches to her War in Ireland feature in *Our Corner*; she had so far forgiven the Avelings that she lectured under the auspices of the Socialist League and, in March was the only non-League speaker at a gathering to commemorate the Paris Commune.[4]

As the year advanced, everywhere, but especially in London, the unemployed began to demonstrate. No settled policy of dealing with their gatherings could be detected in the conduct of the police, but in September, the *Commonweal*, the journal of the Socialist League, accused the Government of adopting Continental ideas, in that the police were being organized as a military force. They were to be placed over and above the people, the *Commonweal* alleged; they were even then asking to know the names of speakers in advance of meetings, and demanding copies of resolutions to be proposed.[5]

Socialists and radicals were further dismayed when, on 20 September, it became known that the Supreme Court of the state of Illinois had confirmed the verdict on the Chicago anarchists. They were seven men sentenced to death as a result of a bomb incident following a general strike the previous year. Seven policemen had been killed. The issue was exceptionally fraught because none of the accused had thrown the bomb, or could be proved to have been directly implicated. They were arrested because they were known anarchist leaders and were on record as advocating violence against servants of the public. In London, as in the United States, their sympathizers said that if the verdict was allowed to stand, it would virtually destroy the right of public meeting. On 14 October there was a gathering at South Place to protest. Addresses were delivered by William Morris, Stepniak, Shaw, and Annie Besant, who was particularily vehement about preserving the right of free speech in England too.[6] As for a reprieve she was not hopeful; she remembered only too well that, after the Manchester trial in 1867, when the accused were acknowledged to be innocent of firing the fatal shot, Allen, Larkin, and O'Brien went to the gallows all the same.

According to the *Commonweal*, four socialists returning from this meeting passed through Trafalgar Square and were moved to indignation by the misery of the unemployed and destitute loitering there. They decided that expression must at once be given to their need. The first speech was made the following day from a seat in the Square—'the finest public site in Europe'—under a banner reading: 'we will have work or bread.' The banner was black with white letters to signify the dark prospects of the unemployed, but it was soon joined by red flags, 'the emblem of sturdy revolt'.[7]

The habit of speeches and discussions directed by a Chairman for the Day (who was sometimes William Morris, or Herbert Burrows, often an ordinary working man) caught on, and for the next three weeks or so the Square was in constant turmoil. Ordinary people were prevented from going about their business; access to the stations was impeded; hotels lost trade; ladies had to refrain from shopping on foot in case they were jostled; the National Gallery was all but isolated; and, in the view of conservative newspapers at least, the vicar of St Martin-in-the-Fields, the Revd Mr Kitto, became a national hero for refreshing gallant constables with sandwiches and coffee.[8]

Annie's activities figured under a headline *The Times* kept permanently set up in type: 'The Police and the Mob'. On Monday 7 November, a day of very heavy rain, with Shaw to keep an eye on her, she went down to the Square to speak from the plinth of Nelson's Column. Her advice to her dripping audience was practical; she told them she and her colleagues had asked the Local Government Board to direct guardians in every part of the country to find work for them. There must *be* work: the country was in dire need of 'clearing and improving', Annie said, and proposed that, for a start, the Post Office be asked to take on extra men for Christmas.[9] Her words were enthusiastically received; she was already a heroine as a result of her work as Hon. Secretary of the Socialist Defence Association.

Numbers of working men caught up in the agitation in the Square had been arrested in the past weeks and charged with offences that carried sentences of hard labour. As Morris explained in the *Commonweal*, 'No one can appeal from a magistrate's decision unless he finds two sureties to bind themselves to pay the cost of the appeal if it is rejected. Where can a poor man find £50?'[10] Annie Besant took it upon herself to find it, summoning bail pledges by telegram. The experience was vastly more disturbing than it had been in 1885 for she held the police to be guilty of beating men in custody, and of interfering with witnesses. She did not flinch from publishing her accusations, and enlisted Lewis & Lewis to defend the most difficult cases. When one of the four original organizers of the protest meetings in the Square (James Allmann) was whisked away from court before bail could be arranged, she went after him to Holloway Gaol and secured his release; eventually the charge against him was dropped for lack of evidence.[11] Her intervention saved at least a dozen men from what Morris said would have been 'summary court martial, unheard, undefended, and on police evidence alone'. He conceived the warmest admiration for her energy; it sprang from such a passionate desire to be of use that, at times, her actions were almost involuntary. 'That Bow Street thing was funny', Annie wrote to one prisoner's wife; 'I had not the slightest right to speak [in court] but I felt that if I did not poor Hicks would be in prison.'[12]

On Tuesday 8 November the Metropolitan Police Commissioner, Sir Charles Warren, banned gatherings in the streets along the route of the Lord Mayor's Show which was to take place the following day; that included Trafalgar Square. This was welcomed by, among others, the 6,000 people who had signed a petition calling for the Square to be cleared.[13] But it dismayed members of the Metropolitan Radical Federation, the alliance of radicals, socialists, and Irish who had already called a meeting in the Square for Sunday 13 November. This was to protest against the treatment in Tullamore Gaol of William O'Brien, MP, who had been arrested for 'conspiracy' when speaking on behalf of the Irish Land League. Denied the status of political prisoner, he was refusing to wear prison clothes, or clean his cell. On Wednesday it became known that his state of health had obliged the authorities to remove him to the infirmary. On Wednesday too it was reported that the Chicago anarchists would be executed in two days' time.[14]

The National Secular Society was affiliated to the MRF and, on Thursday, Charles Bradlaugh was asked to conduct Sunday's meeting in the Square. He agreed, on condition that he had control of the arrangements, and that the meeting be postponed for at least a week since on 13 November he was engaged to lecture out of London.[15] His presence would lend credibility to the demonstration which was now as much in defence of free speech as it was for William O'Brien. Parliament had yet to debate Warren's conduct of public order; the more MPs could be got to criticize him, the better. Bradlaugh had successfully challenged a previous attempt to ban a demonstration. In 1866 the then Police Commissioner had closed the gates of Hyde Park against a meeting called by him in favour of parliamentary reform. Bradlaugh announced that it would go ahead, and without violence. Brilliant organization, which relied on a large number of well-briefed stewards, gave him virtually complete control of an enormous crowd, save for one group that, ostensibly against his wish, tore down the Park railings.[16]

His contribution to the present crisis had been a thoughtful article on the legal implications of a Trafalgar Square ban which appeared in the *Pall Mall Gazette*. Stead, its editor, was a ferocious anti-Warren man. 'To the Square', the PMG declared, and would hear of no delay. 'I never thrust myself on meetings where I am not wanted', Bradlaugh observed, and departed to lecture in the North.[17]

On Friday, Annie took part in a deputation of the organizers of Sunday's meeting to the Home Secretary, Henry Matthews. He rejected the argument that their purpose was legitimate; any assembly was unlawful if it threatened to disturb the peace, he told them; he did not believe they would be able to keep control. Annie told Matthews she would attend, come what may; the numerical strength of those present would, 'by mere pushing', effect an entrance to the Square. Another delegate, Edward Aveling, said they would go 'with the deliberate intention of showing force if necessary'.[18]

The Saturday morning papers were full of sickening descriptions of the last hours of four of the Chicago prisoners hanged the day before. Then came news that Sir Charles Warren had banned all meetings in the Square until further notice. That evening Annie attended a meeting of the MRF at the Patriotic Club in Clerkenwell. The decision was to go to the Square the next day, and 'Sunday after Sunday'.

The approach was to be made from several directions. Annie's contingent assembled at Clerkenwell Green carrying banners—one enormous red one surmounted by the cap of liberty, others bearing slogans: 'Starving for old England', and 'Educate, agitate, organize', the motto of the SDF. Radicals and socialists identified themselves with red ribbons, the Irish with red and green. Before the march began Shaw, William Morris, and Annie Besant were called upon to address it. Annie was vehement in her defiance of the ban, which she described as not worth the paper it was printed on. Her feelings betrayed her into making an assertion for which she had no authority. 'She said that Mr Bradlaugh had told her that the law was on the side of those who wished to hold the demonstration, and next week he would be there to tell them so himself.' She also told the marchers she meant, if not prevented by physical force, to reach the south side of Trafalgar Square where they must follow her, 'as brave men should'.[19]

The march, which attracted huge numbers of spectators, was without incident until it reached Seven Dials at the top of Shaftesbury Avenue where the police fell on it with truncheons flailing. The column scattered at the sight of broken heads. Shaw, who, at some stage, was beside Annie among the leaders, took refuge in the crowd.[20] She rushed on towards the Square which was encircled by a thick black line of police, captured a waggonette, and tried to persuade its driver to pull it across the street to form a barricade. He showed no taste for that, so she jumped down and went to get arrested—as a lady should. 'I pressed quietly against their line and was seized by five of them, whereupon I said I was one of the speakers and would submit to arrest. A superior officer asked if I had committed an assault and would not take the pushing as a technical assault. I confess that my sense of the ludicrous prevented me from going up to a burly constable and giving him a slap.[21] It was getting dark; there was nothing for it but to go to the Hall of Science, and then to Farringdon Street, where she chaired Shaw's lecture on 'Practical Socialism'.

One hundred and fifty people were taken to hospital with injuries received as a result of the confrontation in the Square, and 300 odd were arrested; the incident quickly became known as 'Bloody Sunday'. The PMG's headline over $8\frac{1}{2}$ pages of reports was 'At the Point of a Bayonet'. Stead described the action of the police, backed up by the Life and Horse Guards, as a *coup d'état*: when the country realized what had been done in its name against innocent people the Government would fall.[22] On the other hand *The Times* rejoiced at the utter rout of the marchers; no meeting had been held in the Square; no procession had got near it; it had not even been necessary to read the Riot Act.[23]

In Hull to lecture, Bradlaugh was inundated with telegrams asking for advice. 'Shocked beyond reason' he urged working men not to be made the victims of ill-judged conflict against the armed force of the Tory ministry, and warned them that men who deemed it right to measure themselves against a government must be prepared to replace it.[24] On 15 November the newspapers printed a message from him: 'Protesting indignantly against the prohibition of the Trafalgar Square meetings and denouncing the brutal conduct of the Government, I appeal to London workmen to leave the right of meeting for decision in the law courts and for vindication in the House of Commons when Parliament meets.'

His efforts at maintaining order drew a rare letter of acknowledgement from Gladstone.[25] The Liberal leader's own exhortation to the people appeared on the same day. It was the duty of all citizens to refrain from resistance to the executive Government until the law was clear; to try to claim a public right by means of attacking the police was utterly disgraceful, Gladstone thundered. Since there was grave disquiet in high places about continued unrest among the 'roughs' and 'watermen' and other 'desperate characters' supposed to have formed the majority of the marchers, special constables were hastily sworn in.

On Wednesday 16 November, 100 members of the Metropolitan Radical Federation met once more at the Patriotic Club to decide what to do the following Sunday. A preliminary resolution had already been drawn up recommending that there should be a gathering in Hyde Park, and that Trafalgar Square should be left alone until the right of meeting there had been legally established. This drew support from those who feared the consequences of another confrontation. But Annie Besant stood up in the body of the hall and put a challenging amendment. 'That there should be a meeting in Trafalgar Square at 3 p.m. on Sunday, 20 November; that no procession be made and no conflict with the police be intended; that no arms... be carried and that when the soldiers are called out the crowd disperse.' This was acclaimed by a part of the audience, some of whom were heard to say that hundreds would go armed, and give the police what they deserved. Appalled at the prospect of Annie getting her way, Shaw got to his feet to urge that there was nothing to gain by going; if they were not beaten by police they would be beaten by the military. Others, frightened by vivid images of fixed bayonets, supported him, and Annie's amendment was lost by a large majority.[26]

On her way home from the meeting Annie called on Stead. A disapproving Shaw accompanied her.[27] 'If we persist in a defiant policy we shall be eaten bit by bit like an artichoke', he told William Morris. 'I wish generally that our journals would keep their tempers. If Stead had not forced us to march on the Square a week too soon by his "not one Sunday must be allowed to pass" nonsense, we should have been there by now.'[28]

Stead gave Annie money to pay the fines of arrested men who, for lack of it, had been committed to Millbank Gaol. She went down to get them out. 'We stopped and bought hats to throw an air of respectability over our *cortège*', she wrote, 'and we kept together until I saw the men on to train and omnibus lest with the bitter feelings now roused conflict would again arise.'[29]

On 18 November she attended the inaugural meeting of the Law and Liberty League called into being to preserve the impartial administration of the law regarding free speech. Besides activists like herself, William Morris, Hyndman, John Burns the trade union leader, Stuart Headlam, and Charles Bradlaugh, the acute nature of the threat posed by the 'conspiracy' of magistrates and police against the public persuaded moderates like Dr Richard Pankhurst to join, and Jacob Bright, MP, to agree to be its Chairman. Stead enthusiastically promoted it in the *Pall Mall Gazette*.[30]

On 18 December the LLL gave a spectacular funeral to Alfred Linnell, who had died as a result of injuries received when a police horse had run him down in Northumberland Avenue on

20 November. Annie Besant directed the arrangements with an eye to ceremonial worthy of the Earl Marshal. To the solemn music of the Dead March from *Saul*, fifty wand bearers, veterans of the Chartist agitation, preceded the coffin. Annie paced slowly on its left side with Stead, Herbert Burrows, and Robert Cunninghame Graham, MP. On the right walked William Morris, Robert Darling of the Irish Land League, Frank Scott of the Salvation Army, and James Seddon, Secretary of the LLL. Following in a coach with Linnell's family were Dr Pankhurst and his wife Emmeline, and John Burns. According to Annie the steps of St Paul's were black with spectators; 'the chimney pot hats stayed on but all others came off as the coffin with its escutcheon, "Killed in Trafalgar Square" went by.'[31] She thought close on 100,000 people followed to Bow Cemetery where the service was read by Stuart Headlam, and the address given by William Morris—in characteristically gentle terms.

> Our friend who lies there has had a hard life, and met with a hard death; and if society had been differently constituted his life might have been a delightful one... We are engaged in a most holy war, trying to prevent our rulers making this great town of London nothing more than a prison. I cannot help thinking that the immense procession in which we have walked this day will have the effect of teaching a great lesson.[32]

Annie's next care was for the wives of John Burns and Robert Cunninghame Graham, who were sent to prison for six weeks in January 1888 for the part they had played on Bloody Sunday. 'I am so very much more sorry for you than for your husband', Annie wrote to Mrs Burns, 'because he has the pleasure of the fight, and you the waiting which is much harder.' She invited Mrs Burns to supper at Avenue Road, and to have a chat, and to stay the night. Thoughtfully she enclosed a postal order, 'lest you should be short'.[33] Mrs Cunninghame Graham, in the Gladstonian language now adopted by the Fabians, was a member of the 'classes', not the 'masses'—no postal orders for her. Protocol demanded she be called upon. 'She was so glad to see me', Annie told Stead, 'poor little thing.'[34] Mrs Cunninghame Graham had watched as her husband, with John Burns, had led a charge into Trafalgar Square. 'Where are all the other MPs?', she is supposed to have asked as constables dragged him off by the hair.[35]

Where especially was Charles Bradlaugh, MP? His conduct in staying away from the demonstration on 13 November upset many of his colleagues, while Burns described it as 'simply dishonourable'.[36] Bradlaugh strongly defended his decision in the *National Reformer* and, in so doing, took the opportunity of paying tribute to the only aspect of Annie's behaviour he could approve—and in which he had himself assisted—the effort to secure bail for arrested men. He would not have singled it out as women's work, Bradlaugh told his readers, but 'it has been bravely done, well done, and most usefully done... and I wish to mark this the more explicitly as my views and those of Mrs Besant seem wider apart than I could have deemed possible on many of the points of principle underlying what is every day growing into a most serious struggle'.[37]

Privately he was in a furious temper with her for having said that he would join the attempt to carry Trafalgar Square. Her recklessness had undone months of patient effort in the House of Commons when he had sought to reassure his fellow MPs of his soundness on constitutional issues. Now they were withholding from him vital evidence of police misconduct, evidence he needed to assist in the LLL's declared aim of prosecuting cases when the police were guilty of outrages on individuals.[38] Bradlaugh's morale, already undermined by the shock of finding that he was no longer physically equal to the rigours of provincial tours, was severely shaken.[39] Alice and Hypatia were so incensed at the effect upon him of this latest of Annie's rash endeavours that they sent her to Coventry.[40]

When the Bradlaugh family turned on her, Annie, who depended so much on supportive friends, found herself suddenly bereft. Lizzie Cracknell had gone away to Burma; John Robertson to Germany. Bernard Shaw had withdrawn from the close comradeship she had found delightful. Moreover he had physically deserted her at the crucial moment of the march on Trafalgar Square, then frustrated her intention to return. She looked about her for reassurance and thought she had found it in her 'dear but eccentric colleague' William Stead.[41]

His championing of the cause of free speech, so dear to Annie, was the latest in a series of attempted coups to be brought about through the 'new' journalism he had created, first as editor of the Darlington paper the *Northern Echo*; since 1883, as editor of the *Pall Mall Gazette*. Its success earned him popular acclaim but the influence he wielded as a result was the source of much resentment. Londoners did not like being told that between 1884 and 1888 Stead came nearer to governing Great Britain than any other one man, the New York *Sun* remarked, yet very soon after he took over the *PMG* his personality became one of the controlling forces in English public life.[42]

It was a bizarre personality, aspects of which made his contemporaries nervous. Stead was kind, generous, compassionate, a crusader in the fight against social evil: 'this chivalrous journalist', Annie called him when they first met. He was also turbulent, brash, hyperactive, melancholic, and at times embarrassing—he wept in public.[43] The son of a northern manse, married with six children, his mind ran on a subject that tormented him with guilt. 'He was a puritan who brooded so intensely on sex his brain had gone maggoty; sex and sin being in his view synonymous';[44] among men his conversation was notoriously Rabelaisian. Shaw said he never trusted Stead again after the episode of 'The Maiden Tribute of Modern Babylon'.[45] This was a series of articles in the *Pall Mall Gazette* in July 1885 describing its editor's attempt to expose the scandal of juvenile prostitution. (Stead paid £3 for a little girl, ostensibly for sex, and took her to a brothel where she was examined and certified to be a virgin. She was then left alone in bed until Stead woke her when he gave her into the charge of a woman Salvation Army officer who had accompanied him for the purpose.) One result of the immense uproar his revelations caused was the passing of the Criminal Law Amendment Act of that same year which raised the age of consent from 13 to 16. Another was the sentence of three months in Holloway Gaol on Stead, who was suspected of too much enjoying the escapade. On abducting the child he had carelessly ignored what laws did exist. He served his time in a mood of exaltation; each year on the anniversary of his incarceration he donned prison garb to relive the thrill of sacrifice.[46]

When Annie first knew him Stead was 36, a bearded gangling figure, habitually dressed in a loud, ill-fitting suit and a shabby sealskin hat.[47] She began their correspondence by inviting sympathy for the boycott the Bradlaugh girls had imposed on her.

It is stupid of me but this sort of thing hurts me. I am afraid my mind gets bruised as easily as my skin and it was so pleasant today to see a face that did not look disapproval. See what a baby a woman can be when she placidly walks into a business in which she most certainly thought bludgeon or arrest lay. Queer bundles that we are of strength and weakness!⁴⁸

These letters, which began at Christmas 1887, reveal that, in spite of the remarkable physical and intellectual activity of the past decade, Annie was still tormented by what she conceived as spiritual desire, which could only be assuaged by 'sacrifice'. 'You are brave, good, and kind', she told Stead; 'there is a work that wants doing. I can do much of it but I need a colleague who cares for the salvation of the people more than anything on heaven and earth.' In a phrase that probably was familiar to Edward Aveling, John Robertson, and Shaw: 'I have been looking for "a man" for some time and you seem to be the very one who has head and heart enough for the work.' The mainspring of her life was a belief in the triumph of Right over Wrong, Annie told Stead; work for that end was 'an effective thing, a real thing' and—literally—would usher in a Golden Age.⁴⁹ Her desire to see this Age come about was so intense that, as she told Ellen Conway in her New Year letter, if she thought the work might go for next to nothing, she would commit suicide.⁵⁰

Stead was no stranger to millennial dreams: his took the form of a new Church, founded by himself to bring about a regeneration of the Christian way of life. Annie longed to join in. Atheism had brought peace from the torment of believing in, and hating, an unjust God, she confessed, but it left her 'without a Father'. In an attempt to establish common ground, she told Stead that, by coincidence, she was just then reading the Letters of Oliver Cromwell and had come across a statement of his that seemed to her to bridge the centuries; 'I find this only good; to love the Lord and his poor despised people; to do for them, and be ready to suffer for them.'⁵¹ The Lord Protector, as it happened, was Stead's pattern of a man and leader. Annie had established her credentials to be numbered among the 'saints' of his new order. This was to be set on foot at once by means of the Law and Liberty League, Annie taking the initiative as editor of its journal, the *Link*, while Stead kept in the background.

The *Link* assumed a puritanical directness and austerity of tone. As in *Our Corner* effective use was made of regular headings and special sections; those here included 'The Lion's Mouth', 'The People's Pillory'—Sir Charles Warren was a permanent inhabitant—and 'To be Done, by Order of the Executive of the Law and Liberty League'. What had to be done was to organize the membership into Centres and Circles; the plan being to have at least one Centre in every parliamentary constituency. A Centre would consist of twenty Circles, each composed of eleven 'earnest' men and women. Their task would be to monitor local behaviour—of police, landlords, guardians, ordinary citizens in fact—in order to report whatever did not meet with their approval. In the beginning these cells of the new moral order were called Vigilance Circles; shortly after the name was changed to Ironside Circles.[52]

Centres and Circles all too readily evoked that organization which had most recently made use of them—the Fenian Brotherhood. Its leader had been called the Head Centre, and it was as 'My dear Head Centre', that Annie now privately addressed William Stead. Carelessly she let slip that they were working together to develop the new system and had to make a humble apology to an angry Stead. His lightest word would make her careful not to commit him further than he wished, she wrote: 'in fidelity to what you called our political and spiritual marriage lies the only chance of this special movement succeeding.'[53]

Delegates from more than forty radical and socialist groups, including the Fabian Society, the Social Democratic Federation, the Socialist League, and the National Secular Society, attended a meeting of the Law and Liberty League in January at which the Centres were discussed. On 14 January *Justice* printed a resolution urging its members to have nothing to do with Ironside Circles because they were in effect secret societies whose adherents had to pledge themselves to carry out instructions without limit or control. When Annie invited her colleagues in the National Secular Society to join, Bradlaugh forced the postponement of the decision until he had examined every detail of the scheme. He suspected that the underlying purpose of the Ironside Circles resembled that of the Fenian Brotherhood—to meet force with force. That was how John Robertson interpreted them, and wrote to remonstrate with Annie. 'Have had a hard, horrid, and everything nasty that begins with H, letter from Germany', she told Stead on 19 January.[54]

In advising the NSS to distance itself from Ironside Circles, Bradlaugh confined his comment to the potentially dangerous effect their anti-democratic structure might have upon the lives of their members, exposing them to surveillance, blackmail, and conspiracy. Annie either could not see this, or she thought the risk worthwhile in a situation where, as she believed, the Government would extend coercion from Ireland into England. She was so furious at Bradlaugh's rebuff she announced that she had considered resigning as a Vice-President of the NSS. When her fellow Fabian Graham Wallas also declined to join, she made it clear that henceforth she would consider him one of the weaker brethren: Ironside Circles were for those prepared to take hard knocks, she told him.[55]

But the battle was proving too much, even for her, and she called for quarter. 'I was glad of the drive with you, and you were kind', she told Stead whom she now addressed as Sir Galahad. She would like to sit down on a hearth rug near someone she loved and not talk, only rest, she told him. Detected in the aftermath of a fit of passionate crying, she excused herself obliquely: 'when a woman is very tired and very lonely and cares much for anyone she is apt to be a baby.'[56] When Stead failed to take the hint, Annie declared she was in love with him. Just as Shaw perceived himself to have led her on into a closer relationship than he intended, so Stead apparently did likewise with his reference to their 'political and spiritual marriage'. Annie professed humiliation at the discovery that she was no longer proof against such emotion; her pride and self-respect revolted at being at the mercy of a man to whom she was only one of many, she complained; 'while I have stupidly let myself slip into allowing you to be different from anyone else to me and so to monopolize me that all the rest of my friends have become more or less shadows. That is what hurts me.'[57]

When Stead still did not respond in the way she wanted, Annie continued to write beseeching letters. How exquisitely kind and sympathetic he had been to her, she wrote after one meeting. If he had been at all hard or contemptuous he would have driven her into a somewhat dangerous frame of mind. A little later on she wrote, 'you want to be a Christ so you should be glad to be one to me', and told Canon Shuttleworth that Stead reminded her of him.[58] It was a disturbing echo of the 'half angelic being' of her adolescent fantasy.[59]

Stead shortly took the easiest way out, and perhaps the kindest. On 20 April he told Annie that he was leaving for St Petersburg the following week. She already knew, 'though I was a little surprised to hear of your movements from a stranger'.[60]

Stead's ambition, 'to see the Tsar and find what sort of man he was, and what policy he intended to pursue', was about to be fulfilled through the good offices of one of the many other women in his life, Madame Olga Novikoff.[61] A Russian patriot, an unofficial ambassadress, she deployed her charm in the manner of her predecessor, the Princess de Lieven, to manipulate English politicians; Gladstone was one of those most infatuated with her. Stead had been an intimate friend of hers since the time of the Russo-Turkish War, when her diplomatic effort was said to have been worth 100,000 men to her country. When Stead asked Annie if he could deliver any messages in Russia, she declined. As the Jews had no dealings with the Samaritans, so the Kropotkins and the Stepniaks—who were Annie's friends—held no commerce with Madame Novikoff, whom they regarded as the zealous emissary of the Government they longed to destroy.[62]

On Stead's return to London in July Annie sent him a note: 'When, oh when am I going to have a quiet hour with you? I *do* so want it.'[63] Though they remained in touch with one another, there are no more of her letters in the bundle Stead preserved.

CHAPTER 18

ANNIE MILITANT

THE daily meetings for the unemployed in Trafalgar Square created a new spirit of self-help among them. In the months that followed Bloody Sunday no one worked harder than Annie Besant to preserve and build on that spirit. Her means of influencing working-class opinion were considerable; besides controlling what went in to the *Link* and *Our Corner*, she was still responsible for Daybreak in the *National Reformer*.

Everything she wrote or published at the time was informed by an acute sense of the injustice prevailing not only on her own doorstep, but in Ireland, and in Scotland, where the crofters were up in arms against the system of land tenure.[1] It was Besant who gave Michael Davitt a platform in the first number of the *Link* for his appeal to the working classes of both islands. In 'How to Organize London' he drew a comparison between London and Ireland; the problems they endured were similar: high rents, unscrupulous landlords, and, not least, a harshly repressive government. Among other measures he proposed that the unemployed be organized.[2] This was something those concerned in the Trafalgar Square campaign had already set on foot.

If relief was to do any good it had to be based on sound information. Among the groups affiliated to the Metropolitan Radical Federation, which co-ordinated a drive to establish registers of the unemployed, the National Secular Society had a head start by virtue of its many branches. Annie's writ ran here too; Daybreak exhorted its countrywide readership to co-operate in the scheme. By signing on, men signified their availability for public works which, it was hoped, local authorities would then undertake according to need. The enterprise was a gallant failure: by 17 December sixty registration offices had been opened and 17,112 men signed on. But Annie suspected they were only a quarter of those eligible.[3]

Davitt returned to his proposals in his speech as chairman of the huge and over-crowded meeting in Allen's Riding School on 18 February, to welcome Burns and Cunninghame Graham home from prison. Socialists who had been pleased to claim him for their own were disturbed by what *Justice* described as his 'milk and water' attitude—his was a moderate programme aimed at winning a seat in Parliament.[4] It was left to Besant, who, as one of the organizers, had been rushing about with her hat off, to set the hall on fire with a speech condemning the verdict on Burns and Graham.

They had been convicted of unlawful assembly in Trafalgar Square. The judge at their trial held that the purpose for which Crown Property—for so he deemed the Square to be—was dedicated to public use was for the public to pass by; such places were not to be used for any other purpose. 'If the Square were held to be a place of public resort, then it would be analogous to a public thoroughfare and the public would have no right to hold meetings for discussions on social, political, or religious subjects in the place.'[5]

Besant rejected this smothering pronouncement, holding that the right of free speech was fundamental to the hope of social progress; to her that right was symbolized by unimpeded access to the Square. Almost alone among her colleagues (who had turned their attention to organizing meetings in Hyde Park), she pursued ways of claiming it. In April she proposed that the MRF call an early meeting there, provided that MPs who sat for metropolitan constituencies would attend; there were no takers.[6] In June she asked the Council of the Law and Liberty League whether, supposing she saw her way to challenge Sir Charles Warren's prohibition by leading a procession to the Square, they would approve her action? 'She did not ask them to take part in it but asked to know they would not disapprove.' According to the *Link*, some agreement was expressed; but nothing came of it.[7]

She regularly challenged the Metropolitan Board of Works' by-law prohibiting collections at meetings in the open air. At Mile End Waste, on Clapham Common, and in Victoria Park, after Annie spoke, the hat was ostentatiously passed round. Her name and address were taken only once—probably by mistake—for the authorities were conspiring to ignore her.[8]

They could not do so when she hit on the idea of a 'democratic *conversazione*' in Trafalgar Square. From 4 to 5 p.m on several summer afternoons those invited walked about and chatted in the middle of the Square, 'London's open air Town Hall', the theory being that if they kept on the move they were not infringing the law. When the clock on St Martin-in-the-Fields struck the quarter they were asked to chant; at 4.15 p.m.: 'What is wanted for London is Home Rule. What is wanted for Ireland is Home Rule.' At 4.30 p.m.: 'What is wanted for London is the abolition of landlordism. What is wanted for Ireland is the abolition of landlordism.' At 4.45 p.m.: 'What is wanted for London is the right of public meeting. What is wanted for Ireland is the right of public meeting.' Crowds of spectators lined the northern parapet; at the first sign of indecorum police executed a sweeping movement at the heels of Annie's 'guests'.[9]

Diverting though this was, there was serious work to be done in the field of employment. The working conditions of women and children in particular clamoured to be investigated. They were most vulnerable to unreasonable hours of work performed in unsalubrious conditions for very low wages which constituted sweating, and which a House of Lords' Committee was examining at the time.[10] Herbert Burrows described the practice of sweating as 'robbery by the monopolizer from the worker of a large portion of the product of his trade'.[11] Information as to these practices, much of which appeared in the *Link*, *Our Corner*, *Justice*, and the *Commonweal*, was the concern of the factory inspectorate, whose members were seriously overworked.[12] That was only to be expected, since the Truck Amendment Act, which Charles Bradlaugh had steered through Parliament the previous year, charged the inspectors with prosecuting employers who broke the rules governing the payment of wages.[13] One of the most important of these rules was that wages must be paid in coin; there must be no deductions, particularly not for fines.[14]

The *Link* drew attention to this provision, asking its readers (to whom it still referred as Centres) to collect evidence of offences for the editor to pass on to Bradlaugh, who would put questions in the House.[15] By this means Annie was able partly to resume a working relationship with her old friend. They had resolved their differences over Trafalgar Square in the course of a long conversation in March at Newcastle, where they met while lecturing,[16] and had resumed a degree of domestic intimacy. But the Irish nationalist flavour of Annie's politics was too extreme for Bradlaugh, while he was rapidly making himself *persona non grata* to those (including Besant and Davitt) who were campaigning for an eight-hour working day. Bradlaugh opposed legislation in the House on the idiosyncratic grounds that no limit ought to be put by government to an individual's right to manage his own affairs.

Notwithstanding on 6 July he put a question to the Home Secretary:

Whether complaints have been investigated by the Inspector of Factories in the East London District of breaches of the Truck Act by illegal deductions from wages for fines, and what has been the result? ... Whether [the Home Secretary] is aware that certain *employés* have been dismissed immediately after giving information of breaches of the law, and if this be true, whether he will give directions to the Public Prosecutor to take proceedings in order to punish such an attempt to defeat the ends of justice by terrorizing witnesses?[17]

This question was based on information given to Bradlaugh by Annie Besant; it concerned the *cause célèbre* by which she is most frequently remembered—the strike of the girl match workers at the firm of Bryant & May.

The affair became public when the Revd Adamson, vicar of Old Ford, next door to Bow and Bromley where the company had its factories, gave evidence to the Lords' Committee alleging that the match industry paid scandalously low wages.[18] This provoked a spate of indignant letters from members of the Bryant family (the original Mr May was dead), and shareholders in the firm, which was ahead of its time in the attention it paid to its public image.

'To say that there is any sweating is a gross libel and such men as the Revd. Adamson ought to be locked up for three months for making random assertions simply based on hearsay', A. C. Bryant wrote privately to Lord Rothschild, Chairman of the Lords' Committee, on 4 June. There was a problem of over-supply, especially in the summer when demand was low. Most European countries prohibited the import of English matches while dumping theirs. It was a question of whether some reduction of wages might not be inevitable in the near future, Bryant told Rothschild. The firm imported a large number of foreign matches. '[Foreign competition] does not matter to Bryant and May as they buy and sell these foreign matches, but I can clearly see it is likely seriously to affect those employed in this country.' If there was to be 'a distress', it would not be owing to wage rates, but for want of work, he concluded ominously.[19]

On 26 May the printer H. W. Hobart, a leading member of the SDF, who was in charge of industrial news at home and abroad for *Justice*, had reported that in three Swedish match factories recently acquired by an English firm, wages had been reduced to the lowest level previously paid. Workers' committees were pressing for an increase, but the English management had hinted that if there was a strike they would stop production until new hands could be found to agree to their terms.[20] (The firm in question was Bryant & May.) On 15 June Annie attended a Fabian Society meeting at which Clementina Black, Secretary of the Women's Provident and Protection League, reviewed the unfair practices afflicting female labour, and proposed that a Consumers' Union be formed whose members would pledge themselves to buy only from shops dealing in goods whose manufacturers had a clean bill in the matter of wages. Two of the non-Fabians present, H. H. Champion and Herbert Burrows of the SDF, then proposed and seconded a resolution noting that the shareholders of Bryant & May were receiving over 20 per cent in dividends while paying their workers $2\frac{1}{4}d.$ per gross for making match boxes, and called for a boycott.[21] It was further suggested that the *Link* post Bryant & May in the Lion's Mouth. Besant complied, with a note in support of the boycott.[22]

She and Burrows then went down to Bromley by Bow where their requests for information caused apprehension among the girls. As journalists they were obliged to give an assurance that they would make themselves personally responsible for paying the wage of any girl dismissed as a result of talking to them.[23] The account of conditions at Bryant & May probably came as no great surprise to Annie; the *National Reformer* had reported the occasion in October 1885 when the girls had staged an unsuccessful strike. This time she was moved to a fury of compassion and contempt which she relieved in the *Link* of 23 June. With a grimace in the direction of the Chairman of the Company, who rejoiced in the name of William Wilberforce Bryant, her article was called 'White Slavery in London'.

A typical case is that of a girl of 16, a piece worker; she earns 4s a week and lives with a sister employed by the same firm who earns 'good money, as much as 8 or 9s a week'. Out of the earnings 2s a week is paid for the rent of one room. The child lives on only bread and butter and tea alike for breakfast and dinner, but related with dancing eyes that once a month she went to a meal where she got, 'coffee and bread and butter and jam and marmalade and lots of it'.

Annie's emphasis on the lack of self-pity among the girls was a powerful incentive to public sympathy which counted for a great deal in the affair. 'White Slavery in London' continued:

The splendid salary of 4s is subject to deductions in the shape of fines: if the feet are dirty, or the ground under the bench is left untidy a fine of 3d is inflicted; for putting 'burnts'—matches that have caught fire during the work—on the bench 1s has been forfeited ... If a girl is late she is shut out for half a day ... and 5d is deducted out of her day's 8d.

But the most appalling thing they had to endure was the factories themselves.

These female hands eat their food in the rooms in which they work so that the fumes of the phosphorus mix with their poor meal and they eat disease as seasoning to their bread. Disease I say; for the 'phossy jaw' that they talk about means caries of the jaw, and the phosphorus poison works on them as they chew their food, and rots away the bone.

As if that were not enough, girls of 15 were obliged to carry great stacks of boxes on their heads 'to the great detriment of their hair and spines'.[24]

When shareholders' names were printed in the *Link*, a number of them were conspicuous by the prefix 'Reverend'. 'Shame on you who so long as silence can be kept, do not care whence comes your gold; shame most of all on you who preach love and purity from your pulpits', Annie raged. 'Country clergymen with shares in Bryant and May draw down on your knee your fifteen year old daughter. Pass your hands tenderly over the silky clustering curls, rejoice in the dainty beauty of the thick shiny tresses...', and buy matches from Wilson & Palmer who treated their workers more humanely.[25]

This sort of thing was what Bryant & May feared most. He was sorry their customers were refusing their matches, William Wilberforce Bryant told George Dutton & Sons. His firm had always had good relations with its work-force until the socialist agitation began. The girls earned better wages than in any other place. 'It has happened however, that during the months of June and July this year that in consequence of the very wet weather round London, a large number of our hands who are usually permitted to leave for fruit picking etc. have been kept at home, and this at a time when we are passing through our quietest period.' The firm had been faced with three alternatives, William Wilberforce Bryant explained: to dismiss a third of the girls so that the others might earn more; to close the works on Mondays and Saturdays; to keep on all hands and spread the work. 'It seems to us that those good folks who think it right to discontinue the use of our matches—no doubt because they have not given the matter sufficient thought, inflict serious injury on our work girls for it is, of course, impossible that they can earn such good wages when we require fewer matches.'[26]

Behind the scenes, the Managing Director, G. P. Bartholemew, cast about for evidence to vindicate the firm and halt the slide in the price of its shares. He wrote to manufacturers in neighbouring Hackney and Stratford asking in confidence to know the average wage they paid. He also asked what section of the Truck Act prohibited fines being deducted from wages. He put in an advance order for thousands of copies of newspapers on the day they might be expected to print the Company's annual report, and tried to refuse payment when the item failed to appear or, as in the case of the *Daily Telegraph*, was unsympathetically presented.[27] 'The Tissue of lies' in Besant's article brought a threat of libel action which Annie received with ostentatious calm.[28]

On 26 June she and Burrows and another SDF member, John Williams, distributed copies of the articles outside the factory in Fairfield Road. The following day she was told the girls were being frightened into telling which of them had given her the information. Shortly afterwards three were dismissed, and Besant and Burrows were called upon to redeem their pledge. The sum involved was pathetically small—18*s.* a week for all three—but Besant declared that she and Burrows could not meet it; 'we are very poor' she told readers of *Justice* in appealing for their help.[29] It was this incident that prompted Bradlaugh's question to the Home Secretary. In his reply, without naming Bryant & May, Matthews said that a factory inspector enquiring into alleged breaches of the Act had examined thirty hands. He found that fines had been deducted for carelessness in work but 'not very recently'. On hearing of the dismissals the inspector returned to the factory on 5 July when he found 'the whole of the females out on strike'. The employers denied that the dismissal had anything to do with giving information, and the inspector found that the girls he had previously examined were entered in the wage book as being at work up to the time of the strike. (They had been reinstated after going before the directors who found that there had been a 'misunderstanding'.[30]) The Home Secretary told Bradlaugh he was unable to discover any reason for giving directions to the Public Prosecutor in the matter. Although Bradlaugh and Cunninghame Graham offered to produce girls who would swear that fines had been exacted up to the time of the strike, as far as the Home Secretary was concerned, the matter rested there.[31]

Meanwhile, rumours of the girls' dismissal had reached Harry Hobart who went down to Bow on 5 July, where he found groups of agitated match workers standing about outside the factory. He persuaded them to march to Mile End Waste, where they decided to come out on strike, and Hobart suggested they send a deputation to the House of Commons. A further meeting was arranged for Sunday 8 July to draw up a plan of action. This was attended by Annie Besant, Herbert Burrows, Cunninghame Graham, Clementina Black, and Hobart. The match girls—over 1,400 were present—voted for a resolution affirming the truth of Besant's article; protesting against the 'falsehoods' Bryant & May had been spreading in the press; condemning the shamefully low wages; calling on the Home Secretary to enforce the law against Bryant & May, and to receive a deputation; and 'that a union be formed, to be called the Matchmakers' Union'.[32]

This version of events is taken from *Justice* and rather conflicts with that given in the *Link* which has become part of Annie's legend. According to this the match girls, not knowing what to do, came straight to her for help.[33] The SDF annual report awarded the honours evenly; to Besant and Burrows for spectacularly publicizing the girls' grievances; to Hobart for organizing their first meetings.[34] The latter was taken aback by the match girls' appearance. Addressing them on Mile End Waste he had not particularly noticed how they looked. In passing by St Paul's Churchyard where there were smart shops he suddenly saw how unkempt the members of the deputation to the Commons were; draggled skirts and broken boots contrasting strangely with their beloved hats.[35] These were of black velvet with feathers famous for their size, worn 'clapped on the head like a clam shell'.[36] The public stared and laughed and applauded as they marched to Westminster. Some of them called on Annie at the offices of the Law and Liberty League, where she gave them tea.

They had to be kept from starving until the dispute was settled. Over £700 was collected from well-wishers. Besant, Burrows, and Hobart undertook to register all those eligible for a share, a task that took six hours as there were 672 of them.[37] They were assembled in a hall, the free use of which was donated by Frederick Charrington,[38] one of Stead's most obsessive supporters in the drive for 'social purity' in the East End. Among the Fabians who attended at this rather late stage in a strike that lasted less than two weeks were Graham Wallas, Sidney Webb, Sydney Olivier, and Bernard Shaw. John Mackinnon Robertson who had returned from Germany, was also present.

Various parties were anxious to act as mediators, including the Revd Samuel Barnett, Warden of the University Settlement at Toynbee Hall. He led a group of serious young men to see the directors of Bryant & May. It was after this meeting that on 17 July two members of the London Trades Council,[39] with the match girls' strike committee, which included Besant and Burrows, met representatives of the firm. After a protracted discussion the terms drawn up to end hostilities were submitted to the girls assembled in Charrington's hall, who called out 'yes' to each proposal. Among these were: the abolition of all fines, and of deductions from wages for any purpose; an increase in the rate for piece work; grievances to be reported to the Managing Director without intervention of a foreman; all strikers to be taken back without exception; barrows for transporting boxes formerly carried on the head. 'The firm further said that they would as soon as possible provide a breakfast room so that the girls would not be obliged to eat in the room where they worked, and also expressed a strong wish that the girls would organize themselves into a union so that further disputes may be officially laid before the firm', the *Commonweal* reported.[40] According to notes made by G. P. Bartholemew for circulation at the annual meeting of Bryant & May shareholders on 31 July, 'when the delegation from the work girls came before us with the London Trades Council it was found that they had absolutely no grievances'.⁴¹

The rules for the new union were drawn up at a meeting attended by the same two members of the Trades Council, Clementina Black, Annie Besant, and Herbert Burrows, the two last becoming Secretary and Treasurer respectively. Power to take business decisions was vested in a President with a committee of twelve. At the first quarterly meeting, in November 1888, it was decided to admit men as members. Because of the low rates of pay contributions towards sickness and other benefit could be either 1d. or 2d. a week.⁴²

As soon as she took office as Secretary, Annie set about raising money for a members' club in Bow, a match girls' drawing-room which would have a piano, tables for papers, for games, and for light literature, 'so that it may offer a bright homelike refuge to these girls who now have no real homes, no playground save the street'. Friction between the workers and the company continued for a time but, under the influence of common sense, it gradually disappeared. 'We have found the manager ready to consider any just grievance and to endeavour to remove it, while the Company have been liberal supporters of the Women's Club at Bow', Annie was to write in 1893.⁴³

The fame that came to her as a result of the success at Bryant & May brought appeals for help from many other workers; among them shop assistants, tin box makers, boot finishers, shirt makers. She was always ready to spring into action as, for example, in the case of the fur pullers of Bermondsey. They were paid 1s. 6d. per sixty rabbits; when their employer refused to pay more for the much heavier animals imported from Australia, the women came out and went to Besant. She got together a few pounds for the most needy, assembled the strikers in Flat Iron Square, and told them to organize, which they promptly did. When Beatrice and Sidney Webb came to write the history of the trade union movement they acknowledged the importance of Besant's contribution to the 'new Unionism' which swept the country after 1888,⁴⁵ when men were organized by the industries that employed them rather than by craft. It was through her work for the match girls and other unskilled workers, rather than, as she had intended, by her membership of the Fabian Society or the SDF, that Besant came closest to realizing her ambition of creating a new political force.⁴⁶

Annie's achievement was tremendous, the more so since it required patience in negotiating, something for which she was not famous. She became a respected figure at trade union gatherings and, in November 1888, with John Burns, Tom Mann, and the young Keir Hardie, attended the International Trades Union Congress in London which brought together delegates from England, France, Belgium, Holland, Denmark, and Italy.[47]

Some weeks before, Bernard Shaw, who, in spite of the upheaval in their personal relationship, was much in Annie's company at the time, noted in his diary: 'At School Board meeting I learn that Mrs Besant has joined the Social Democratic Federation.'[48] The news probably came as no surprise to her colleagues. As a result of the events Annie had lived through since Bloody Sunday, the SDF, which she had found so unacceptably militant at the beginning of her socialist career, was more in sympathy with her mood than the Fabian Society. Though she remained an active member of the latter she had lost faith in its capacity to deliver political results. In the spring of 1888 the Fabian Parliamentary League, on which she had placed much hope, was merged with the Political Committee, depriving Annie of an outlet for her initiatives.[49] By that time she was aware that she would never be a member of the inner group but must remain, as Shaw so neatly put it, 'fifth wheel to the coach'.[50] 'Her position among the Socialists is uncertain', Beatrice Webb wrote. 'Socialists are not characterized by devotion to leaders and men are jealous of women who assume leadership.'[51] When Herbert Burrows became Annie's inseparable companion, her translation to the SDF was only a matter of time.

It was Burrows who taught Annie about the class war and introduced her to the language of revolutionary socialism. She interpreted his commitment in her own way as a desire to sacrifice himself in the service of the people, whose respect he had laboriously won through years of working among them. She was encouraged to express her own aspirations in terms that Burrows used. 'I love the poor—these rough coarse people who have paid their lives for our culture and refinement, and I feel that the devotion to them of the abilities cultivated at their cost is the mere bare debt that I owe, for my class, to them.'[52] 'A Socialist every inch of him from top to bottom', in the age of stiff collars and starched shirt fronts, Burrows wore a sagging tweed suit which emphasized his gauntness, a brown flannel shirt, and a strawberry coloured tie.[53] Annie, who had always enhanced her roles by the appropriate costume, now put on laced boots, short skirts, and a red tam-o'-shanter, which gave the caricaturists an easy target but which, to be fair, were better suited to the lanes and alleys and rubbish-strewn yards that she and Burrows frequented when they spoke at workers' meetings.

When on 1 September *Justice* hailed her first appearance in its columns, it also expressed the hope that she would have time to stand as a candidate for the London School Board election in November. But, while Annie had been urging her readers in Daybreak, *Our Corner*, and the *Link* to run candidates, she refused to throw her own hat in the ring until just before the formal adoption date. She had set her heart on standing for a newly created body that would have wider powers than any School Board. The first elections to the London Country Council, brought into being with other local elective authorities by an Act of 1888, were due to be held at the beginning of 1889. Annie's prevarication was in case, at the last moment, women might be pronounced eligible to stand. They were not; so, on 27 October, she announced that she would contest the Tower Hamlets district of the London School Board, taking the place of Burrows who had stood at the last election in 1885.[54]

Women had been eligible for School Boards since their creation under the 1870 Act, and some had already served, including Helen Taylor, Elizabeth Garrett, Henrietta Müller, and Florence Fenwick Miller, the last a staunch supporter of Bradlaugh and Besant's publication of the *Fruits of Philosophy*.[55]

First as sub-editor, then as co-proprietor, of the *National Reformer*, over the years Besant had written countless reports on the activities of School Boards up and down the country, and urged upon her readers their duty as free-thinkers to claim the right, embodied in the Education Act of 1870, to withhold their children from religious teaching.[56] In 1882 her interest became closer and more personal when Edward Aveling became the first secularist to win a place on any School Board, when he was elected to serve as member for Marylebone.

Even in 1888 the London School Board had not yet caught up with the tremendous problems that had faced it at its creation eighteen years before. Then more than half the children in the capital were without any form of education; there were no buildings in which to teach them, hardly any teachers, and a great aversion on the part of parents to the loss of income when a child was not at work. According to the final report of the Board, which was absorbed by the London County Council in 1904, it took twenty-five years after its creation in 1870 to reach a state of equilibrium. These were years of rapid building, when 9,000 children a year were added to the roll. Attendance was compulsory; fees ranged from 1d. to 9d. a week per child. Parents of absconding children were liable to be fined. The system operated under stress: there was too much to do, too many children in a state of extreme poverty, and the continuing hostility of parents was an added burden.[57] Discipline was harsh, classes overcrowded, learning was by rote. The educational process which Annie was to confront in Tower Hamlets was as far removed from her own experience of practical observation and rational enquiry under Ellen Marryat as it was possible to be. And there was an aspect of it that socialists in general, and the SDF in particular, detested: Board school children were all of one caste. In the scathing words of R. H. Tawney, the system was intended to produce 'an orderly and obedient population with sufficient education to understand a command'.[58]

The parliamentary manifesto of the SDF required all its candidates, for whatever post, to subscribe to a declaration: that they admitted the existence of a class war between the earners and owners of wealth; and that this antagonism would not be healed until the means of production, including the land, passed into public ownership, and every adult member of the community was called upon to do their share of work.[59] Besant not only endorsed these principles, she held on to them in the teeth of opposition from various moderate groups—liberal committees, radical organizations—who supported her because they were anxious to make headway against the Tory domination of the Board: her chance of winning was always considered to be good. Unusually, her address was to the electors *and non-electors* of Tower Hamlets. She promised them closer attention than they were accustomed to receive from their MP; he was a representative, acting at his discretion; she would be a delegate, the people's mouthpiece on the Board.

Her programme embodied ideas long put forward by her colleagues in the socialist movement before she came among them, but it was distinguished by her verve. 'Free secular compulsory, technical education for *all*. Education should be for all classes alike, so that it may break down class divisions and lay a basis for real equality', Besant wrote.

> The Common School system is the foundation of true democracy. Extension of Education and School management in favour of Evening Continuation Schools and the opening of higher education to the poor. I shall never consent to starve either the Schools or the teachers; teaching is arduous and exhausting work, and those engaged in it should always be treated with consideration and sympathy. I am in favour of putting the Board schools at the service of the public ... and the throwing open of the playgrounds to all children during the hours of daylight after school time, and during the holidays.
>
> No more *hungry children*. Children sent to school without breakfast and given but little dinner cannot profit from the education you parents and ratepayers pay for ... they break down under what is called overpressure but what is really underfeeding. If we force them to learn we must make them *able* to learn. Until Parliament gives us power to provide a free meal we must be content to organize voluntary effort ... Not a penny will be added to the rates but to the misapplied City Charities and taking these we shall use them for filling the stomachs of thin children instead of gorging the paunches of fat aldermen.
>
> I shall if elected, steadily oppose any work of the Board being given to firms who treat their workpeople unfairly or who pay less than trade union rates of pay. I am strongly in favour of the appointment of working men and women as managers of our Board Schools.
>
> Lastly I ask the electors to vote for me, and the non-electors to work for me because women are wanted on the Board and there are too few women candidates. Women have proved themselves useful, especially in connection with administrative details whenever they have been elected to offices of public trust and, as a woman, I ask you to elect me.[60]

Annie campaigned with her customary panache, driving about in a dog cart with a red ribbon in her hair: 'there was in her brown and tanned face and in the firm thin lips and knotted brows [something] which unmistakeably told of a woman of courage and determination', the *East London Observer* remarked.[61] The proceedings were enlivened by the vociferous opposition of clergymen, vestrymen, representatives of hastily summoned mothers' meetings, who raked up Annie's undeniably eventful past. The vicar of Stepney, the Revd Edwyn Hoskyns, issued 30,000 leaflets of unexampled virulence harping on themes put forward in *The Elements of Social Science*—the abuse of chastity, the glorification of sexual licence—'themes I loathe', Annie told

Stead.⁶² Wearily she issued a writ for libel. But her supporters were rather encouraged than otherwise, since at least a high turn-out was ensured.

Tower Hamlets—for which Annie's girlhood hero Captain Frederick Marryat was briefly a parliamentary candidate—was one of the largest and poorest districts of the School Board. Thirty-five per cent of its inhabitants were more or less permanently in want. The figure was higher in certain areas like Whitechapel, where, in the autumn of 1888, as far as anyone knew, Jack the Ripper was still at large.⁶³ As for the voters, Annie could count on all those Irish dockers in Limehouse who put politics above religion. The Jewish immigrant vote was large enough to make a difference; thousands had recently taken refuge here from pogroms in Russia and eastern Europe. There was a strong conservative candidate in Claude Montefiore, a member of the old Jewish aristocracy. That Annie prevailed over him was partly owing to her connection with the Russian exile group, mostly because of a campaign in Yiddish on her behalf by the militant tailor Lewis Lyons, who owed her a debt for finding him bail in 1885, when he had been arrested while demonstrating.⁶⁴

When the votes were counted Annie came top of the poll with a staggering 15,296—four times the number Burrows had received three years before. Stuart Headlam was second; he won Hackney with 13,000 votes. 'Ten years ago under a cruel law ... Christian bigotry robbed me of my little child', Annie wrote in the *National Reformer*,⁶⁵ 'now the care of the 763,680 children of London is placed partly in my hands, 98,890 of these being in my own constituency.' Of the members elected in the eleven districts of the Board she reported that thirty-two were conservatives under the leadership of the arch-Tory the Revd Diggle: they were known as the moderate party. The opposition, the progressives, a loose alliance of radicals, liberals, Nonconformists, and representatives of the left (a word just coming into use) numbered twenty-three. Out of the fifty-five members of the Board sixteen were clergymen of various persuasions, causing *Justice* to remark that the most important issue to be settled was whether the schools belonged to the parsons or to the people.⁶²

The reputation of the formidably victorious Mrs Besant must have given the longer-serving male members of the Board some anxious moments. Helen Taylor and Henrietta Müller had conducted themselves aggressively, Müller uncomfortably so, for she was permanently at war with those who were insufficiently supportive of the feminist cause.[66] Annie was at pains to dispel rumours that she was 'difficult' if she did not get her way. 'If you ask anyone I work with, I have the character of being the most workable with', she pleaded to William Stead, 'unless a point of principle arises and, I must admit, then I am hard.'[67] The amount she accomplished in the three years she served bears witness not only to the fact that her programme was relevant but to the sense and skill of her approach. 'Her speeches ... radiated a sunniness, moral generosity and sincerity that was clearly infectious and which brought out the best in otherwise sour and dour colleagues.'[68] Lyulph Stanley, leader of the progressives, was charmed by her dash and courtesy. But each time he was forced to recognize her from the Chair, the Revd Diggle looked as if her name would choke him.

Besant obtained the committees of her choice, three of the most influential—works, by-laws and school management. By the end of January 1889, with the help of Stuart Headlam, her effort in the first of these produced the rule that all firms tendering for a Board contract must sign a declaration that they paid trade union rates. Thus a famous principle was established, to be adopted shortly afterwards by the London County Council. Next Besant and a fellow free-thinker, Mrs Ashton Dilke, together waged a furious battle against Eyre & Spottiswoode, the firm which had the contract for printing religious material, and which they accused of sweating. 'There is something especially revolting in cheap labour at work on cheap Bibles', Besant forthrightly declared to the horror and indignation of the clergymen around the table and, in spite of a writ that arrived in the middle of the meeting, went on to remark that Eyre & Spottiswoode's wages were so low (6s. and 7s. a week) that their women workers would be forced to take to the streets.[69]

'Diggle and Co.', 'Diggleism', stood for the hegemony of the Church of England in all fields of education, especially in Board Schools. Besant won a notable victory when, after a protracted struggle with the Chairman and his associates, she prevented the dismissal of Arthur Moss, a school visitor whose offence was that he devoted his spare time to lecturing in free-thought.[70] She also pounced on the opportunity for propaganda in the cause of educational equality by drawing to the attention of the school management committee a complaint that the well-to-do were sending their children to Board schools. After a wrangle over wording, when her colleagues amended Besant's robust approach, the Committee went on record as to the equal right of citizens to the use of schools which they built and supported through their rates.[71]

It was the by-laws committee's task to issue summonses against parents whose children, for whatever reason, failed to attend school. Besant proposed to spare them the strain of attending police courts by the appointment of an itinerant magistrate to hear offences against the Compulsory Schools Act. The idea found favour but was shot down when it got to the Home Secretary.[72] Case by case Besant fought to have children allowed to continue in school during the lengthy, and sometimes hopeless, business of recovering their fees. Little by little the desirability of making elementary education free was borne in upon her colleagues, until she and Headlam gained support for a resolution, moved against the full force of the Revd Diggle's opposition, that the Board petition Parliament to make such schools free. The Church was fiercely opposed to this since it feared the effect upon the finances of its voluntary schools. However, pressure from many sources resulted in an Act of 1891, making it a matter of discretion for individual Boards whether fees were charged, or not. The Final Report of the London School Board was quietly congratulatory on this point. 'The Board, even before the 1891 Bill was passed, decided that, should Parliament give them the power, they would abolish the fees in the whole of their schools and make education free so far as their schools were concerned.'[73]

Even Annie Besant's power of persuasion could not shake the opposition to free school meals which came from all sides, including Mrs Dilke. The best she could do was to persuade the school management committee to enquire into the scale of the need. It found that, in the year to March 1889, 12.8 per cent of all children in the London Board schools were in want of food; no provision at all being made to feed half of these, 24,739 to be exact. 'Oh to gather all those 24,739 children into one piteous throng', Besant exclaimed, 'to fill with it Palace Yard and the approaches to Westminster.' She had to be content with chivvying the voluntary London Schools' Dinner Association which, by December 1889, had given one free meal a day to 36,000 children.[74]

Her duties at the Board were at the expense of time spent lecturing. Recognizing this, a group of well-wishers undertook to pay her £150 a year while she remained on the Board. Her income from lectures was falling; the result, she told Stead, of embracing socialism. *Our Corner* was the first casualty. When Edward Carpenter offered her a paper in July she replied that she was always glad to print anything of his, 'but before accepting I must say that my poor *Corner* is on its last legs and is not able to pay contributors. I am out of pocket every month and have spent on it all I could spare.'[75] Its demise was followed by that of the *Link*, which ceased publication after the School Board elections. It had never really recovered from the fiasco of the Ironside Circles, and its fate was sealed when the places where it could be obtained—Kelmscott House for example—gave up taking it. Annie perceived that she was almost poor; fretful remarks about her rent indicated that the ample style of life at 19 Avenue Road was in jeopardy. The money that mysteriously appeared in 1876 to enable her to lease her first house seemed to have vanished just as mysteriously. There is a possible explanation. Mrs Maria Wood, 'Aunt Ben', who may have been Annie's benefactress twelve years before, was still alive, aged 90. We know that Annie expected a legacy. But by then, various branches of the Wood family were aware that Aunt Ben intended to leave everything to Kitty O'Shea. Annie's sudden insecurity may have owed something to the abrupt end of her expectations.[76] Be that as it may, when Alice Bradlaugh died, Annie proposed that her father take over the lease of 19 Avenue Road. Bradlaugh refused.

Alice died on 2 December at Avenue Road, where she had been moved because the eccentrically arranged accommodation in Bradlaugh's lodging made nursing onerous. During the weeks of Alice's fever and delirium from meningitis, and in the very middle of the School Board campaign, Hypatia became aware that Annie was avid to observe the moment of her sister's death. Annie had begun to investigate the claims of spiritualism and wanted to discover whether the life force made itself known at the moment of its departure into another world.[77] She was in the early stages of a journey into a new belief; one that would absorb her until the end of her own life. Beatrice Webb, who well understood why Annie was attracted to it, and herself saw value in it, described this new belief, Theosophy, as a 'wonderful fairy tale'.[78]

CHAPTER 19

'SUPERHUMAN, SPIRITUAL REALITIES'

As atheists Annie Besant and Charles Bradlaugh did not contend that there was no God, but that no evidence of his existence had been discovered so far. One of their functions as co-editors of the *National Reformer* was to examine and report on beliefs claiming supernatural origin. Because of its immense popularity, spiritualism, which swept the United States in the 1850s and England a little later on, received a good deal of attention. In 1869 Bradlaugh had served with two distinguished spiritualists, Alfred Russel Wallace and Sergeant Cox, as a member of a committee of the Dialectical Society to examine its claims. Broadly these were that there was life after death, and that the dead were in communication with the living through mediums. After taking evidence of phenomena, the committee's conclusion was that 'the subject was worthy of a more serious investigation than it had hitherto received', a form of words that reflected Bradlaugh's reservations.[1] While the *Reformer*'s coverage continued to be extensive throughout the 1870s and 1880s, its tone (which Annie apparently had no difficulty in reflecting) was markedly sceptical.

In 1888 Annie's desire to look more closely at what she called 'the obscure side of consciousness—dreams, hallucinations, illusions, insanity',[2] may have owed something to the enthusiasm of her Fabian colleagues Pease, Podmore, and Shaw, who were members of the Society of Psychical Research which met in Podmore's rooms in Dean's Yard, Westminster, as, from time to time, the Fabians did. 'Thus I have attended a Fabian meeting, gone on to hear the end of a Psychical Research one, and finished by sleeping in a haunted house with a committee of ghost hunters', Shaw remarked.[3] Annie herself ascribed it to a conviction that her philosophy was not enough; this had been growing upon her since 1886—when Hypatia had observed her fits of brooding.⁴ At their first meeting the following year, Beatrice Webb was struck by her obvious unhappiness. 'We met and [I] felt interested in that powerful woman with her blighted wifehood and motherhood and her thirst for power and defiance of the world', she noted in her diary.

> I heard her speak, the only woman I have ever known who is a real orator, who has the gift of public persuasion. But to *see* her speak made me shudder. It is not womanly to thrust yourself before the world. A woman in all relations of life should be sought. It is only on great occasions when religious feeling or morality demand it that a woman has a right to lift up her voice and call aloud to her fellow mortals.⁵

To Annie the duty her gift imposed on her was precisely that; at every opportunity to address her fellow human beings on great themes. By 1889 she despaired of the fact that, while she knew she had the power to move her audiences, its effect scarcely lasted beyond the emptying of the halls. Otherwise the Fabian Society would have fallen in behind her in her march to Parliament, socialists would have worked in harmony with other socialists, and the Ironside Circles would have obtained a grip upon the land. She was forced to conclude that the message was at fault. The socialist position sufficed on the economic side, she remarked, but where was the human material for the nobler social order? Where—as she put it—were the hewn stones for the Temple of Man?⁶ Her disillusionment was echoed by Herbert Burrows in a series of articles for *Justice* lamenting the loss of idealism by social democrats.⁷

According to Hypatia, there was a period of transition lasting several months before Annie announced her conversion to Theosophy. During it she read and pondered as nervously as she had at Sibsey, on the eve of her departure from the Church. Burrows and his wife accompanied her to seances, and sat in on her experiments with clairvoyance and thought reading. The phenomena produced seemed to Annie 'indubitable, but the spiritualistic explanation of them incredible'.⁸ Probably the doubts she had expressed often enough in the *National Reformer* were reinforced by the public outcry in October 1888 when Kate and Margaretta Fox confessed that, as children, they had contrived the famous rappings in their parents' house in upstate New York.⁹

In 1848 the sensation that the Fox sisters' spirit messages caused swept the United States whence it was imported into

England by American mediums. By the early 1850s table-turning, levitation, lights, bells, and manifestations were also understood to be communications from the dead. Much depended on the medium, who was vulnerable to suspicion of fraud. Some were exposed, but others—among the greatest was Daniel Dunglass Home[10]—produced phenomena which defied the most rigorous investigation. On the one hand the new discovery acquired the spiritual and moral values that accompany religious faith; on the other its potential for conjuring tricks was exploited by Phineas Taylor Barnum in America; in England by J. N. Maskelyne.

By her own account (written after her conversion to Theosophy), having experimented in various ways suggested in the books she read, Annie got some curious results. 'I finally convinced myself that there was some hidden thing, some hidden power... and by the early spring of 1889 I had grown desperately determined to find at all costs what I sought.'[11] A hint of the direction she was taking is contained in the letters she wrote to the Revd Williams Ashman, an enthusiast for telepathy who, on hearing of her interest, offered to introduce her to his system. He was conversant with books that were hardly required reading for a Church of England parson. 'I have *studied* nothing in occult science', Annie told him on 14 February 1889, 'Only read anything that came my way—two books of Sinnett's, some stray pamphlets... But I am quite ready to study carefully any works throwing light on the matter.[12]

Occultism embraced all kinds of magic, besides astrology, alchemy, and forms of mysticism. However, the books by A. P. Sinnett[13] dealt with a particular and recent development by which a movement founded in America in 1875 by Helena Petrovna Blavatsky and Henry Steele Olcott, which had its origins in spiritualism, acquired many of the attributes of oriental religion, a connection with the ancient past of India, and a belief in the perfectibility of man; all of which was calculated to appeal to the contemporary taste for strange happenings in exotic places.

That is to put it crudely. The aims of the Theosophical Society were both subtle and far-reaching—to establish a brotherhood of man; to promote the study of comparative religion and philosophy; to investigate the mystic powers of life and matter.[14] Those who gave themselves to its service often found that their lives were transformed, while the potential for doing good in India, which became the movement's headquarters, was vast.

As their respective autobiographies show, Annie Besant and Beatrice Webb were alike in their intellectual development.[15] Webb's commentaries on her experience, delivered with detachment and clarity of thought, provide a useful gloss on the metaphysical ideas that attracted both of them. While Besant suffered more acutely in her passage out of Christianity (Webb describes her own hold on it as feeble), both were swept along by currents of thought that Webb describes as stirring the minds of those who frequented the outer and more unconventional and cultivated circles of society[16]—Besant's mentors being Charles Voysey, Thomas Scott, and Moncure Conway. These currents were Eastern religion, which Webb described as the most immediately subversive; and the religion of science, the most deeply influential, because it promised to solve the great preoccupation of the time—man's relation to man, and to the universe—by purely scientific methods. The fact that these two currents were made to come together in Theosophy—a word originally used to describe any system of speculation which based a knowledge of the phenomena of the natural world upon a special understanding of the divine nature, as, for example, in the works of the seventeenth-century mystic Jakob Boehme—attracted cautious approval from Beatrice Webb. 'As a working hypothesis for the conduct of life the whole theosophical concept of the universe and Man's destiny has many points in its favour. If it were true and held to be true the world would gain in goodness, capacity and charm.' The trouble was, Webb went on, there was no verification for its claims and so she had to conclude that it was 'a daring piece of intellectual gambling'.[17]

Let Josephine Ransom, a historian of the Theosophical Society, who was also a member, provide the introduction to it.

Like all truly great movements the sources of the Theosophical Society are not to be sought in mundane regions and activities, but in the heights of superhuman spiritual realities, whence emerge those forces which direct evolutionary destiny...

It has been consistently affirmed by all who have exercised the highest authority in the Society that these origins are to be found in a group of Superhuman Men, Teachers, Masters, Adepts, Whose universal knowledge of evolution and its laws constitutes them the wise Initiators and Guides of all movements designed to influence profoundly the growth of the world, and Whose directions the Theosophical Society has through its leaders with considerable success striven to follow.

Mrs Ransom then summarized the nature and work of these Superhuman Men from 'what accessible information' existed. The purpose of Those interested in promoting the Theosophical Society was;

1. To assist in showing to men that such a thing as Theosophy exists, and to help them ascend towards it by studying and assimilating its eternal verities. This department of the work... was entrusted to Madame Helena Petrovna Blavatsky (HPB) because of her rare and precious gifts.
2. To promote a genuine practical Brotherhood of Humanity where all will become co-workers with Nature. In this department Colonel Henry Steele Olcott proved eminently practical and efficient.[18]

The concept of secret knowledge, conferring on those within the circle powers beyond those of ordinary human beings, must have recalled to Annie Besant the rapt attention she had devoted in her adolescence to accounts of magi in the Fathers of the Church. Later, through Moncure Conway's introduction to Hinduism and other Eastern faiths, she became familiar with the idea of Masters, Teachers, Adepts. At the same time her scientific studies under Edward Aveling prompted her to look for a natural extension of her own capacity, the sense of which was so strong it drove her to wander on the borders of imagination and belief. Again, this was a territory that attracted Beatrice Webb when she encountered it in the works of the 'mysterious' William Winwood Reade, whose short life as an African explorer, joined to his books, which had titles like *The Veil of Isis*, and *The Martyrdom of Man*, brought him the kind of attention later to be enjoyed by Henry Rider Haggard.[19] Beatrice was fascinated by Reade's vision of the paradise on earth that would come about once Nature's laws were comprehensible. Man would become a superhuman being, able to subdue the powers of evil without and within; to eradicate disease and overcome death. The earth being small, man would cross endless Saharas separating planet from planet and sun from sun, and, as master of the forces of nature, create his own systems and manufacture worlds. The earth would become a Holy Land where man would worship his own divinity.[20]

While Webb read *The Martyrdom of Man* as a parable of the religion of science; others proposed a more literal interpretation for the kind of supernatural changes Reade foresaw. In sending

'SUPERHUMAN, SPIRITUAL REALITIES'

Alfred Russel Wallace a copy of *Isis Unveiled*, 'my first ever literary work', Madame Blavatsky apologized for its misleading title. 'Needless to tell *you*, who has lived in the East, that the final mysteries and secrets of inspiration are never given to the general public', she told Wallace. But Winwood Reade had pre-empted her preferred title, *The Veil of Isis*. There were facts in her work that should be known alike to those who advocated, and those who opposed, the study of spiritualism, Madame Blavatsky declared. If it did not soon develop into a philosophy and a science the adversaries of spiritualism would drag it down and it would fail to satisfy the reasonable expectations of a public which was outgrowing the tyranny of theology and of materialistic science. 'Behind the veil lies the key to modern spiritualistic phenomena through which can be discovered the secret of secrets; what is man, his origin, his power, and destiny', Wallace was told.[21]

Isis Unveiled,[22] which was published in 1877 when Blavatsky and Olcott were in New York and the Theosophical Society was barely two years old, was a hectic jumble of extracts and paraphrases from the ancient writings that scholars attribute to many different hands; among them the third-century, and later, treatises known by the name of Hermes Trismegistus; Pythagoras; Paracelsus, and the authors of the Cabbala. According to Madame Blavatsky, the book appeared 'by order' of the Masters, who showed her what to write. On a more mundane level, while it demonstrated the result of compulsive reading on an unusually imaginative mind, it must also have owed something to the fact that Madame's rooms were only a step away from 42nd Street and the great glories of the New York Public Library.[23]

The first meeting between the founders of the Theosophical Society came in 1874 at Chittenden in Vermont, where Olcott had gone to report on spiritualist manifestations in the Eddy brothers' house, for a New York newspaper. Also present, he noted for his readers, was 'a stout and remarkable looking woman, wearing a perky hat with plumes, a *grande toilette* satin dress with much trimming, a long heavy golden chain attached to a blue enamelled watch with a monogram ... in cheap diamonds, and on her lovely hands, a dozen or 15 rings'.[24] This was the rumbustious and enigmatic Helena Petrovna Blavatsky, a Russian who sometimes hinted that she was as old as time but, for everyday purposes, was probably 43. She had come to rest in America after wandering the world in pursuit of occult knowledge, in the course of which she had had adventures, never more than half-explained, resembling tales told at the court of Louis XVI by Cagliostro, to whose strangely compelling personality hers seemed the natural successor. 'She was much more a man than a woman', Annie Besant was to write after HPB's death, 'outspoken, decided, prompt, strong willed, genial, humourous, free from pettiness, and without malignity, she was wholly different from the average female type.'[25]

HPB's origins, as she told them to her biographer A. P. Sinnett,[26] were equally innocent of any connection with the average. According to her she was born in 1831, in Ekaterinoslav, the daughter of General von Hahn, whose family originated in Mecklenburg. When her mother died, Helena, aged 11, was sent to live in Saratov, where her maternal grandfather was Civil Governor. By far the most influential figure in her childhood was her grandmother Elena Fadeef, born a princess of the ancient Dolgorouki family. She was a famous scholar whose studies in geology, archaeology, and numismatics were said to fill twenty published volumes. Either she was indifferent to the point of eccentricity about the welfare of her granddaughter, or she felt it wiser not to check the notoriously hysterical Dolgorouki temperament. Helena was left to roam the furthest corners of the huge and gloomy mansion, to handle the rocks and shells and coins and skulls of the private collection, and, like many another bright and solitary child, to play and talk with companions only she could see. 'Fairy tales and old legends told to her by the servants were Helena's special delight', Josephine Ransom wrote; 'she avoided as far as she could the dry lessons of the governesses though she had brilliant abilities.' And all the time she read. Her family took it for granted that she was psychic; that by holding objects she could 'see' their origin. They knew that wherever Helena was phenomena occurred.[27]

When she was 18 she became the wife of Nicephore Blavatsky, a high civil servant who was apparently about 40 at the time. But the new Madame Blavatsky, who seems to have embarked on the marriage in a fit of bravado, always spoke of 'old whistle breeches' as in his dotage. After three months she ran away. Her account of what happened during the next ten years of her novitiate was deliberately vague. She claimed that she entered Tibet where she sat at the feet of the Masters—or Mahatmas—who directed her to found the Theosophical Society. She met her own Master, a Rajput prince, in London in 1851. In 1858 she returned to Russia, where she was reunited with her family and practised as a medium.[28]

A different account of these years is given by writers considered hostile by the Theosophical Society. They speak of time spent with a travelling circus; of a liaison with an Italian opera singer; of a spiritualist society she founded in Egypt among whose members were camp followers of the engineers building the Suez Canal; even of a child she had by a Russian nobleman.[29] During these 'lost' years she came in contact with the most famous medium of all, Daniel Dunglass Home. It was as a result of an attack on them by Home that Madame Blavatsky and Colonel Olcott left the United States for India.

Home's *Lights and Shadows of Spiritualism*, which was published in the same year as *Isis Unveiled*, condemned fraudulent phenomena and corrupt mediums for bringing the subject into disrepute. Three years previously, spiritualism had suffered a most serious reverse when its doyen, Robert Dale Owen, was obliged to confess publicly that he had been deceived in his endorsement of a spirit called Katie King who appeared at seances conducted in Philadelphia by the mediums Mr and Mrs Nelson Holmes. Owen was warned that the Holmeses' honesty had been called in question, but persisted in his support for so long that, when it came, his recantation provoked spectacular derision.[30] In 1874 Madame Blavatsky and Colonel Olcott attended the Holmeses' seances. Although she declared that she could get in touch with Katie King whenever she pleased, the Owen débâcle seems to have persuaded HPB to abandon elementary mediumship for the more complex and romantic philosophy of the Theosophical Society which she and Olcott and a small group of followers founded the following year. Josephine Ransom relates how, on 7 September 1875, after a lecture given in HPB's rooms on 'The Lost Canon of Proportion of the Egyptians', there was an animated discussion among the seventeen or so people present.

Olcott then spoke of the philosophic character of ancient Theosophies and their sufficiency to reconcile all existing antagonism . . . He proposed to form a society for the investigation of science and religion; the society to entirely eclectic, the friend of true religion and the enemy of scientific materialism. It would be a nucleus around which might gather those willing to work together to organise a society of occultists, begin to collect a library and to diffuse information concerning those secret laws of nature which were so familiar to the Chaldeans and Egyptians but were unknown to the modern world of science.

On 8 September Colonel Olcott was elected Chairman and William Quan Judge, a young law clerk, as Secretary. Later on Olcott's title was changed to President.[31]

In *Lights and Shadows of Spiritualism*, Home deplored the kind of sensational publicity the Theosophical Society received in its early years. Home also poured scorn on Olcott's claims in the press for HPB's powers of mediumship and described the Colonel's own observations as worthless.[32] HPB was vulnerable to Home's ridicule. Not only was he much respected internationally, he was in touch with spiritualists in Russia who knew something of her history.[33] Yet so wildly unconventional were her appearance, behaviour, and language, people might have been forgiven for supposing her indifferent to what they thought of her. As the demands of her strange calling weighed more heavily on her, and her health declined, perky plumed hats and satin dresses gave way to a voluminous wrapper which she wore all day, in the depths of which she kept cigarette papers and tobacco. She was restless and peremptory, issuing orders in a ringing voice, prodigal with oaths, threats, denunciations, whenever she was crossed, or felt at a disadvantage, which, in view of her temperament, was often. 'She never could learn to estimate occurrences at their proper relative value', Sinnett lamented.[34] The Irish poet W. B. Yeats, who knew her at the same time Annie Besant did, described her as a female Dr Johnson and, alternatively, as a kind of old peasant woman with an air of humour and audacious power.[35]

This audacity also struck Moncure Conway, whose first encounter with her was in December 1878 on her journey from the United States to India. He considered her a genius in her own way, a moral phenomenon to be studied, but one consumed by ambition, 'with a morbid desire to sway men'.[36] It followed that if publicity was important to her, bad publicity was disastrous. She was on record as saying that Daniel Dunglass Home's book destroyed her in Europe—as it did in America.[37]

In this her implication was that her partner was at fault, for Home's attack was largely directed not at her, but at Henry Olcott's claims on her behalf. In the early years of the Theosophical Society his attitude towards publicity, as befitted a former journalist, was robust; the only weapon to fear was silence, he is reported to have said, and: 'Abuse however truculent is almost as beneficial as praise',[38] a remark he might just as well have applied to himself, considering how HPB spoke of him. 'Colonel', as Theosophists learned to call him—a title derived from the American Civil War—was the first to admit that in occult matters he was a novice, even though his interest dated from the 1850s when he was farming in Ohio, a state that bent to the first phase of spiritualism as to a prairie wind. Olcott was convinced that the Masters had chosen him for the task of directing the Theosophical Society under Madame Blavatsky. 'First by her testimony and next through her instrumentality I came to know of the existence of the Mahatmas, the nature of Their exalted powers and the system of training by which They may be evoked.' HPB was a woman of great faults but also of the rarest endowments, intellectual and psychical, he told the celebrated oriental scholar Max Müller: 'her passionate desire was to oppose ignorance and superstition and to promote knowledge, especially that of the Ancients.'[39]

HPB treated Olcott with loud and affectionate disapproval. According to her there were only a dozen real Theosophists in the world, 'one of them stupid (Olcott I imagine), the rest flapdoodles',[40] Yeats wrote, borrowing the word that conveys the essence of HPB's boisterous self-assurance. In one of her most revealing phrases Olcott was 'a psychologized baby'.[41]

It was one of his charms that he retained a schoolboy's sense of humour—and of wonder—into old age. An enterprising and inquisitive man, in his youth (he was born in 1832) he followed his nose into strange corners.[42] This taste for adventure was matched by a ready pen, so when his attempts at farming failed, he became Associate Agricultural Editor of the New York *Tribune*. In the troubled years before the Civil War the paper was a leading advocate of the abolition of slavery, a cause which Olcott took to his heart, to the extent of reporting John Brown's execution at the risk of lynching. When hostilities broke out between North and South, Olcott served in the War Department as an investigator into profiteering, and afterwards performed the same function for the Navy, where the headlong methods and blustering manner which had served him well in the Army were not so well appreciated.[43] By the time of Lincoln's assassination Olcott's experience in what were essentially undercover operations earned him a place as one of three detectives (he liked to refer to himself as a commissioner) charged with hunting down the accomplices of John Wilkes Booth. Emerging from the war with a colonelcy, and testimonials which he carried with him wherever he went, Olcott qualified at the New York Bar as an expert in marine law and insurance. Whenever business was slow he exercised his investigative powers on the claims of spiritualism. By 1878 his close association with Madame Blavatsky proved too much for his marriage, though he always maintained that he and HPB were 'simply chums'.

In 1877 Olcott began a correspondence with members in Bombay of a devout and puritanical movement, the Arya Somaj, which was becoming influential among Hindus in northern India. The message of its Luther, Swami Dayanand Saraswati, was that national regeneration, of which India stood in desperate need, would be the result of a return to the ancient Vedic religion. With the agreement of his colleagues in New York Olcott affiliated their group to the Indian movement, renaming it the Theosophical Society of the Arya Somaj.

The Society's propaganda was aggressively anti-Christian at the time, a stance that seemed to its founders likely to fit well with the inclination of the Arya Somaj. In a letter to Swami Dayanand, Olcott discussed his belief in a monotheistic being as the originator of the universe. 'The Supreme One whom you teach your disciples to contemplate . . . is the very same Eternal Divine Essence whom we have been pointing the Christians to as the proper object of their adoration instead of their own cruel and remorseless . . . Moloch—Jehovah.' Olcott went on to denounce Christian worship as self-indulgent, its churches as ostentatious, and its priests as rapacious and hypocritical. For her part, Madame Blavatsky hailed the Swami as a Master. Preparations were made for the founders to go to India, which they regarded as the cradle of the present race; the source of ancient wisdom.[44]

When HPB and Olcott landed at Bombay in January 1879 they had very little money; their only asset was their connection with the Arya Somaj, which in itself sounded a note of warning to the Anglo-Indian community. Thirty-two years after the bloody chaos of the Indian Mutiny the system of government designed to prevent such a thing ever happening again was firmly in place: the heyday of the Raj was near. The pattern of living for its administrators and their families was well established as separate from, and superior to, that of the indigenous population which so hugely outnumbered them. While it was never openly admitted that the viceregal Government ruled the vast and teeming country by consent, it was definitely 'bad form' to do or say anything to 'upset the natives'. In particular, every recruit to the Indian Civil Service had it dinned into him that the Raj must hold itself aloof from popular religion. Madame Blavatsky and Colonel Olcott were different in that respect, alarmingly so it seemed, for the newspapers (including the *National Reformer* [45]) were soon reporting Olcott's speeches to large and astonished gatherings. 'It was new to Indians to hear a foreigner extol their religions and urge them to evolve their own leaders', Ransom wrote. Olcott told his audiences to expect the spiritual teacher who would help India to 'awaken'.[46] While the various groups of missionaries received this challenge with equanimity, Madame Blavatsky shortly insulted them in terms they never forgave. In a letter to the *Indian Spectator* complaining of the lack of respect shown by certain Anglo-Indian papers to her colleague, HPB blamed Christianity for the outbreak of the Indian Mutiny, a remark that quickly found its way into police files.[47]

For obvious reasons a watch was kept upon the press in India, on foreigners, and, in view of the vulnerable state of the North-West Frontier, especially on Russian nationals. Olcott and Blavatsky found themselves under surveillance; it was suggested that HPB was a tsarist spy, an accusation that she vehemently denied. Whether she was or not—there is no proof either way—it has to be said that the cumulative effect of her teaching was profoundly destabilizing to British rule, in Ireland as well as in India. To evoke the pride of race, of former hegemony, of old religion, proved to be the surest way of promoting united resistance to government by aliens, resistance which, though years were to pass before it came about, triumphed in the end. Besides, though it may have been coincidence, it is curious that, at their first appearance in the country, Blavatsky and Olcott happened upon precisely those people—very few at that period—who were already launched upon agitation; in religion, Swami Dayanand; in politics, among others, S. K. Ghose, one of the founders of the nationalist paper *Amrita Bazaar Patrika* which was more than once threatened with prosecution for sedition; and the 22-year-old Krishnavarma Shyamaji whose later exploits against the Raj led Maxim Gorky to dub him the Mazzini of India.[48]

HPB's introduction to Anglo-Indian society came about through the interest in spiritualism of the editor of the *Pioneer* of Allahabad, a paper only slightly less prestigious than the *Times of India*. A. P. Sinnett gave the Theosophical Society extensive coverage and invited correspondence. He and his wife, who were well established in the Anglo-Indian social pecking order, took up the founders (when addressed as Countess, Madame Blavatsky rarely bothered to object). Viceregal society in its summer retreat at Simla was pleasantly diverted by the phenomena which occurred at HPB's behest, while Sinnett's commitment to Theosophy was firmly engaged by a series of letters addressed to him, delivered by an unseen hand, purporting to be written by the Master Koot Humi and the Master Morya. His account of this period was published in 1881 as *The Occult World*, one of the first books by him that Annie Besant read.[49]

According to Allan Octavian Hume, who joined the Theosophical Society after meeting the founders at Allahabad, many of the phenomena described in *The Occult World* were genuine. He was a most distinguished civil servant, who had been relegated to the Board of Revenue in Allahabad to serve out the remaining years of a career which had erred on the side of too much sympathy with the Indians he governed.[50] Among Charles Bradlaugh's papers is a letter written by Hume in 1884, in which he said that he found Madame Blavatsky to have so much that was false in her that he withdrew from open support of the Society in 1882, while remaining convinced that it was one of the channels for the dissemination of occult knowledge by which means mankind would be regenerated.[51]

Bradlaugh's connection with all this—which, of course, was known to Annie Besant—arose for two separate reasons. The first, which she whole-heartedly supported, was his involvement with the slowly emerging movement for Indian self-determination, which A. O. Hume did so much to encourage after his retirement from the ICS. Hume arranged the first gathering of politically minded Indians in 1883 and in 1885 it became the Indian National Congress; at which time it began to look to London for support. When Bradlaugh entered Parliament the following year, he assumed the mantle of the late Henry Fawcett, as 'the Member for India', to whom Indians brought their grievances against the Raj. Bradlaugh therefore got to know leading nationalists, among them S. K. Ghose. Like A. O. Hume, he was a member of the British Committee of the Indian National Congress.[52]

Meanwhile, in 1882, Olcott and Blavatsky moved from Bombay to Madras, where the writ of the Arya Somaj did not run. That connection did not long survive a meeting between the founders and Swami Dayanand: each side found discrepancies in the other's belief, as well as other drawbacks. The Swami accused Olcott and Madame of pretending to be disciples while really setting themselves up as religious leaders, and publicly condemned HPB's phenomena as 'cheating jugglery'. The founders of the Theosophical Society riposted that their only interest in the Swami had been that he would demonstrate the ancient brahminical doctrine of Yoga Vidya, but 'he was no Adept', they declared.[53] Olcott also blundered into hot political water in Bombay. His strenuous complaints about police surveillance finally drew a letter signed by an under-secretary. This laid down guide-lines by which the Theosophical Society operated in India for years to come: so long as the members confined themselves to the pursuit of philosophical and scientific studies wholly unconnected with politics, they would no longer be troubled by the police. Olcott apparently interpreted this as an official endorsement of the Society; when, somehow, the newspapers got hold of his remarks, the Government of India was obliged to issue a crossly worded refutation.[54]

Madras was a sea-washed, sleepy, trading town remote from the centre of British administration at Calcutta. Its people were Dravidians, a separate race from the Aryans of the north, among the most deprived of the whole subcontinent. Though poor they were independently minded, as was evident from the influence enjoyed by free-thought publications, whose most read, most respected authors were Charles Bradlaugh and Annie Besant.[55] Publishing was to become a most important function of the Theosophical Society after its headquarters were established on a large estate bought by a well-wisher at the mouth of the Adyar

River seven or eight miles from the centre of Madras. Publishing was the means by which Theosophy gave its message to the world and the founders kept in touch with the widely scattered branches. Publishing was vital to sustain the influence of Madame Blavatsky. As the years went by her vast bulk, poor health, and uncertain temperament made it almost impossible for her to move about in India. Nor was it worth the effort in a male-dominated society. This female incapacity seems to have infuriated HPB to the point where she marked down a surrogate, Mrs Annie Besant, the one woman who was demonstrably capable of moving thousands with a message that in its hostility to the hated Christian creed and—who knows?—its support for Indian self-government, its robust denunciation of British policy in Ireland, was of particular significance to her.

This was a characteristically audacious move, because the object of her attention was hostile at the time. Annie Besant was disturbed by what she saw as Theosophy's attempt to suborn Indian free-thinkers. From 1879, when it reported Olcott's speeches in Bombay, the *National Reformer* was besieged by comment and enquiry from its own readers. Bradlaugh was astonished to discover that his potential rival was the same Olcott he had met in New York in 1875, and even more astonished when one of his hosts on that lecture tour, his old friend D. M. Bennett, the distinguished American free-thinker, called on him on his way home from India to say that he had joined the Theosophical Society.[56] When a Hindu member of the National Secular Society, who also belonged to the Madras Freethought Union, gave a favourable account of the Adyar group in the *Reformer*, 'Can a Secularist be a Theosophist?', Besant privately urged him to have nothing to do with it. At the same time she used her Daybreak column to warn readers to beware of this 'vaguely worded' belief which had connections with spiritualism.[57]

HPB judged the time was ripe to intervene by means of her monthly journal, the *Theosophist*, with an observation shrewdly calculated to flatter. 'For one so highly intellectual and keen as the renowned writer [Besant] to dogmatise and utter autocratic ukases after she had herself suffered so cruelly and undeservedly at the hands of blind bigotry and social prejudice in her lifelong struggle for the freedom of thought seems absurdly inconsistent',

HPB wrote in August 1882.[58] Thereafter the *Theosophist* kept in close touch with the *National Reformer* and carried its advertisements, as it did those for works by Bradlaugh and Besant—always excepting *The Fruits of Philosophy* and *The Law of Population*, whose subject the Masters condemned. Nevertheless, at the same time, 'They' apparently made overtures to the champion of neo-Malthusianism by means of 'Their' letters to A. P. Sinnett, some of which were not made public until the twentieth century. In 1883 Sinnett was dismissed as editor of the *Pioneer*, having exhausted the patience of the proprietors by his undiscriminating coverage of HPB's phenomena. (Behind the scenes the *Pioneer*'s standing with the Government of India may have been compromised to the point where it risked losing certain privileges.[59]) Sinnett and his wife went home to England, where he founded the London Lodge of the Theosophical Society. Communications from the Masters continued to arrive, encouraging Sinnett in his desire to help Theosophy win the hearts and minds of the multitude. Towards the end of 1883 the Master Koot Humi secretly told Sinnett to 'use every effort to develop such relations with A. Besant that your work may run on parallel lines and in full sympathy; an easier request than some of mine with which you have ever loyally complied. You may, if you see fit—show this note to her *only*.' An inscription on the letter (which is among the Mahatma Papers deposited in the British Library) indicates that the contents were shown to 'A. Besant'. There is no hint of her response.[60]

Equally there is no evidence that, when Blavatsky and Olcott arrived in London at the beginning of 1884 for a stay of some months, they had any contact with Bradlaugh or Besant. But efforts were made to enlist the sympathy of the *National Reformer* on behalf of HPB when, in her absence from Madras, she was subjected to a devastating attack upon her integrity. In September 1884 an article entitled 'The Collapse of Koot Hoomi' appeared in the Christian College Magazine. This was based on allegations by Monsieur and Madame Coulomb, friends and dependants who had been left in charge at Adyar during the founders' absence. They described how certain phenomena, including the mysterious appearance of letters from the Masters, had been contrived by use of a special cupboard, called the shrine, in HPB's room.[61] Just how large an uproar the revelations caused emerges from a confidential letter addressed to Madame Olga Novikoff by the Governor of Madras Sir Mountstuart Grant Duff, who thereby redeemed the faith she had placed in him during her progress through London society as the 'MP for Russia'. Grant Duff was obviously exasperated by the raging controversy over Madame Blavatsky; he detested the whole subject of Mahatmas and astral bodies; he thought if Madame Blavatsky did not sue her accusers she would be excluded from the company of 'persons of good character'. But he gave Madame Novikoff the same uncompromising declaration that Olcott had extracted from the viceregal government: the Government of Madras, of which he was the head, had issued 'the most peremptory orders against interference with the religious or philosophical views of the members of the Theosophical Society, or of anyone else'.[62]

One of the many pamphlets which supported HPB in the furious debate that followed her 'exposure' was *Madame Blavatsky and Her Slanderers* by Dr Franz Hartmann, an American Theosophist who had been at Adyar for some time. On 2 November 1884 the *National Reformer* referred to this in terms that must have made the doctor rue the day he had sent Bradlaugh and Besant a copy.

> To an untutored mind the bulk of the last two or three years' declarations of Madame Blavatsky and Colonel Olcott are so startlingly at variance with the possible that we hardly read the *Theosophist* as a serious publication, nor can we treat members of the Theosophical Society as serious persons. Many of them are very good, very respectable and very mad. Some of them are less mad and less good,

the *Reformer* commented. It continued,

> a Mr and Mrs Coulomb say they have been parties to frauds and cheats in astral manifestations and that Madame Blavatsky is a prime mover in the imposture. The story of the Coulombs, if true, shows very clumsy fraud, and the reply of the theosophists shows the marvellous force of superstition. We are not inclined too readily to accept the evidence of the Coulombs who ... continued to connive at fraud until a quarrel arose, but we can only acquit Madame Blavatsky by regarding her as an enthusiastic semi-spiritualist who has managed to get fairly crazed in dabbling with the wonderful in company with many simpletons.[63]

It seems likely that, in composing this crushing review, Bradlaugh took account of A. O. Hume's opinion, for the letter by Hume about Blavatsky that is among the Bradlaugh Papers is dated shortly after 'The Collapse of Koot Hoomi' burst upon the world. Bradlaugh's opinion did not alter, which is why, when Annie was in the throes of her approach to Theosophy, she did her best to conceal the fact from her former partner. When Bradlaugh tried to discuss it with her, he 'only got a smile'.[64]

As Grant Duff had foreseen, the question of suing the Coulombs was crucial to HPB's reputation. As soon as she returned to Adyar in 1885 she apparently announced her intention of starting proceedings. With the support of Theosophists in conference, Olcott vetoed this. HPB fell ill and, in March 1885, sailed for Europe, never to return. For the next two years she moved about on the Continent (one place she chose to stay was Würzburg because it was the birthplace of Paracelsus). In May 1887, gravely weakened by disease, she arrived in London, carrying with her the enormous bulk of papers that, after much editing by Bertram Keightley, Secretary of the London Lodge, was to become her last and most celebrated work, *The Secret Doctrine*. The first volume came out in October 1888, the second in early 1889.[65] Their appearance was the signal for the London press to rake up all the old stories about the author and a few more. In 1885 the Society for Psychical Research sent a special investigator, Richard Hodgson, to Adyar, to examine the Coulombs' allegations on the spot. The conclusion he came to then once more was widely quoted: 'we regard [Madame Blavatsky] neither as the mouthpiece of hidden seers, nor as a mere vulgar adventuress: we think she has achieved a title to permanent remembrance as one of the most accomplished, ingenious, and interesting impostors in history.' Hodgson was also reported as saying that HPB was a Russian spy.[66]

On 3 January 1889 the *Pall Mall Gazette* printed an article linking Madame Blavatsky to Madame Olga Novikoff; both were Russian lady spies, it told its readers. Novikoff, who occasionally translated letters for Blavatsky, told William Stead that HPB was 'a great Russian patriot', and dragged him off to visit her. 'I was delighted with and at the same time repelled by Madame Blavatsky', Stead wrote. 'Power was there, rude and massive, but she had the manners of a man, and a very unconventional man, rather than those of a lady.'

Because he could not bear to read it himself Stead gave Annie a copy of *The Secret Doctrine* to review, and, when she was bowled over by it, an introduction to the author.[67] Considering the antagonism shown to Madame Blavatsky over the years by the *National Reformer*, Annie must have thought this to be essential. Stead was the natural choice as go-between; although he had dashed her greater expectations he and Annie remained on friendly terms, while his interest in spiritualism and related matters, already great, was destined to obsess him. One soft spring evening Annie and Herbert Burrows, who thought as she did, went along to the substantial house in Holland Park which had been placed at HPB's disposal by Bertram Keightley, where they were greeted very cordially. According to Annie, HPB talked easily and brilliantly of travels, of various countries; 'Nothing special to record, no word of Occultism, nothing mysterious; a woman of the world chatting to her evening visitors.' But when they rose to go the veil was lifted; two piercing eyes met Annie's and with a 'yearning throb' in the voice—'Oh my dear Mrs Besant if you would only come among us!'[68] Annie felt a well-nigh uncontrollable desire to bend down and kiss her but pride forbade, and she said a commonplace goodbye. But the outcome was not in doubt and, to Annie's friends and colleagues, it was devastating. A note among the Bradlaugh papers laments the perverse infatuation she conceived for the adroit and enigmatic Russian.[69]

CHAPTER 20

GONE TO THEOSOPHY

MADAME BLAVATSKY, who considered Annie Besant to be dangerously naïve, took precautions to guard against the consequences. Her new disciple was never to hear irreverent talk about occultism or Theosophy, and exaggerated claims must be avoided, she warned her American colleague William Quan Judge, telling him that Annie was the soul of honour and uncompromisingly truthful; not psychic or spiritual but 'all intellect'.[1] In this assessment HPB ignored the practicality which contributed so much to Annie's effectiveness in public life, and was to be of great service to Theosophy. But Annie's correspondence shows that, if only in the beginning of their relationship, she applied it to the strange new personality that had at last made contact with her.

Madame Blavatsky's physical condition was the exact opposite of what was supposed to be the result of occult training, Annie told the Revd Ashman. Her life, instead of being maintained at a high point of efficiency, was obviously wasting away. How was that associated with the training she was supposed to have undergone? 'It seems to me that—for the moment accepting the Mahatmas hypothesis—they have been working through her, not training her to work for herself and so have exhausted her vitality . . . does it not seem rather cruel if they have worn her out and thrown her away? . . . I am much puzzled', she wrote in March 1889.[2] We do not know how HPB explained this anomaly, nor did Annie pursue it, once she had convinced herself that Theosophy could resolve the moral and intellectual dilemma that had oppressed her for so long.

Reincarnation—barely mentioned in *Isis Unveiled*—played a significant role in *The Secret Doctrine*, an edifice of Babylonian proportions, raised upon themes which appeared in many Eastern faiths, with the interesting exception of Islam. Students of oriental religion like Moncure Conway and Edward Carpenter dismissed it as simply ridiculous. 'No words can describe the general rot and confusion of Blavatsky's "Secret Doctrine"', the latter wrote.[3] But to Annie Besant, as to many others over the years, the book was as valuable as it was enthralling—'natural', 'coherent', 'subtle yet intelligible', Annie called it. Significantly, she recognized in it 'the means of realizing the dreams of childhood on the higher plane of intellectual womanhood'.[4] On reading it, disjointed facts appeared to her as part of a mighty whole, and puzzles, riddles, problems, disappeared. All at once Annie perceived that the possibility of immense happiness lay within her grasp. After half a lifetime of strenuous enquiry and painful disappointment she had been granted a revelation. She was to teach that reincarnation, together with the doctrine of karma, was the true theory of life and immortality. Reincarnation proposed that people were born over and over again. Their karma dictated in what circumstances each successive birth would take place. As the individual acted and reacted, well or ill, to events in the course of one life, so his or her prospects in the next would improve or deteriorate. Thus, as Annie came to interpret it, karma was not fate to be hopelessly endured (as Hindus, for instance, might think of it), it was the natural law by which inequality of opportunity and of capacity would eventually be resolved. This belief satisfied both Annie's intellect and her deepest feeling, since upward progress was to be achieved by practising magnanimity towards one's fellow men. 'If the more stupid or more vicious is merely a younger soul with less experience ... than the more clever or more virtuous; if sickness and deformity are the result of past evil and of cruelty [and not, as she had been taught to believe by the Church, the sins of the father visited on innocent children], then Justice is replaced on her pedestal', Annie wrote. As justice was austere, so the doctrine of karma was an austere concept of salvation, but how much more satisfying than a vicarious Atonement![5]

The result of Annie's 'conversion' was not wholly positive for, with the knowledge that all ills were the result of human error, capable of correction given the will—and a cosmic period of time—came a certain falling off in the magnificent strength of her compassion, while the conviction that she had been vouchsafed the ultimate truth would betray her into arrogance.[6] But her first emotion was happiness at her release from doubt, as Bernard Shaw remarked when detecting her in a joke.

Long afterwards he related how, shocked by news of her decision, he rushed to her office in Fleet Street where he denounced Theosophy in general, and Madame Blavatksy in particular. As a member of the Society for Psychical Research, Shaw had attended the meeting when Hodgson presented his report on the Coulomb affair, and had read a number of Theosophical books in order to review them.[7] Annie listened with amusement and remarked that since she had become a vegetarian—a cause next to Shaw's heart—the regime had probably affected her powers of reasoning.

'In short she was for the first time able to play with me' Shaw conceded; 'she was no longer in the grip of her pride; she had after many explorations found her path and come to see the universe and herself in their real perspective.'[8] He always insisted, somewhat testily, that it was he, not Stead, who gave Annie the copy of *The Secret Doctrine* that precipitated her decision.[9] Whether he did, or both of them did, is immaterial. More to the point is how she dealt with the strictures in the Hodgson Report. Annie went on record as denying that she had ever heard of it until HPB herself drew her attention to it, when she read it and dismissed the allegations.[10] That is hard to reconcile with the running reports in the *National Reformer*; with her almost daily contact with Shaw; and with Stead, who kept a fascinated eye on Madame Blavatsky in the *Pall Mall Gazette*. But, more especially, it does not consort with the fact that Annie was in touch with Moncure Conway, who visited Adyar in 1884 a few months before the Coulomb storm broke, with the specific purpose of observing Theosophy and its leaders on their home ground.

Conway, who was in Australia, on leave from South Place Chapel, found that the talk among his hosts ran on the 'Countess', the colonel, and the Mahatmas. Since he was due to visit India, Conway, who was very disturbed by their susceptibility, decided that it was his duty to investigate. He spent six hours at Adyar but could not carry out his intention to the full since the Mahatmas were neither sending nor receiving letters. HPB, who had been forewarned of his arrival, engaged him in witty social niceties. She was on her guard. Conway's high standing among religious and secular leaders, his American connections, and his scholarship made him a formidable adversary. After a while she went on the offensive. 'You ought to know the truth, it [her power] is all glamour; people think they see what they do not see; that is the whole of it', she told him. Conway perceived that she had spiked his guns: her confession, made without witnesses, could easily be repudiated if he published it. Glamour, moreover, was a vague word which did not give away the reality of her pretension. He was disturbed by the possible effect of her propaganda. He thought she had no real interest in the moral and spiritual regeneration of India. Yet Westerners could not imagine the effect when there appeared from America this company of people who had abandoned every form of Christianity to lead in the work of rehabilitating the ancient system of religion, and pointed Hindus and Buddhists to their own prophets and scriptures as symbols of faith and hope. No wonder they moved and attracted Indians more than Christian missionaries did.[12]

Five years later Conway laid these misgivings before Annie Besant after she had joined the Theosophical Society and her Master had appeared to her for the first time. What if HPB said it was all glamour? That implied a good deal, Annie argued. To make one see a person in one's room when there was no person there was a marvellous power. Much to Conway's distress, she told him he must have been mistaken in thinking that Madame Blavatsky had ever said, 'that is the whole of it'.[13]

G. W. Foote, who had objected to Annie preaching socialism from the platforms of the National Secular Society, now accused her of leading free-thinkers astray by lecturing to them on Theosophy from the same platforms. She hit back in terms nicely judged to wound. 'To set up a new infallibility is to do what Churches...have done', unblushingly she told him; 'to set up their own petty fences round the field of truth—that is to be false to the freethinkers' creed.'[14] She was going *forward*, having discovered new evidence, convinced that the President of the NSS, Charles Bradlaugh, would approve. She certainly did not imagine that her new belief would come between them. According to Hypatia, blind ignorance led her to this view. 'She never examined herself... Why should she? People praised and flattered her, or grossly abused her. Such abuse amounted to a kind of flattery, while the men she charmed never criticized her, or only gently.'[15] There was nothing gentle about the statement Bradlaugh printed in the *National Reformer* on 30 June: it was a savage denunciation of Theosophy which did not spare his colleague. He had not read *The Secret Doctrine*, but during the past ten years he had read much else from the pen of Blavatsky, Olcott, and other Theosophists.

> They appear to me to have sought to rehabilitate a kind of Spiritualism in Eastern phraseology. I think many of their allegations utterly erroneous and their reasoning wholly unsound. I very deeply regret indeed that my colleague and co-worker has with somewhat of suddeness, and without any interchange of ideas with myself, adopted as facts matters which seem to me to be as unreal as it is possible for any fiction to be.

Under this a brief statement by Besant defined the three objects of Theosophy as: to found a universal brotherhood without distinction of race or creed; to forward the study of Aryan literature and philosophy; to investigate unexplained laws of nature and the powers latent in man. The founders of the Society denied a personal God and a 'somewhat subtle' form of Pantheism was taught. This appeared to Besant to promise a solution of some problems, especially problems in psychology which atheism left untouched, she explained to the patient readers of the *National Reformer*.[16]

But if Theosophy claimed her thoughts, she had no intention of letting it interfere with her career. In 1889 the major part of her activity was concerned with politics in one form or another. In April she was re-elected as a member of the Fabian executive. It was decided to publish a collection of essays with contributions by Bernard Shaw, Sidney Webb, William Clarke, Hubert Bland, Sydney Olivier, Graham Wallas, and Annie Besant. This was the first concerted attempt in England to explain the meaning and direction of a movement which, in whatever country it had taken root, was, as yet, sparsely documented. What literature did exist was revolutionary rather than constructive.

The first edition of *Fabian Essays in Socialism*, which came out in December 1889, 'went off like smoke', according to its editor, Bernard Shaw.[17] He arranged the book in three sections: I: 'The Basis of Socialism'; II: 'The Organisation of Society'; III: 'The Transition to Social Democracy'. Graham Wallas and Annie Besant figured in II; he on Property, she, on Industry, under Socialism. It was a subject she did her best to avoid and, when she could not, complained that she and Wallas were set to the task by the 'Pharaohs' of the executive.[18] At the very moment when her thoughts were hovering in the Himalayas, Annie had to sit down with—of all people—the irreverent Shaw, and submit to his editing an essay on a mundane theme. 'I am only trying to work out the changes practicable among men and women as we know them', she explained in the course of her essay, 'always seeking to lay down not what is ideally best but what is practical.'[19]

Her obvious irritation at having to ignore the 'ideally best' was matched by a certain bossiness which betrayed itself in words like 'draft', 'send', 'take-over' when rearranging the future of these disappointingly ordinary men and women. Her excuse was that she believed drastic measures would continue to be necessary to preserve workers from unemployment. She found the answer to hand in the Local Government Act of 1888. By dividing England into districts ruled by County Councils she felt the Act had virtually created the Commune which would make socialism possible. Now it remained to fill the gaps—to give all adults a vote in County Council elections; to restrict the term to one year; to pay Councillors so their time was entirely at the public's disposal; to enable Councils to take and hold land. Then these bodies would act as managers, entrepreneurs, industrial organizers on behalf of the workers—operating an eight-hour day, paying trade union rates, making the continued existence of private capitalists increasingly difficult.

The circumstances which attended the appearance of the *Essays* meant that, while Besant was still the only well-known name among the essayists, and the role of industry a crucial issue, Annie's contribution was the least convincing. In his preface to the 1920 edition Sidney Webb had the grace to assume at least partial responsibility for the way 'Industry under Socialism' had missed the target. In 1889 he and his colleagues had yet to discover the proper place of local authorities in the social organism, Webb remarked. Their mistake at first was to think of them as associations of producers on whose behalf jobs had to be organized in competition with private capitalism. In fact they soon realized that Parish, Vestry, Town, and County Councils were associations of *consumers* with largely autonomous—and powerful—local governing bodies.[20] Here Webb was referring to what came to be known, fondly, as 'gas and water socialism'.

In July Annie went to Paris as one of the SDF delegates to the hundredth anniversary of the French Revolution. She found the ceremonies at the Hôtel de Ville, the Place de la Bastille, and Père Lachaise deeply moving for they also commemorated more recent martyrs; the Communards and the Chicago anarchists.[21] So it was with irritation that she was forced to contemplate the deep division among her comrades which meant that, instead of there being one great conference of workers in Paris on 14 July to demonstrate the movement's strength and progress to the world, there had to be two: the International Workers' Congress arranged by the French trade union and workers' party—the Possibilists; and the International Socialist Labour Congress, supported by the Marxists—which came to be known as the Second International. In England the division was reflected by the fact that *Justice* advertised the Possibilists' Congress, while the *Commonweal* and a new journal, the *Labour Elector*, advertised the Marxists'. The latter was edited by H. H. Champion, former Secretary of the SDF who had now fallen out with Hyndman. He and his allies, among whom were John Burns and Tom Mann, supported the Marxists, as did the Avelings. Annie Besant and Herbert Burrows took their places among the Possibilists.

Burrows's was the decisive voice in Annie's joining the Theosophical Society, whither he had preceded her in the aftermath of his wife's death, which left him distraught and ill. In her *Autobiography* Annie, who took him in and nursed him through his collapse, described him, without naming him, as a person of remorseless energy who broke his heart over wrongs he could not remedy. It was characteristic of him that, when he thought he was dying, he said to Annie, 'tell the people how I have loved them always'.[22]

After the ceremonies were over Annie and Burrows went to join Madame Blavatsky, who was taking a holiday just outside Paris in the Forest of Fontainebleau. There she enlarged upon the source of her powers, though she never committed the folly—Annie's word—of claiming certainty for her philosophy on the grounds that she could work wonders. Instead she framed her explanation in scientific terms calculated to appeal especially to Annie Besant. 'Constantly she would remind us that there were no such thing as miracles, that all the phenomena she had produced were by a knowledge of nature deeper than that of average people, and by the force of a well tuned mind and will.' Some of these were what HPB was bold enough to call 'psychological tricks'—the creation of images by suggestion. Annie accepted this teaching gratefully, not because Madame Blavatsky thereby claimed authority over her, but because it woke like powers in her.[23] The question was what use could she make of these powers once she had attained them? HPB had the answer to this too in a book she produced during their stay at Fontainebleau: *The Voice of the Silence*. This introduced the figure of the Bodhisattva, the person who postponed his own spiritual enlightenment in order to be of service to others in the world—a course that was of particular appeal to Annie Besant. 'To reach Nirvana one must reach Self Knowledge, and Self Knowledge is, of loving deeds, the child', the Voice of the Silence proclaimed.[24]

One night at Fontainebleau, Annie woke from a deep sleep to find 'the air ... thrown into pulsating waves and there appeared the radiant astral figure of the Master visible to my physical eyes'. Annie regarded these occasions, which she recorded by a special sign in her diary, as the mark of great favour and a source of renewed strength.[25] Of the Masters, or Mahatmas, charged with the task of guiding Theosophists to the higher knowledge, Morya and Koot Humi made the most frequent appearances[26]—apparently this first encounter was with the Master Morya, often referred to as the Master M. There were others: Serapis, Ilarion, Master the Count. Opinion as to their nature has varied over the years and has proved a fertile source of difficulty. Annie seems to have regarded them as manifestations deriving from the mind and will of an adept—Madame Blavatsky being the prime example—whose skill she hoped one day to emulate. More recently some Theosoophists have preferred to see them as wholly metaphorical, the means of making a complex spiritual idea more readily understood.

The experience drew Annie closer to HPB, who responded with endearments. Letters addressed to 19 Avenue Road began 'My darling Penelope', and were signed by Madame Blavatsky as 'Your ... female Ulysses'.[27] A question that occurred to many of those who saw her was, to what degree of femininity did HPB belong? Her appearance was androgynous, her behaviour that of a hermaphrodite. There was no conspiracy of silence over this.

Henry Olcott, for example, referred to her as *her* merely through force of habit. 'Dear lord, boys', he admonished some male friends, 'in my opinion she is no more a *she* than you or I. Putting aside her actions, habits of thought, masculine ways, her constant asseverations of the fact... I have pumped enough out of her to satisfy me that she is a man... a very old man... a most learned and wonderful man... a Hindu man.'[28]

Needless to say, the suggestion was made that the relationship between Annie Besant and Madame Blavatsky was lesbian. As with Annie's other friendships after her marriage ended, it is impossible to know the truth. Madame Blavatsky insisted that she was a virgin, the state that best supported occult powers. And the moment was fast approaching when Annie would publicly endorse celibacy as a condition of her Theosophical novitiate. What no one could doubt was her fondness for HPB, who combined the self-assurance which so much attracted Annie, with vociferous approval of her conduct, and great warmth. Moreover she was visibly in need of the kind of help Annie was best able to give. A young disciple—the word Theosophists used was *chela*—described the pathos HPB conveyed in her last years.

> She could not express her thoughts as she wished and would call on Annie Besant to explain and answer questions. Some very foolish questions were asked; some sensible. HPB would start up impulsively and endeavour to satisfy the questioner... then turning to her friend would say, 'Annie, you explain.' Annie would stand up and very wonderfully answer the questions and HPB would sink back in her chair with a deep sigh and roll her cigarettes thoughtfully, her keen yet tired old eyes roving over the audience.[29]

During all this Annie's work at the School Board went on, as did her efforts on behalf of unskilled workers. The full Board met once a month; committees much more frequently. Whenever she was able Annie sat in on lessons. Parents had to be seen, teachers and managers interviewed, and much attention given to the extra curricular activity of feeding and clothing the poor children of Tower Hamlets.[30] Annie used the lengthy journey to and from her 'constituency' to catch up with her reading, which otherwise had to be done in the small hours. Even then she might be called out to attend meetings, like those on behalf of shorter hours for the bus and tram workers which, of necessity, had to begin after midnight.

Her *Autobiography* is curiously silent on the subject of the great dock strike which began in August 1889 and was for an extra 6*d*. on wages—'the dockers' tanner'—as well as fairer methods of hiring men. We know from other sources that Besant was one of those (Burrows was another) who attended at dock gates in the early mornings to speak to the pickets whose action brought the nation's commerce to a near standstill; and that she and Bradlaugh helped one of their leaders, Ben Tillett, draw up the rules for the Dock, Wharf, Riverside and Labourers' Union which, in time, grew into the mighty Transport and General Workers' Union.[31] Her silence probably reflected the breach that had arisen between her and some of the most prominent leaders of the new trade union movement.

On her return from France at the end of July Annie was the victim of a stab in the back. When he left the SDF Henry Champion did not scruple to attack his former colleagues. The *Labour Elector* of 30 July accused Besant of trying to obtain details shown on the credentials of Russian delegates to the Paris congresses in order to betray them. What seems to have prompted this devastating smear was Annie's association with those two 'notorious' supporters of the tsarist government Madame Blavatsky and W. T. Stead. John Burns and Cunninghame Graham were on the Board of the *Labour Elector*. Annie had already fallen out with Burns over Bradlaugh's absence from the events of Bloody Sunday; that left Cunninghame Graham. 'No one cares what Messrs Burns and Champion may say, but as you are put forward as responsible for the *Elector* the matter is more serious', she told him. 'I ask you as a man of honour to distinctly repudiate... any part of this vile charge against me and I take it for granted that you will at once repudiate this harmful and libellous attack.' Anyone who knew the arrangements for verifying credentials would know the charge was groundless, she went on; each nation verified its own, so only a Russian could betray a Russian.[32]

She had barely recovered from the unfairness of the *Elector*'s charge when a further example was visited upon her. In November her libel action against the vicar of Stepney for circulating as her views 'vile passages' from *The Elements of Social Science* at the School Board elections came to trial. Annie observed that she had against her not only the Solicitor-General, who was defending the Revd Hoskyns, but the judge whose bias against her resembled that of the Master of the Rolls in her action to recover custody of Mabel. Annie was cross-examined for five hours in an attempt to make her admit that, by advocating limitation of the family, she condemned chastity. When she remained unshaken the judge informed the jury that Hoskyns's statements might be regarded as privileged since he was a clergyman defending his faith against an unbeliever. The action failed since the jury could not agree. Wholly disheartened, Annie could not bring herself to press for a new trial.[33]

She would not have given in so easily in the days when she and Bradlaugh published the Knowlton pamphlet. But her belief in family limitation was wavering, having come under attack from two different directions. On becoming a socialist Annie had been made aware of the disapproval of her male colleagues. Hyndman, Morris, and Burrows, for example, reproached her with the thought that neo-Malthusianism diverted the attention of the proletariat away from the real cause of poverty, the monopoly of land and capital by one privileged section of the community.[34] Nevertheless, Annie persisted in active membership of the Malthusian League. But, when Theosophy claimed her, Madame Blavatsky intimated that this was one aspect of her life she would be required to renounce.

Annie swiftly came to see that artificially to interfere with the act of procreation was incompatible with the idea of reincarnation. The difficulty was not so much that the numbers of those eligible for rebirth might dwindle to a point of near extinction, but, as she had reminded the jury in the Knowlton trial, that contraception would be practised by the more intelligent, those further along the upward path who, in failing to reproduce themselves, would bring about a deterioration in the race. She became convinced that her works had been the means whereby thousands of children of more advanced parentage had failed to enter the world.[35]

With direct knowledge of the misery of the poor, and a personal distaste for the tribulations of childbirth, Annie pleaded with HPB to allow her to recommend family limitation, for a time at least, as a palliative and 'as a defence in the hands of a woman against intolerable oppression and enforced suffering'. But Blavatsky taught her that the sexual instinct had been so excessively indulged that it was the source of misery in every generation. The only remedy was to reduce it to its natural limit by self-denial; to develop the intellect at the expense of animal passion, so to raise the whole man. 'Theosophists should sound the note of self restraint in marriage', Besant wrote, 'and the gradual—for with the masses it cannot be sudden—restriction of the sexual relation to the perpetuation of the race.' In 1890 she bought up all existing copies of her hugely successful *Law of Population* and had the plates destroyed.[36]

This was a shock to her friends and associates. Shaw sometimes gave the impression long afterwards that his connection with Annie Besant came to a dramatic end at this time. But his *Diary* continued to record their meetings, including his attendance at 19 Avenue Road for lectures on Theosophy. He also made a point of noting in it the occasion when a mutual acquaintance, Walter Besant, introduced him to her son.[37] Digby, then aged 22, had been fortunate in that his Uncle Walter had seen to his education and welcomed him to his home. 'I had long made up my mind that as soon as I came of age I should go and see my mother and judge matters for myself', Digby wrote. Frank Besant replied to this news in a long letter warning Digby against any such venture. When he persisted, Frank sent his personal belongings after him and broke off all further communication. A patriarchal figure, he lived on as vicar of Sibsey until 1917 when he was found dead in his study, a volume of Plautus in translation fallen from his hand.[38]

Like Annie before her, Mabel Besant was less privileged than her brother. Left to grow up in a girls' boarding school, she demonstrated her mother's spirit when, at the age of 18, she began making clandestine visits to Avenue Road. Before she came of age she went to live there permanently. The law was on Frank's side had he chosen to claim her. Instead he banished her as he had his son.[39] An impression arose that Mabel was 'difficult' and caused Annie much disquiet before she married and went off to Australia. Her behaviour may have been influenced by the unusual nature of the household that, in the summer of 1890, took over 19 Avenue Road.

When Madame Blavatsky and her disciples had to find other accommodation, after the lease of the house they occupied had expired, Annie offered them hers. Rooms on the ground floor were set aside for HPB. An inner sanctuary was constructed for meetings of the Esoteric Section. As its name implies, this was a group of initiates who met in seclusion to study occult subjects. The ES—Theosophists were addicted to initials—had been founded by Blavatsky on her arrival in London, much against the wish of the President. Colonel Olcott was extremely anxious to dissociate the Society from further connection with 'phenomena' following the Coulomb affair.[40]

According to W. B. Yeats, who became a member of the ES shortly after it was founded, its members, including Annie Besant, engaged in the study and practice of magic.[41] When he encountered Annie Yeats was already convinced that he would have a part to play in the spiritual regeneration that was imminent in Ireland. This idea naturally appealed to Annie: 'the ancient land once inhabited by mighty men of Wisdom...the Island of Saints...shall come again to be the Island of Sages',[42] she observed, in Theosophical language. According to Blavatsky, an avatar, a teacher, would appear to show how Ireland might be freed from Anglo-Saxon rule (as India would be freed).[43] Yeats, whose predilection for mysticism, secret knowledge, and Irish legend was well established before his connection with Blavatsky, took a more detached view of her than did Annie. He approved *The Secret Doctrine*'s synthesization of religion, and he was intrigued by the proposition that scientific discoveries might be explained by reference to occultism.

But he was undecided whether the Masters were living beings, as Blavatksy claimed, or dramatizations of her unconscious mind. His poetic imagination visualized them as living in the Himalayas, amid the snows, meditating through the centuries.[44] Yeats liked HPB for much the same reasons as he liked William Morris, whose Sunday lecture evenings at Kelmscott House he attended, as did Annie. Both were unpredictable, illogical, restless, and incomprehensible people. Yeats found HPB humorous and unfanatical, with a surprising honesty, though he waited in vain for any explanation of the Coulomb affair. Of Annie Besant, who shared the incantations at 19 Avenue Road, he said nothing more than that she was a very courteous and charming woman.[45]

Annie rejoiced at living in the midst of a colony of like-minded people whom she described as earnest, loyal, self-sacrificing, and studious. HPB insisted on regularity within the household. They breakfasted at eight, worked until lunch at one, then again until dinner at seven. After dinner they gathered in HPB's study to talk over plans, receive instructions, and listen to her explanation of 'Knotty points'. Annie regarded her enforced absences as something of a deprivation.[46] Her friends thought otherwise; relegated to a distance they read significance into small details. Hypatia, for example, deplored the attitude that banished Annie's beloved creatures from her presence; animals were no longer allowed at Avenue Road, apparently because they disturbed the 'aura'.[47]

Sidney Webb was one of Annie's colleagues who noted that she was on the point of resigning all participation in political life. 'Her reason to us, and to herself, is her growing distaste for all the methods of politics and her wish to devote herself to Theosophy', he wrote in October 1890. 'We have been fearing this for some time.' To Webb's mind Annie was being over-fastidious. Just as he was not mindful of soiling his boots or of inhaling foul air when lecturing to a working men's club, so he could not afford to be over-careful of mental contamination in doing other necessary work. He thought Annie would go on to the end of her term on the School Board but refuse to stand again. 'It is a pity. If all the good people go out of politics, politics will be run only by the bad.'[48]

Annie's last service to the Fabian Society was to take part in the 'Lancashire' campaign, a series of lectures by leading Fabians which led to the establishment of provincial branches—something she had urged on first joining the Society. Pease's *History* stressed the importance of these branches, which were succeeded by branches of the Independent Labour Party and, in due course, by those of the present Labour Party.[49]

The Lancashire campaign was still under way when Annie arrived in Ireland for a hectic programme of lectures on three separate subjects—Theosophy, secularism, socialism. At the age of 43, and for the first time in her life, she set foot in the country she regarded as her true birthplace; to whose political aspirations she had devoted an immense amount of time, energy, and her own particular talent. As evidence of her effectiveness she claimed that the authorities at Dublin Castle had her followed wherever she went.[50]

But while Theosophists gave her the warmest welcome and secularists stood up to argue about her new belief, there was no great response from politically minded people. In fact, though Michael Davitt's new paper, the *Labour World*, faithfully recorded her appearance in Dublin's Antient Concert Rooms, it was not able to report that she was received as one of Ireland's own. Yet, in lecturing on the class war, Annie delivered a rousing speech, attacking the most sensitive issue of the time. Landlordism was robbery, she proclaimed, echoing Davitt; excessive rent was theft even when it was sanctioned by law. Law did not make a thing right. In Ireland they would not change the law while landlords had soldiers and police to force their tenants to pay up.[51] The reason for the lack of fire in the audience's response may have been that Annie came too obviously late to a situation that had caused great suffering; that—as Shaw seemed to feel[52]—her claims to be one of them were overdone. The Irish had their own champions.

On her return to London Annie submitted her resignation to the Fabian Society. Though her colleagues on the executive directed Pease, as Secretary, to discover whether this was due to any cause they could remove, it took effect on 21 November. Although Annie stated that, from first to last, her relations with the Society had been most cordial, there is evidence that it was, at least in part, provoked by her anger over the Society's dealings with her as co-partner with Charles Bradlaugh in the Freethought Publishing Company. They had handled Fabian tracts and, when the Society decided to put the third, and cheap, edition of *Fabian Essays in Socialism* out to an independent publisher, they naturally hoped for the contract. They were refused, probably because the FPC was known to be in financial difficulty. Annie's anger over this, and over what she regarded as the Society's amateurishness in business, was increased when, in the course of 1890, she received 'dunning' letters seeking payment for the Fabian publications the FPC had handled.[53] To the relief of the members of the executive her letter of resignation did not mention this. 'Evidently it is Theosophy, though she does not say so', Shaw informed a friend.[54] On the other hand, Pease did not doubt what it was. He drew a red line through her name on the list of members and—with a sense of satisfaction?—wrote beside it: 'Gone to Theosophy.'[55]

The parlous state of the Freethought Publishing Company was only one of the worries by which Charles Bradlaugh was beset.

Another was his concern at Annie's 'subservience' to Blavatsky.[56] His health was now rapidly declining, a reproach to all who saw him, which was intensified by the anguish he continually expressed at his inability to continue work he felt that only he could do. His first collapse came at the end of 1889. The doctors held out hope of a recovery if he undertook that universal panacea, a long sea voyage. His was in the direction of Bombay, where he addressed the annual meeting of the Indian National Congress, which received him with the warmest gratitude for his work on their behalf. On his return to England, in order to embark on one more parliamentary year, he was obliged to lay down other burdens. In February 1890 he presided for the last time over the National Secular Society's gathering. It was a very sad occasion for all concerned, and embarrassing for Annie, who was alone in thinking that Bradlaugh must name her as his successor. Instead she had to watch her arch-rival, G. W. Foote, take up the presidential gavel.[57] Two weeks later she resigned from the NSS. 'The one thing that never changed with her was a desire to hold an important place in any party with which she was associated', Hypatia remarked, but nevertheless paid tribute to Annie's absolute sincerity.[58]

It was difficult for Annie to be much with Bradlaugh in the last crisis of his illness which overtook him in January 1891, because Hypatia feared her presence would exhaust him. He died on 30 January, shortly after the Speaker, W. H. Smith, did him the honour and kindness of waiting on his sick-bed to inform him that the Commons had voted to strike from the record all mention of his expulsion from the House in 1881. That countless others shared the desire to make amends for the exceptional difficulties Bradlaugh had encountered in the course of his career produced a tidal wave of obituaries. According to Bernard Shaw these cited the usual string of qualities: eloquence, determination, integrity, strong common sense, and so on, with the result that, if names had been omitted, the reader would have been unable to say precisely who the subject was—Gladstone, William Stead, or anyone else, 'no more like Bradlaugh than Garibaldi or the later [sic] Cardinal Newman'.[59]

To Shaw Bradlaugh was 'quite simply a hero', a man of tremendous personal force who found his role in life as a destroyer of idols. Lest that be thought too straightforward a judgement, Shaw improved it with a paradox.

Though Charles Bradlaugh preached the gospel of rationalism, he acted throughout life in the most irrational manner. Instead of choosing the line of least resistance—by which Darwin... taught... the order of the whole world had been produced—he chose the line of greatest resistance. When he met one of these idols, instead of taking off his hat and filling his pockets—which was the sensible, rationalistic thing to do—he hit the idol as hard as he could, and very often knocked it down.[60]

Annie Besant's memorial of the man with whom she had sustained the longest and closest, if sometimes tempestuous, relationship of her life so far was as good an example of the defiant attitude towards society and its conventions that Beatrice Webb had divined in her as anything she ever wrote. It was partly a statement, as much on her own behalf as Bradlaugh's, of the mission to reform society they had pursued together; partly a celebration of the private and domestic man whom she claimed to have known more intimately than anyone else. She too used the word hero and, as befitted her new-found confidence in the way the universe was ordered, looked to the future to provide a vindication of the man whose contemporaries had so grievously maltreated him.[61] At his funeral her feelings could be deduced from her attire. Bradlaugh's wish had been for the simplest ceremony. Among a sea of billycock hats—the vast majority of the crowd were men, and of the working class—her long black veil set her apart even from Hypatia, who was chief mourner.[62] There were many Indians among the crowd who watched the papier mâché coffin lowered into the grave. One of them was a young law student, Mohandas Gandhi, whom Annie already knew. He was interested in Theosophy, and had visited Madame Blavatsky at 19 Avenue Road.[63]

In the spring that followed Bradlaugh's death Annie had ample opportunity to reflect on karma. Hers at that moment mingled irony with tragedy. She had to watch as her new mentor succumbed to the same illness—Bright's disease—that had condemned Bradlaugh to his premature decline. In her feeble state HPB placed increasing trust in her, and wrote to William Judge describing her as a wonderful woman, her right hand; her successor, 'when I will be forced to leave you'. In the sixteen years since his co-founders had departed to the East, Judge, who remained in New York, had devoted himself to building up the Theosophical Society and was now General Secretary of the American Section. On 1 April Annie sailed to the United States as Blavatsky's personal representative to the convention at Boston of American Theosophists which Judge organized. He and Annie got on famously; he was, for example, versed in occult knowledge, a leading member of the Esoteric Section.[64] After the convention ended Annie went on to Washington and New York, and a bewildering number of small towns in between, where she lectured, sometimes on Theosophy, at others about socialism. She had to cope for the first time with American newspaper reporters, who buzzed in swarms around one who combined charm, clarity of expression, and ease of manner with sensational copy.

It was May before Annie sailed for home. When the ship put in at Queenstown Herbert Burrows came on board with dreadful news. Madame Blavatsky had died on 8 May. Influenza—unusually for the time of year there was an epidemic—had proved too much for her weakened state. Faithful to the hieratic nature of her extraordinary calling, before she died, she issued orders, one of which read as follows: 'I hereby appoint in the name of the Master, Annie Besant, Chief Secretary of the Inner Group of the Esoteric Section, and Recorder of the Teachings.'[65]

'Do nothing until I come', Judge cabled.

CHAPTER 21

'MRS BESANT, I SUPPOSE?'

THE further it receded the more devastating Bradlaugh's death appeared to Annie. 'The foundation is destroyed',[1] she wrote, in desolation at the knowledge that there could be no going back. Three months later, her grief at the loss of HPB was tempered by determination to refute the slanders in the many obituaries of the new Sphinx of the age. Her use of secularist platforms for this purpose aroused the antagonism of Bradlaugh's successor, G. W. Foote. To him *The Secret Doctrine* was a terrible jumble of second-hand knowledge and first-hand pretence. He expressed alarm at the effect upon Annie Besant of what he called the two 'evil angels' of Theosophy—spiritism and celibacy; more especially the latter, whose practice he thought was attended with the greatest physical, moral, and mental danger.[2]

Foote complained that, while Annie had used free-thought platforms to press her views on socialism, she had never advocated secularism at socialist gatherings. In the summer of 1891 he issued instructions that evening lectures in the Hall of Science must be confined to free-thought subjects. Besant took this as the final challenge. She announced that she would give a last lecture from the platform hallowed by Bradlaugh's memory on 30 August, when the title would be '1847–1891: A Fragment of Autobiography'. In a life that was remarkable for dramatic incident, like her speech on Clerkenwell Green, this was an occasion when the excitement of the crowd, her own emotion, betrayed her into an assertion that would have disastrous consequences.

That Sunday evening the Hall of Science was crowded to the point of suffocation when Annie, accompanied by her daughter Mabel and Herbert Burrows, mounted the platform that had been the focus of her power in the free-thought movement. Foote and John Mackinnon Robertson were in the audience, as was Henry Olcott, who, on hearing of Blavatsky's death, had made haste to

London, arriving shortly after Judge. As a result of the well-publicized differences between Besant and the new leaders of the National Secular Society, the mood was tense. 'When she rose to speak someone shouted "Shame"; others cried out, "No, No"; and then the audience cheered wildly for several minutes', Esther Bright, a friend of Annie's, who was in the audience, wrote.[3] Besant launched into an attack upon the attempt to restrict her that drew hisses and boos as well as cheers.

Now I shall never speak under such conditions. I did not break with the great Church of England and ruin my social position in order that I might come to this platform and be told what I should say. Our late leader [Bradlaugh] would never have done it. I do not challenge the right of your Society [which had until so recently been her Society] to make any conditions you like. But, my friends and brothers is it wise?

The audience fell silent when Besant, with obvious sincerity, denied the dreadful rumour that she had apostasized to the extent of believing in a personal God. Her belief had indeed changed, she told them, as a result of her fortunate encounter with Helena Petrovna Blavatsky, whom she warmly defended against the charge of fraud and charlatanry. That raised the crucial issue of the Masters. Without the Masters, the Theosophical Society was an absurdity, Besant was to write, and there would be no use keeping it up. On this highly charged occasion, in her desire to vindicate Blavatsky, she surrendered to impulse. Her audience of free-thinkers had been taught by Bradlaugh and herself always to call for evidence. She would give them evidence:

You have known me in this hall for sixteen and a half years [cheers] You have never known me tell a lie to you [No never, loud cheers] My worst public enemy has never cast a slur upon my integrity [Never, and cheers] I tell you that since Madame Blavatsky left I have had letters in the same handwriting as the letters which she received [Shock, sensation] . . . I do not ask you to believe me but I tell you it is so.[4]

The letters to which she drew attention in this sensational manner were written on the same kind of rice paper as the Master's missives to HPB. Their phraseology was enlivened by the odd Americanism: for example: 'Judge leads right. Follow him and stick.'[5] The explosion of public interest that greeted her declaration was awe-inspiring; Theosophists could only congratulate themselves that she had come among them. The new hall in the garden of 19 Avenue Road could not cope with the 'monster' audiences for her lectures; extra ones had to be hastily arranged. The correspondence columns of the newspapers were overwhelmed with letters pro and con Besant's Theosophy, and the leader writers gave her a formidable amount of space. Their analysis of her character and motive was largely sympathetic, for no one who heard her speak could doubt her dedication. Her claim to have been turned away from materialism by experiments in hypnotism, an imperfectly understood phenomenon which was attracting serious attention among the medical profession, led the cautious at least to suspend judgement. However, it provoked her former mentor, the Revd Charles Voysey (who had turned against her over the Knowlton pamphlet), to dismiss the whole affair simply as the result of her susceptibility to hypnotic suggestion.[6]

But those who had worked beside her in the turbulent 1880s recognized that she was responding to a need. It had long been predicted that she would end by becoming a Catholic. Sidney Webb foresaw her destiny as the convent in some form or another.[7] That was to misread her temperament: the convent implied seclusion, passivity; it is not too much to say such things frightened Annie. William Stead fell into the same error. Describing himself as an old friend who knew more than he could publish, he paid her the compliment of a long analysis of her life and career in the October issue of his very popular *Review of Reviews*. If she had been born a Catholic she would have become a nun, he remarked; since she was an Anglican, she married a curate. More seriously he described her as profoundly religious: one of three remarkable women of the century who were propagandist, militant, apostolic. The others were Josephine Butler and Catherine Booth, wife of the founder of the Salvation Army. Booth was dead and Butler had retired from the fray but Besant was still young enough (she was 44) to live to take her seat in the House of Commons, where Stead expected Millicent Garrett Fawcett would join her. His prediction reflected the current state of optimism among supporters of the women's cause, who had not yet taken the full measure of the hostility that existed to any such enlargement of the franchise.

Bernard Shaw's assessment was detached, unsentimental, and shrewd. 'Like all great public speakers she was a born actress. She was successively a Puseyite Evangelical and Atheist Bible smasher, a Darwinian secularist, a Fabian socialist, a strike leader, and finally a Theosophist exactly as Mrs Siddons was a Lady Macbeth, Lady Randolph, Beatrice, Rosamund and Volumnia. She "saw herself" as a priestess above all. That was how Theosophy held her to the end.'[8]

According to Colonel Olcott, Annie Besant was the greatest gain to the Theosophical Society since A. P. Sinnett. He was impatient for her to join him in India where, as he grew older, he found the task of holding the Society together much too arduous. The Hindu temperament called for constant overlooking yet the members were widely scattered, he complained. He was tired: of days spent in suffocating trains; of colds caught by falling asleep by the open window, a wet handkerchief on his head against sunstroke; of cramp from hours in a jolting bullock cart too small for his long American legs. He had to have an assistant; ideally he or she would be a person of good education, of eloquence as a speaker, force as a writer, and familiar with Indian history, its religious and caste observations. Besant appeared to him just such a person, 'a natural Theosophist', he rejoiced.[9]

But for the next two years Annie resisted every invitation to the East. She went instead to the United States where, like Bradlaugh before her, she restored her fortunes by lecturing. As Judge pointed out, 'she can make plenty of money in America: in India she can't make a cent.'[10] Several times in 1891 and 1892 she crossed the continent by train, journeys in the course of which she passed from scenes of industrial squalor like those at Pittsburgh—'pale, lined faces, sad looking men and women' which reminded her of England—to the vast forests and fire-blackened clearings of the rapidly retreating frontier.[11]

But she knew that India waited upon her coming. It was of the greatest significance. Theosophy's headquarters were there, the President Founder was there. Most important of all the future was there. Blavatsky taught that India, the place of origin of the ancient wisdom, like Ireland, a country that had lost its way under British rule, was shortly to experience a spiritual regeneration. Of course the hope of a better religious, social, and moral order was not confined to India, but the magnitude of the task excluded everything else. Annie was under no illusion that by abandoning her other responsibilities her public reputation would suffer. There is a hint of self-pity in the letter informing the London School Board that she would not stand in 1891. 'I elect to leave the more popular work in other hands and devote myself to the less attractive duty of pressing the claims of a spiritual philosophy on a public largely dominated by materialism', she wrote.[12] The inhabitants of Tower Hamlets, unaffectedly sorry to see her go, threw an enormous party in Poplar Town Hall to express their appreciation of her work on their children's behalf.

Because of her antipathy to the Christian Church Besant had to force herself—not always successfully—to observe the central tenet of Theosophy: that all religions were equally entitled to respect. This doctrine especially appealed to Indians. Long before Blavatsky landed at Bombay it formed part of the message of certain Hindu holy men, the most famous of whom was Ramakrishna Paramahansa, whose ashram near Calcutta became a place of pilgrimage. Ramakrishna died in 1886 but his message was conveyed to a wider audience by a young disciple, Swami Vivekananda, who became an object of devotion in his turn.[13]

In 1893, upon impulse, Vivekananda decided to attend the Parliament of Religions to be held in September in conjunction with the Chicago World's Fair. He applied to Colonel Olcott in Madras for money to help pay his expenses: he was refused.[14] The Theosophical Society at Adyar was sending representatives of its own; one a distinguished Buddhist, the other a Hindu. They broke their journey in London, where the latter, Gyanandra Nath Chakravarti, captured the imagination of Annie Besant, who hailed him as her guru. Chakravarti was a brahmin of Allahabad, aged about 50 at the time. By profession he was a lecturer in mathematics, but his attraction for Besant lay in the fact that he was a mystic: a spiritual teacher, with a profound knowledge of the ancient Hindu scriptures. He was Secretary of the Yogi Somaj (loosely translated as the Indian Psychical Society), and a long-standing member of the Theosophical branch at Allahabad. In his person Chakravarti linked West to East; science to religion; ancient knowledge to new practices and theories. He was also— and probably this was not known to Besant—the delegate to the Parliament of Religions of a number of brahminical societies in the most holy city of Allahabad whose influence was greater than that of the foreigners at Adyar. As a brahmin Chakravarti was exposed to loss of caste by crossing the sea and through contact with unbelievers. That he undertook the mission to Chicago suggests that he went with the blessing of his religious leaders.[15]

As soon as he arrived at 19 Avenue Road, Chakravarti created a sensation with his determined participation in discussions and at lectures. His English was fluent, and he spoke brilliantly. His remarkable singsong voice, 'a fascinating monotone' according to Besant, came, it was said, from years of Sanskrit chanting. From the moment she met him Besant's aspect and behaviour underwent a change and she became extremely anxious to visit India. Observers suggested that Chakravarti hypnotized her.[16] Certainly his power over her was plain for all to see. On 26 August she sailed with him for the United States, where she was to be Olcott's personal representative at the Chicago Parliament of Religions. The purpose of these great exhibitions and fairs was, by their displays of inventions, new processes, machinery, art, and scenes from history, to explain, to illustrate, to encourage progress, and to foster national pride; so highly regarded was this mission that Annie's brother Henry Wood had just been knighted for his work as a Commissioner at the 1889 Paris Exhibition.[17] At the Chicago World's Fair in 1893 comparative religion came before the public for the first time on this grand scale. Years of literary and biblical scholarship, of tales by travellers of the calibre of Richard Burton and Laurence Oliphant, of popular poetry like Arnold's 'Light of Asia', of romances like Rider Haggard's *She*, were beginning to transform Western attitudes to non-Christian religion; ignorance and suspicion were giving way to sympathetic interest.

Hinduism took this Parliament of Religions by storm; the public became aware, as never before, of how much it had to offer. The herald of this new enthusiasm was not Chakravarti, though his discourses were popular, but Vivekananda, conspicuous in ochre-coloured robe and turban. Besant's own commanding position as a female lecturer was challenged by the appearance at Chicago of celebrities like Susan B. Antony and Julia Ward Howe.[18] Having already covered Besant's exploits on behalf of Theosophy in her American lectures of the past two years, the newspapers were keen to discover a different 'angle', and thought they had found it in her association with Chakravarti. Their allegation that he slept on the threshold of her bedroom was strenuously denied.[19] The brahmin's influence over her was a source of anxiety to William Judge: his advice to his good friend Annie to steer clear of India was about to be ignored. As soon as the Parliament of

Religions was over she was to sail for Colombo in Ceylon, where Olcott was to meet her and conduct her to Adyar.

Judge was concerned to hold the balance between East and West in the Theosophical Society; he complained that Olcott was giving influential Indians the impression that the Society's real aim was to propagate Buddhism. If the brahmins, as guardians of Hindu faith and tradition, were given a false idea of the Theosophical movement they would preach against it and 'we shall have a silent, subtle, untouchable influence negating all our work'.[20]

There was some foundation for his anxiety. Of all the religions that Olcott examined in the spirit of tolerant enquiry he took to be the mark of a true Theosophist, Buddhism held the greatest attraction for him. Since about 1882 he had worked to restore a way of life that had fallen into decay, and had shown himself to possess revivalist fervour, tact, and, where necessary, cunning. Thus he had founded schools, composed a Buddhist catechism which he published in a cheap edition, and taken the initiative in preserving Buddhist shrines—at Gaya for example.[21] He began to assemble ancient manuscripts, of every faith, including some extremely rare Tibetan ones which formed the nucleus of an important library at Adyar. He was invited to lecture to Buddhists in Burma and Japan, and conceived the desire to bring about a world-wide Buddhist Union. His prestige stood highest in Ceylon, where by careful diplomacy he had persuaded the British colonial administration to make allowances for the special standing of Buddhism among the Sinhalese.[22]

Deprecating his own abilities, Olcott sought out experts; from E. B. Havell, head of the Madras School of Art, who designed the great carved doors of the Adyar Library,[23] to the most eminent Professor of Comparative Philology at Oxford University. Max Müller, whom Olcott respectfully badgered for help in his publishing enterprises, was anxious not to endorse a connection between the Vedic scriptures he had been one of the first Western scholars to investigate, and Theosophy. He was wary of the 'theosophic Russian Countess'. Blavatsky's name and prestige were doing real mischief among people who were honestly striving for higher religious views and were willing to recognize all that was true and beautiful and good in other religions, he told Olcott, whom he also rebuked for propaganda about 'esoteric' Buddhism.

There was nothing 'esoteric' about it, Max Müller remarked: it was the religion of the people, pure and simple. In May 1893 he aired these views in the *Nineteenth Century*, causing Besant to spring at him: Max Müller was no more than 'a narrow priggish scholar,'[24] she remarked.

As soon as she committed herself to visit India the Theosophical pendulum swung in the opposite direction to that pursued by Henry Olcott. Hinduism was revealed to be Mrs Besant's strong preference. The message she preached to India's multitudes in the forty years that remained to her was essentially the same she offered to her fellow passengers on board the *Kaisar i Hind* in October 1893. The passage out was rough. 'Lectured hanging on to a corner of the table', her diary recorded—she was a bad sailor.[25] However, those members of the Indian Civil Service returning from home leave who listened to what she had to say in the first-class saloon were probably more shocked than queasy.

There were two views of India, Besant told them. 'You know her ... by taking part in the foreign government by which she is subjugated and therefore you are very largely shut out from the real thought and real life of the people ... Whereas to me she is ... the Holy land ... whose polity was built by King Initiates, whose religion was moulded by Divine Men.' The brahmin families of the north and north-west of India—Aryavarta she loved to call it, the heartland of the Aryan race, the people who in prehistoric times had introduced the Vedic religion—represented the noblest physical, mental, and spiritual type ever produced on earth. Their civilization was unique because it was formed solely for a spiritual purpose. It was founded on a system of caste, she went on, not, as in the present day, depending on endless subdivisions, but on four great ones: those given in the Book of Manu who was the Lawgiver in Vedic times. The brahmins were the spiritual teachers; the Kshatriyas were the warriors, the 'royal and ruler' class; the Vaisyas, the merchants and agriculturalists; lastly the Shudras, the 'submerged people' Annie called them, were the serving class. These forms had given stability to Indian life; they had preserved her civilization in spite of conquest. What was needed now was to sweep away the memory of the dark times when the brahmins had oppressed instead of serving the people, and to restore all castes to their proper function.

Linking this firmly to her belief in reincarnation Besant explained that the four castes were the four great schools of the evolving soul: 'the Brahmin caste, pure in its blood, developing the most delicate organism, the subtlest brain, the most perfect mental mechanism should be inhabited by the most advanced souls.'[26] This ordered system pleased Besant, but it was highly controversial. As a good American, Olcott, for example, detested caste, and devoted a great deal of time and money to alleviating the condition of the Untouchables (whom he called panchamas). Those Western-educated Indians who were just beginning to contemplate what India's future ought to be saw caste as perhaps the greatest barrier to democratic progress. In the difficult years to come, Besant's persistent advocacy of a reformed but essentially unaltered system exposed the Theosophical Society to bitter criticism.

At Colombo in Ceylon Olcott was waiting to greet Annie and her companions: Miss Willson, who shared her Indian adventures for many years, and the Countess Wachtmeister, who had been HPB's devoted friend.[27] They disembarked to a reception fit for a proconsul. Not for Annie Besant the shock of noise and indifferent scurrying crowds and unfamiliar heat. Greeted with magnificent courtesy she was garlanded, sprinkled with rosewater, and applauded by admiring crowds that had been assembled on the quayside hours before the *Kaisar i Hind* came into view. From the most eminent of the Buddhist priests to the Governor himself the whole island of Ceylon appeared enraptured by her presence. The triumph owed everything to her sponsor, Colonel Olcott. When she observed the deference and affection shown to him on the island—which he always scrupulously referred to as Lanka—Annie had to revise her opinion of him. This was the man of whom HPB had spoken, with affection certainly, but also with contempt, as destined to be gulled, whose *naïveté* she had used while despising it. After her death an attitude grew up towards Olcott as the distant figurehead, a 'papal authority' whose orders were irksome to the transalpine branches of his empire. Now she was forced to acknowledge a presence that commanded crowds and moved as of right among real imperial magnificence.

'Mrs Besant put her foot on Indian soil for the first time at the hour of 10.24 a.m. on 16 November 1893', Olcott's diary recorded. This was at Tuticorin near the southern tip of India after a short crossing of the narrow choppy Gulf of Mannar. Her progress up the remote peninsula resembled that of a messiah. To go from place to place, to be met by curious and friendly crowds who were astonished at her reverence for their faith, to speak and be enthusiastically received, forged a sacred bond between Annie and the Indian people.[28] Privately she noted those occasions on which she received the Master's encouragement before she spoke.[29] The most respected pandits deigned to listen to her. 'It needs a mind like Annie Besant's to grapple with the metaphysical subtleties they bring forward', countess Wachtmeister dotingly remarked.[30]

What Annie always called 'tom toms', and flaring lights, and popular religious symbols greeted her wherever she went. Elephants saluted her by throwing up their trunks, and sacred bulls guarded the flower-bedecked platforms from which she spoke. Details like this caused some amazement among the social democrats in England. 'Oh I say, Mr Hyndman, have you seen that Annie has been walking in procession behind two white bulls, *two white bulls?*'[31] At Bangalore the fever of anticipation was so great that offices were closed early for people to go to her lecture. Here the local brahmins asked her to lead a national Hindu movement for reform. For the moment she returned a non-committal answer.

At Kumbakana Annie chose to lecture on renunciation. 'The audience felt the beauty of such an idea and I was glad', she wrote, 'the Hindus have sought only the Path of Liberation [from the British] for so long that the other idea has faded away.'[32] Renunciation, as proposed by her, was hardly an idea that had faded out of Indian consciousness, but a concept new to most of them—it was derived from the principle of sacrifice of self for others that the Evangelical Miss Marryat had instilled once and for all in the young Annie Wood. One who kept a close—and disapproving—eye on Besant throughout her Indian career, the diplomatist and writer Valentine Chirol, noted the significance of this aspect of her teaching for nationalist aspirations. Hindu philosophy did not tend to develop the sense of social duty of which democracy should be the political expression, he wrote, for it taught the individual to seek, not the relief of human suffering around him, but his own escape from an endless cycle of suffering existence.[33] Besant's odd, apostolic fervour was to be one of the forces that gradually brought about a change.

Shortly after Christmas she and her party reached Madras, where Theosophists from all over the world had gathered for the eighteenth annual convention.[34] In her various accounts of the rise of Indian nationalism,[35] Besant claimed that the Theosophical convention was the model for the Indian National Congress. It is true that among the lawyers and journalists and other professional men who responded to Allan Octavian Hume's invitation to them to confer at Bombay in 1885 were a number of Theosophists.[36]

The headquarters occupied a large estate some seven miles from the centre of Madras, which lay along the south bank of the Adyar River where it entered the Indian Ocean. From November to March an almost constant sea breeze tempered the heat to that of a Mediterranean summer. Trees and a host of flowering shrubs gave shade, and shelter to the numerous birds—chattering parrots and loquacious doves, kingfishers, herons, waders, and gulls come to a temporary haven from the open sea. With its latticed screens and elegant portico, the largest of the buildings on what had been known as Huddleston's Gardens resembled a Southern plantation house. Beyond the grove of coconut palms stretched the flat bright beach and the heavy pounding waves of the Coromandel coast. At low tide buffalo waded through the shallows so deliberately they scarcely caused a ripple. Towards evening the colonel could be seen, comfortably grounded in the water, smoking a cigar and conversing with the little boys who paddled in beside him. Olcott loved Adyar, as he loved all India and its people.

On this first visit, as on all others, Annie Besant held court in the manner of an Eastern potentate. 'Every morning from 8 to 10 I sat in the hall and answered questions; from 3 to 5 the same; from 5 to 7 private interviews; after 8 questions or what came up. During Convention I lectured each morning and there were public lectures besides.'[37] When in January 1894 she and Olcott set out on an extended tour of Theosophical branches, these public lectures delivered to large audiences in a succession of cities turned the journey into a potentially explosive political campaign.

In Calcutta Town Hall Besant faced a packed assembly already stirred by rumours that the 'Englishwoman' who presumed to tell Hindus how to behave risked physical assault. In the event she won a huge ovation. She began by asserting that India had been created by the avatars of the gods, and that the Hindu scriptures, the Vedas, came from the mouths of those who were more than men. As far as the *Indian Mirror* was concerned Besant was herself an avatar (its editor was a Theosophist). 'The messenger of the gods has come to tell Hindus that the gods will return to India if her sons will only... attempt to win back their ancient heritage', it enthused. *Hope* detected a mighty awakening as the educated classes began to distinguish between the solidity of the Hindu system and the hollowness of the West. The most significant, because immediately practical, response came from *Amrita Bazaar Patrika*, among vernacular newspapers a persistent and courageous critic of the Raj. Its editor, Motilal Ghose, had been a regular correspondent of Charles Bradlaugh.[38] Now the *Patrika* hailed his former colleague, well known for her own efforts on behalf of justice for India, as, quite simply, the leader for whom they had all been waiting. This was to take a calculated risk: the *Patrika* was already threatened with prosecution for sedition under the Press Act. In a long first leader the London *Times* castigated Mrs Besant for allegedly urging the youth of India to cease fawning on a foreign power. If they acted in accordance with tradition they would not long remain under a foreign yoke. These remarks were so reactionary and dangerous, *The Times* remarked, it took leave to doubt they had been accurately reported.[39]

Besant's intervention came at an awkward time. Confidence in its capacity to hold the country had to serve the Raj instead of a standing army. Recently confidence in England as well as India had been shaken by a series of communal riots provoked by the Cow Protection Society: as its name implies, a Hindu organization which appeared to threaten Muslims. In the opinion of at least one expert the atmosphere was surcharged with religious dissension.[40] Now here was the notorious Mrs Besant poised to stir the masses further. Her antecedents, especially her support for Michael Davitt, worried English politicians who feared that India would go the way of Ireland.

The comparison was inescapable. In 1888 the then Viceroy, Lord Dufferin, called on the Secretary of State for India to make concessions so as to weaken the radicals. 'Unless some line is definitely taken we shall have something like a Home Rule League established in India on Irish lines under the patronage of Irish and Radical members of Parliament.'[41] As recently as August 1893, his successor, the marquis of Lansdowne, had remarked that the Cow Protection Society was dangerous since it offered a way of connecting the admittedly small number of politicians in India with the great mass of the Hindu population. 'The Home Rule agitation [in Ireland] did not become really formidable until Parnell had taken up the agrarian question', Lansdowne pointed out.[42] It is not clear whether Besant's activities in Calcutta were influenced by her Irish nationalist friends, or proceeded as a matter of course, from her association with Charles Bradlaugh, the 'Member for India'. Earlier in the year she had spoken in support of the Parsee businessman Dadhabhai Naoroji, at the Central Finsbury by-election, at which he became the first Indian to win a seat in Parliament. Previously, when no seat could be found for him, Michael Davitt suggested that he stand for one in Ireland, and that the Irish party at Westminster should make itself responsible for Indian affairs.[43]

Naoroji was a founding member of the Indian National Congress which, when Besant first arrived in India, was so moderate in its demands as to appear apologetic: 'a foolish debating society' was how Lansdowne thought of it.[44] One of its most prominent members, Gopal Krishna Gokhale, remarked that the proper exercise of the political institutions of the West could only be acquired by an Eastern people through puritanical training and experiment.[45] His restraint was echoed by Naoroji and the great majority of educated Indians. One man spoke out, proudly and harshly, linking religion to politics and making a special bid for the allegiance of the young. Bal Gangadhur Tilak was the originator of the Cow Protection Movement; a Hindu scholar, a preacher, determined to make the British quit India, by force if necessary.[46]

As he did, Besant, by her outspokenness, threatened the fragile situation whereby, if most Indians did not subscribe to the convention that they were incapable of governing themselves, they condoned it for the sake of peace and security. Henry Olcott was quick to grasp the danger to the Theosophical Society from her unbridled remarks. It is not known exactly what kind of battle was fought behind the scenes but Annie emerged holding the white flag. On 30 March *The Times* reproduced a letter by her to the editor of *Amrita Bazaar Patrika*. She declined the honour of leadership; she was no longer a politician but a worker in the field of education and religion. Though her association with Charles Bradlaugh had been the proudest of her life, she would never resume his mission. Her task now was to arouse India's sense of self-respect, and pride in the greatness of her religious and cultural traditions.[47]

From Calcutta Besant and Olcott journeyed on, via Lucknow and Benares, to Allahabad where a vast crowd had gathered to celebrate the most important Hindu religious festival, the Khumba Mela. Local newspapers described Besant, dressed as a Hindu, bathing daily with the pilgrims at the sacred spot where the Ganges met the Jumna. This was her first appearance in the striking costume she wore most frequently for the rest of her life. 'A lovable, loving, attractive *white* lady' was how one of her Indian protégés, as a little boy, remembered her: 'white from head to foot with white hair, white clothes, white stockings and shoes.'[48] Many garments went to make up this whole; shawls, stoles, sashes, much fine embroidery. Loose white garments coped with the heat and dust and gave her small, and by now plump, figure an added dignity. At home in England, among the black coats, against the grime of London, white appeared bizarre, theatrical. Ordinary people never wore it. It was understood that Annie Besant did so as a sign of mourning for the wrong Britain had done to India.[49]

According to Olcott's diary, when Besant was reunited with Chakravarti at Allahabad, where she stayed in his compound, the brahmin soothed her nerves which were 'overstrained'.[50] This blandness concealed disaster. The tension did not arise so much from Annie's blunder into politics; it was the result of a suggestion made to her at Adyar: that the letters she had so fervently proclaimed as by the Masters were forgeries, and by none other than William Quan Judge. Invited to sign a document accusing him of this, she held out until she could consult Chakravarti, whom she had not seen since Chicago. The brahmin's decision was that Judge should be arraigned. Olcott, whose own conduct, private and public, had been denounced by Judge, now had the pleasure of requesting his old colleague to resign or the charges would be made public.[51]

There followed a vicious struggle over the office of President in which Besant's credibility was hopelessly compromised. The

Theosophical Society had developed in separate directions, one autarkic, the other democratic; the President was the only link. Olcott had no power over the members of the Esoteric, or Eastern, Section whose affairs were conducted in strict obedience to the Outer Head—or Heads, since Besant and Judge were jointly in control following Blavatsky's death. An oath of secrecy bound the members of the ES. As Yeats's correspondence indicates, the purpose of this inner group was to advance communion with those spirits in which the members believed, in preparation for a higher form of life. How they went about it was considered to be no concern of anyone but themselves. But their statements were revealing. Language is in itself a kind of magic; the words they used were obviously weapons turned against an enemy; they were terms undreamed of by her socialist colleagues, whose disputatious habits Besant declared had virtually destroyed her sympathy for that cause.[52] Those who frustrated members of the ES were evil; were manifestations of the black order of the universe; were, at best, victims of 'glamour'—Blavatsky's word—which distorted their true nature. In this context the reason why Besant postponed her visit to India for so long was not because she needed money, nor, as she claimed, that her health would not stand the climate; it was that Judge told her that if she went Olcott would poison her.[53]

There was another rumour to the effect that Chakravarti proposed to get rid of Judge and Olcott and rule through Besant, who would be President. While he was in Chicago Chakravarti asked Judge to repudiate a letter written in 1881 by HPB, acting on behalf of the Master who, of course, composed it. The Allahabad Theosophists to whom it was addressed found it deeply insulting in its attitude to their religious customs, and in its references to 'you natives'. Besant doubted its authenticity; either it was a forgery, or HPB had taken it down too quickly, she said. Judge stood firm. If this letter were shown to be a fraud, all Blavatsky's claims of communication with, and teaching from, the Masters must fall to the ground.[54]

When he received Olcott's ultimatum Judge reacted coolly. Instead of resigning he demanded a hearing. This took place in London in July 1894, when Olcott presided over a small committee, one of whose members was Judge. Annie Besant need not have been involved, except that while in India she consented to play the part Olcott devised for her: that of the Accuser, in which capacity, with her customary thoroughness, she put together a brief which included confidential papers pertaining to the Eastern Section.

Judge flummoxed his opponents by proclaiming his faith in the Masters who had been in touch with him, and, through him, with others. That was as far as he believed things could go. The Theosophical Society had no dogma as to their existence; to subject them to the deliberations of an official committee would mean that, after nineteen years of unsectarian work, the Society was determined to affix this dogma to the Constitution. This adroit argument apparently frightened Olcott, who was the guardian of the Constitution and had always stressed its open nature. Judge remained, and Annie Besant was obliged to admit that it had been a mistake on her part to bring the charges. That went some way to mollify the members of her lodge who objected to the slur cast on Judge, who was popular. A less recondite explanation was advanced: that Judge, who had been at Adyar during the uproar over the Coulomb affair, and who had disposed of evidence by burning the shrine, had threatened to make public all he knew.[55]

'I do not charge, and have not charged Mr Judge with forgery in the everyday sense of the word', Annie Besant wrote, 'but with giving a misleading form to messages received psychically from the Masters in certain ways...I know...that I believed the messages he gave me were messages directly precipitated or directly written by the Masters.'[56] This recantation was carefully phrased. The 'messages' were the letters she had commended to the Hall of Science audience three years previously. A huge storm was brewing, but it did not break before she departed on her longest sea voyage yet—to Australia, a visit she made at Olcott's request to gather recruits.

In late October, when she was still half a world away, the *Westminster Gazette* produced a scoop that, in colour and interest, rivalled Stead's best effort, and was second to none in satirical effect. For nine consecutive days, under the title 'Isis Very Much Unveiled: The Truth about the Great Mahatma Hoax', it published articles which gave a devastating glimpse of a sect at war: accusations, denials, protests, threats; talk of Black Magic—the private language of the Eastern Section and some of its secret documents were exposed to daylight and laughter. The horrified inhabitants of 19 Avenue Road perceived that the country was being treated to a comprehensive view of the case Besant had prepared against W. Q. Judge.[57] The articles were written by Edward Garrett, 'one of the most... caustic literary experts of the London press', Olcott remarked, no doubt with feeling.[58] Garrett charged Annie Besant with being gulled, 'so as to deprive of value any future evidence of hers on any question calling for the smallest exercise of observation and common sense'—a crushing indictment for a pupil of Miss Marryat.

In Annie's absence, W. T. Stead did his best to mitigate the harm. He did not care how Koot Humi made his wishes known: by post, or dropped from heaven on rice paper. What mattered was that Blavatsky had succeeded where he and others had failed, in convincing Annie Besant that there was an existence beyond the material. In other words the Mahatmas might be signs, symbols, metaphors for something as yet undefined but, for all that, true and natural. 'This generation is behind no other that ever existed in thirsting for a sign.'[59]

A very bitter war broke out among Theosophists. On her return to London Besant found herself in the thick of it, sneered at by opponents, cut by former friends, damned by erstwhile colleagues. Even Herbert Burrows turned against her for condoning acts by HPB, knowing them to have been fraudulent.[60] Furious at the treatment of their General Secretary the Americans seceded in a body. They were supported by Irish Theosophists, led by the poet AE (Yeats's friend G. W. Russell). He never forgave Annie Besant for departing from the path laid down by Madame Blavatsky, and for having believed her capable of fraud.

The Judgeites ever afterwards objected to undue emphasis being given to the teaching of the East. 'The Theosophical Movement was begun by Masters in the West, by western people', Judge himself wrote, 'it is not Their desire... to have us run after the present East.'[61] He died in 1896 from consumption, his demise hastened, it was said, by the Theosophical civil war. How curious it was, Stead wrote, that the Theosophists were splitting into eastern and western sections, very much after the fashion of the Christian Church.[62]

'Theosophistry' was a splendid public joke. Mahatmas turned up everywhere, including the Egyptian Hall, where Maskelyne and Cook put on a burlesque. At Christmas the subject filled out the patter in the pantomimes. Afterwards it was immortalized in *The Story of the Amulet* by E. Nesbit; as the wife of Hubert Bland she knew Annie well. The children are at the British Museum with the Psammead, a magic creature with the power to take them back in time. They encounter a Babylonian queen who takes the opportunity to release objects she considers hers from their glass cases. They fly out of the door and she follows them, whereupon a bystander is moved to remark: ' "Theosophy I suppose. Is she Mrs Besant?" "*Yes*", said Robert recklessly.'[63]

CHAPTER 22

ARYAVARTA

AFTER 1893, and until her death, India was the channel into which Annie Besant chose to direct her superabundant energy. Her life there was a bizarre adventure, a fantasy in the manner of a Rider Haggard novel, made real by the exercise of a tenacious will. At the same time as she ruled as high priestess over a sect whose members were obsessed with ancient teaching, her scarcely concealed objective was to seize the fulcrum of contemporary power and, once it lay within her grasp, to use it to bring about an irreversible shift in India's political direction. By her own account the enterprise was planned. The first stage was to revive confidence in the old religion; the second to combine traditional learning with the best of Western education; only then to enter politics.[1] Her aim was visionary; millennial—it was to bring about a universal, theocratic state under whose firm, wise rule men could not but behave as brothers. Benares, in the north of India, heartland of the Aryan race, was where the first two stages of the plan were put into effect.

Adyar was Olcott's domain. Besant attended the yearly conventions there but did not attempt the challenge of a longer stay. In 1896, ostensibly because Adyar was so remote, Benares was made the headquarters of an independent Indian Section.[2] Money given by Ursula Bright, a close friend of Annie's, enabled a small estate in the district of Kammacha, near the centre of Benares, to be bought. From November to April each year Annie was in residence. There was a hall for meetings; offices; a printing press; and a pharmacy. There was, besides, a house for Chakravarti and his family. Birds, a tame deer, and a number of companionable mongooses shared the garden. Annie's own house in a corner of the compound was low and solid; tall rooms looked on to a courtyard full of flowers. She called it Shanti Kunj, Abode of Peace. Ganesh, the elephant-headed god, looked down from an outside wall. He was the symbol Tilak chose to represent the call to India's freedom; a fact that cannot have escaped Besant however apolitical she had promised to become.

The means to the first stage of her plan began at the gate of the Kammacha compound. Benares, the holy Kashi, was the Hindu Rome. One of the strangest places in the world, it lay in a curve of the Ganges (Ganga to Besant), facing directly into the rising sun, which pilgrims greeted every day. A crescent of river and stone, worn steps, massive towers, parapets, pillars, temples, palaces, and burning pyres over which vultures floated; it was sanctified by time, mysterious; as weird as ancient Kor.[3] Besides the common people who flocked to the sacred river, Benares attracted acolytes from all over India to its monasteries, whose priests were the guardians of brahminism. According to E. B. Havell, who wrote a book about it as it was when Annie Besant lived there, not only were the priests and pandits ignorant of all things modern, they regarded as worthless all knowledge not contained in Hindu sacred writings. Propagandists like Swami Vivekananda were frowned upon.[4] Yet it was in this forbidding city, among people of an alien race, that Annie Besant sought to wield authority in matters of religion and moral conduct, and, for a time, apparently prevailed. When Havell saw what progress she had made he was astonished.

On first coming to India Besant set herself to learn Sanskrit, and as early as 1895 published an English translation of the most beautiful of all the ancient texts, the story of Krishna and Arjuna in that part of the *Mahabharata* called the *Bhagavadgita*. Her version was by no means the first, and she had the help of several Hindu scholars, but her introduction distinguished it from others. According to her, the central lesson of the *Gita* was that spiritual man need not be a recluse: 'that union with the divine life may be achieved and maintained in the midst of worldly affairs.'[5] In other words, the *Gita*'s teaching embraced that sense of social duty Besant was determined to arouse among Indians.

Her next publishing venture was a textbook to be used as the basis for religious instruction at the Central Hindu College for boys she founded in Benares. She took great care in the preparation of the *Sanatama Dharma* (Virtuous Conduct), which was scrutinized by a board of trustees before it was adopted. The new boys' institution was to be Hindu monastery and English public school—Harrow of course—combined. They were intended to emerge as pious Aryan gentlemen able to hold their own with Western citizens. They would be endowed with the ancient Aryan virtues which, according to Besant, were: reverence, self-reliance, freedom, moderation, calmness, gentleness, justice, and courtesy.[6] The school was divided into houses; order was maintained by prefects. The first $1\frac{1}{2}$ hours each day were given to prayer, meditation, and questions on the sacred books. Besant herself took classes in the *Mahabharata* and the *Ramayana*. Though she considered herself well qualified by her studies to do this, not everyone agreed with her. Swami Vivekananda let it be known he thought her knowledge superficial, while one Collegian confessed he thought she talked nonsense on these occasions.[7] The boys were constantly reminded they had reached the third stage of a brahmin's life; that of *Brahmacharya*, the time of celibacy; they were admonished to be pure in thought, word, and deed.

Science lessons were based on practical experiment, using apparatus laboriously constructed by the first Principal, Dr Richardson. Whenever any of Annie's acquaintances from her days at London University came near Benares she begged them for a lecture.[8] Many of the teachers were brought out from England; they were expected to work for the pittance their Indian colleagues were accustomed to receive. Beginning in 1895, it took Besant only three years to raise the money for the CHC, during which time she travelled thousands of miles; sleeping at dak bungalows, or camping at rail junctions. Everywhere she went asking for money for her own project, she urged local inhabitants to start religious instruction in their own schools, whatever their faith. One of her most enthusiastic helpers was a prominent lawyer of Allahabad, a member of the Indian National Congress who had been introduced to Theosophy by Chakravarti. He was Pandit Motilal Nehru, father of Jawaharlal. From 1899 until 1902 the future Prime Minister of an independent India was tutored by a Theosophist, F. T. Brooks, whom Annie Besant recommended at Motilal's request.[9]

The outlook for the Central Hindu College was transformed when the Lieutenant-Governor of the United Provinces denounced Besant's initiative as a disloyal act: by founding it she was using education as a cloak for politics, Sir Antony Macdonald complained.[10] This remark so enraged the maharaja of Kashmir he bestowed a large sum of money on the College. Not to be outdone, the Maharajah of Benares donated one of his spare palaces, a symphony of jumbled spires and twisted cornices, just across the road from the Theosophical compound, to be a permanent home for the school. Macdonald's warning, however, received the warm approval of the Secretary of State for India, Lord George Hamilton. As Minister in charge of education for part of the time Bradlaugh and Besant ran the Hall of Science school, he had been their very bitter adversary.[11]

The CHC grew to house 1,000 boys—'the natural leaders of young Hinduism', Annie liked to call them. That it flourished in spite of her long absences was due to the judicious administration of its Secretary, Babu Bhagavan Das, whom Annie had persuaded to abandon a promising career in government service as a magistrate in order to assist her. He was a brahmin, a solitary man, of independent mind and studious habits, profoundly knowledgeable about the philosophy of religion, and versed in the ancient Hindu writings—he was one of those who helped Annie with her translation of the *Gita*. He was also kind, for he allowed her a degree of intimacy with his orthodox Hindu family that was given to few Europeans. In return she treated his children as if she were a grandmother, bringing them presents every time she returned from Europe, saving them stamps from the wonderful assortment of foreign letters she received.[12]

But, however much she tried to behave as a Hindu, she never lost sight of the fact that her life was dedicated to the Masters, that she was the intermediary between them and Theosophists at a lower stage of progress than herself—a humbling thought. 'I am always so afraid of lowering the standard of discipleship', she told a friend, 'people know I am a disciple and my words and actions give them evidence I am.'[13] But she grew intolerant of criticism, and authoritarian in her conduct of the ES, which she developed as a source of power to rival that of Adyar. She formed an inner group of ES members who took an oath to obey her 'without cavil or delay';[14] their function was to protect her, as a priest within the sanctuary is shielded from the common gaze.

Her capacity for intercession on behalf of weaker brethren arose, she was convinced, by reason of her access to the Masters. In her case, as in Blavatsky's, proof of this authority was required—not least by her own demanding conscience. But the kind of circumstantial evidence the older woman had produced was now out of bounds. Besant's inclination was, in any case, for an intellectual explanation rather than 'phenomena'. This evidence came more easily from another, from a teacher versed in occult matters. Charles Webster Leadbeater was not on Bernard Shaw's list of Annie's leading men; probably Shaw forbore to mention him because his effect upon her reputation was so damaging.

Though he became a Theosophist long before she did, in 1883, Leadbeater's rise to power in the Society was far less rapid than Besant's. For at least the first ten years he was held in such low esteem he was permitted, rather than encouraged, to make himself of use. A curate, unhappy with his lot, Leadbeater took refuge in Theosophy when he abandoned the Church of England. In 1884 Madame Blavatsky took him out to Adyar. The following year Olcott sent him to Ceylon to supervise the schools he had founded. His complaints about the squalid conditions in which he was obliged to live did not bring a reprieve until 1889. Back home in London he scraped a living as a tutor; among the boys he taught were A. P. Sinnett's son and George, nephew of Miss Francesca Arundale, one of the wealthier and more unswerving members of the Society.[15]

Part of the donkey work Leadbeater had to do was to answer the many letters of enquiry about the doctrine of Theosophy, a task for which he developed unusual aptitude. The leadership was pleased, the members flattered by the scale and singularity of the doctrinal edifice he gradually unveiled. His work appeared in the *Theosophist*, which in due course Besant edited. She was intrigued by his ability to express the immaterial in scientific terms. They became collaborators, fellow explorers of an unknown country in search of knowledge. Communication with the Masters, guardians of this treasure, was no longer by courtesy of the GPO, but on the astral plane, journeys which CWL (as Theosophists liked to call him) described in the compelling style of more popular adventures. He became a leading authority in the occult, reflecting prestige upon the Society.

Each year as the heat became intense in India Annie sailed from Bombay into an English spring. Part of her time at home was spent in retreat with Leadbeater and his boys—an unconventional reading party—'bringing through' information about the astral plane. They combined their special interests in a book called

Occult Chemistry, which presented the results they had obtained by using clairvoyance to examine elements and atoms.[16]

When Annie decided to spend part of each year in India, the lease of 19 Avenue Road was given up and she made her home in England in the very different atmosphere of the Bright household. Jacob and his wife Ursula were prominent members of a network of social reformers related by friendship and marriage—Brights, Maclarens, Garretts, Fawcetts, Butlers—these were the very people from whom Besant had formerly held aloof, fearing a rebuff by reason of her notoriety.[17] She had earned their respect by her work on the London School Board. Ursula, moreover, was a Theosophist, as was her daughter Esther, and another member of the circle, Henrietta Müller.[18] Esther conceived a lifelong affection for Annie Besant which was reciprocated. 'AB' wrote to Esther once a week from Benares addressing her as her 'Knight of Friendship'. These letters reveal how great were the demands on her of the ES, which she controlled through an unremitting daily correspondence and frequent visits to the membership, which was scattered from coast to coast in the United States, as well as in Scandinavia and Europe. When they were together Esther did her best to ease the burden on AB, coaxing her away from the writing table (which had given her a pronounced stoop), seeing her off on the interminable sea voyages with rugs, and shawls, and lentil soup. Esther did not doubt Annie's ability to sustain her lofty mission: 'she has so much wisdom in seeing character and making allowances for everything and treating things naturally.'[19] While Ellen Marryat would have been delighted at this testimony to what she had tried to achieve, not many people agreed with Esther. In 1906 a huge storm burst, causing another schism; at the centre of it was this very question of Annie Besant's wisdom; of her judgement—or the lack of it.

For the past few years AB had worked increasingly with CWL: they had begun to bring about a change in the nature of the Society which was far from commending itself to all the members. CWL's occult vision was so precise and confident it left nothing to individual surmise: his interpretation hardened into dogma. This was something Theosophy had originally opposed, but which—in contrast to her free-thinking past—Besant now endorsed. At the same time she rediscovered her taste for ritual, which Leadbeater, who had briefly followed Pusey during his time in the Church, shared. A special kind of Masonry called co-masonry offered the best opportunity for the elaborate ceremonial they both enjoyed. As this was also alien to the Society over which Henry Olcott still presided, they were cautious about when and where they introduced it.[20]

These changes would not have come about if, towards the end of the 1890s, Chakravarti's influence over Besant had not begun to wane. What came between them is not clear; it may be that he reflected a growing disquiet among orthodox Hindus in Benares about her intentions. Certainly he disliked Leadbeater's cast of mind. In a speech in 1900, Chakravarti drew a firm distinction between spirituality and psychism, remarking that dalliance on the astral plane was juvenile.[21] In 1906 Leadbeater was accused of teaching young boys in his care to masturbate, and of enjoining them not to tell their parents. He admitted the advice as a way of relieving sexual tension, but consistently denied other charges involving indecent assault: these were never proved.[22] Besant, who was given confidential information at an early stage by the mother of one of the boys—Helen Dennis, who was Secretary of the ES in America—refused to repudiate him. While she deplored Leadbeater's advice as ill considered, she insisted that one who had stood with her before the Masters was incapable of evil intent. Asked why, in that case, the Masters had not spoken out, she confessed that her occult powers were not sufficiently developed to permit her to retain everything transmitted on the astral plane. It had been the same with HPB in spite of her splendid psychic ability. 'I heard her warmly invite Oscar Wilde to come into the Theosophical Society at the very time when, as afterwards proved, he was practising the nameless abominations that landed him in jail', she claimed. If Wilde had accepted Blavatsky's invitation she would have been condemned, as Annie now was, for condoning immorality.[23]

Her friends seized on this very point.

> Can you not realize [one of them wrote] that if by any chance this affair and your action in it got really into public knowledge then you are ruined. You could never appear on a public platform in England nor could the Theosophical Society *officially* support you. You must know that you have never really lived down your 'Law of Population' and the hostility you raised then would be ten times intensified now by this present line of conduct.[24]

Annie remained unmoved, so a note—partly in cipher—by Leadbeater to one of the boys was put before her. It was hoped this would finally destroy her confidence that Leadbeater's motive was benign. This was the crux of a hidden conflict that was devastating those who knew of it. How far was Besant aware of the significance of CWL's teaching?, they asked. She knew of the claim that the practice of 'self-abuse' led to an increase in physical well-being. Did she know that it was also supposed to intensify the capacity for psychic experience? Was she aware of the homosexual nature of the practices?

It was put to her that, in pursuit of an explanation as to why she claimed Leadbeater's motives were 'good', American members were saying that he had some method of raising and activitating the *Kundalini* (psychic energy) forces which he *and she* understood, while those who opposed it were blinded by ignorance; that far from causing harm, this practice conferred occult powers, clairvoyance, for example.[25] Whose side was Annie on?, a furious Mrs Dennis asked, in what she described as the conflict between Theosophical ideals of thought and sex purity and 'phallicism pure and simple'.[26]

Leadbeater's note was addressed to:

My own darling boy ... I am glad to hear of the rapid growth and of the strength of the results. Twice a week is permissible but you will soon discover what brings the best effect ... The meaning of the sign ☉ is urethra. Spontaneous manifestations are undesirable and should be discouraged. If it comes without help he needs rubbing more often but not too often or he will not come well. Does that happen when you are asleep? Tell me fully. Glad sensation is so pleasant. Thousand kisses darling.[27]

Annie's reaction to this thunderbolt came in a confidential letter to various officals of the ES. She remarked that her first inclination was to proclaim that conventional ideas of morality did not bind the occultist, and, as a disciple, Leadbeater was not amenable to their criticism. 'The old warrior and martyr spirit rose up in me and I could joyfully stand beside him against the public in the pride of the occultist.' Conscience, however, forced her to admit that she had blundered. The kind of teaching Leadbeater had given young boys without their parents' knowledge perverted the sex impulse implanted in man for the preservation of the race; degraded the idea of marriage, of fatherhood and motherhood; befouled the imagination; polluted the emotions;

and undermined health. In short, she said, it was 'earthly, sensual, devilish.'[28]

By the time she had drawn up this capitulation, Colonel Olcott had already been to London, where he presided over a 'judicial enquiry' into the affair. Leadbeater, who appeared before it, freely answered questions but without any expression of regret. It emerged that some 'indicative action', i.e. touch, had also been involved. He was not expelled but his resignation from the Society was accepted.

That concluded the first stage of the affair. Its chief casualty was Henry Olcott, whose spirits were low and whose health was deteriorating. His condition was aggravated by a fall on the journey home. Shortly after reaching Adyar he took to his bed in the expectation that he had not long to live. One of those who watched most patiently beside him was Annie Besant. She was present, therefore, when, shortly before his death on 17 February 1909 at the age of 77, the Masters appeared. According to her testimony, which was corroborated by another person present, they instructed Olcott to nominate her to succeed him.[29] Among the officials and members of the Society who were disturbed by this manifestation was the Vice-President, A. P. Sinnett. He objected to the exoteric and esoteric sections coming under the direction of a single person. However, when it came to the vote by individual members, Mrs Annie Besant was found to have been elected President by a large majority. Shortly after she took office, Sinnett was further alarmed by a report that Besant had attended a meeting in favour of *swadeshi* (the use of only native Indian products) in Bombay Town Hall where, to his horror, she had been greeted by the nationalist slogan of '*Bande Mataram*'. Sinnett could hardly believe his newly elected Chief had lent her support to what he considered only thinly veiled sedition.[30]

Annie had been liberated by Olcott's death from her undertaking not to engage in politics at a moment when the British had good cause for alarm. During the past decade there had been a revolution in the attitude of Indians towards their rulers. In 1908 Tilak had become a national hero when he had been given the longest prison sentence yet of any agitator. His offence was extremely dangerous: he had succeeded in connecting a political demand—freedom for India—to a religious movement. His base was at Poona and in the Bombay hinterland where, besides the

Cow Protection Society, he had introduced celebration of Shivaji, the warrior hero of the old independent Maharashtra state, and threatened to export its fervour to other parts of India.[31] In the Punjab, Lala Lajpat Rai; in Central India, G. S. Khaparde; in Bengal, Bepin Chandra Pal, risked arrest for their attempts to stir up feeling. These so-called Extremists were suspected of secret links to terrorists whose methods included bombs and murder.[32] The effect they desired to achieve was assisted by the policy of the Viceroy who was in office from 1898 to 1905. Annie Besant described Lord Curzon's personal arrogance as the very best encouragement to nationalist aspirations.

Even before Olcott died she joined with Gokhale, Naoroji, and other leaders of the Moderates—so-called—to oppose the Curzon Government's introduction of stricter control over educational establishments. These were suspected, rightly, of being hotbeds of disaffection. However, in 1905 she earned the disapproval of the Extremists by ordering the staff and boys of the Central Hindu College not to join in the day of mourning called to protest against Curzon's decision to partition Bengal. Her aim was to stand alone; to be seen from all sides as an independent force; this required nerve as well as skill. As she said to Esther Bright, 'My old political work is now full of help for me.'[33]

In April 1908 she had to be called to account by the Commissioner of Benares for articles in praise of Lala Lajpat Rai which appeared in the CHC Magazine (circulation 15,000, editor A. Besant), and for inviting him to lecture to the boys. Lajpat Rai wanted what she wanted, Besant explained; a self-governing Indian nation within the British Empire.[34] The incident illustrates the peculiar difficulty the Government of India faced in dealing with a maverick like Besant. The viceregal Government considered Lajpat Rai dangerous enough to warrant deportation but had to rescind the order at the behest of the Secretary of State who was responding to pressure by MPs. 'The weakest thing in the world to do is to repress a little. The House of Commons would never stand the amount of repression necessary to make the present unrest inaudible.'[35] This pessimistic assessment of a situation Besant increasingly exploited for her own ends was by James Dunlop Smith, private secretary to the earl of Minto, Curzon's successor as Viceroy.

Relations between Dunlop Smith and Mrs Annie Besant were extremely cordial. She hoped, through him, to enlist Minto's support for her scheme to turn the CHC into an independent national university of India. And she was angling for the College to receive the accolade of a visit from the Viceroy and Vicereine. That staunch republican Charles Bradlaugh would not have recognized Besant in the ardent royalist she had now become. Her ostensible reason was one shared by many experienced Anglo-Indians who were concerned to see Indians achieve political progress. If that was to happen the multitude had to be given an obligation higher than religion, or their kindred. The monarchy was just such an idea; simple but compelling and, in its Indian presence, splendid.[36]

Before receiving her, Minto took advice from Francis Younghusband, who was both a mystic with a profound knowledge of Eastern faiths, and a high official of the Raj. After his attention had been drawn to the *Sanatama Dharma*, Younghusband kept a fascinated eye on Besant, went to her lectures in places as far apart as Simla, Srinagar, and Indore, and read her books. He told Minto he thought she was sincere but very dangerous. By her praise of their ancient knowledge she was encouraging Hindus to believe themselves superior to Europeans, which, in the circumstances, was an attitude that might endanger peace. Younghusband saw her as a neurotic and partially educated woman, and perceived the real object at the back of all her efforts: 'to turn all her followers into Theosophists'. While this was carefully wrapped up and sugared over, 'she thinks Theosophy is the universal real religion and that she is its high priestess'. He recommended the Viceroy to commit himself to nothing more than kind words of sympathy with her 'very real and genuine desire to improve the lot of Indians'.[37]

Annie was not to be diverted by platitudes. In February 1910 she mounted a sensational attack upon the British for racial prejudice. Her 'Appeal', which was circulated with the Central Hindu College Magazine, was ostensibly provoked by an incident which occurred on a train when a venerable Theosophist was turned out of a first-class carriage by an Englishman. Reminding the Viceroy to whom it was addressed that its author had direct access to 1,000 young men and was known to tens of thousands more, the Appeal called on Minto to check the hatred that was dividing community from community as a result of thoughtless acts. 'Speak strongly as only you can do to these lower English who are destroying your work and undermining the British Empire. Bid your officials to guard your Indian children and shield them from Outrage and from wrong.'[38]

The tremendous uproar that greeted this was prolonged when Besant publicly expressed her fear, on the one hand, that the Lieutenant-Governor would deport her; on the other, that her life was in danger from irate Anglo-Indians. Gokhale was moved to intervene. He was told that, while the official view was that Besant was being very unfair to the great majority of English, no action was contemplated. 'The hearts of millions and millions of Indians have been with you', Gokhale told her; 'the insults and vexations you have had to endure will be repaid a hundredfold by the gratitude of the Indians for the way you have spoken out.'[39] Other equally loyal Indians accused her of incitement to racial hatred. And 'You write as you speak—strongly', the Governor of Madras told her. 'I do not mind how keenly my peccant compatriot is made to smart. I am thinking how your words may be used to stir the impressionable youths of India to revolt.'[40]

In a confidential letter to the Viceroy apologizing for the 'bother' she had caused, Besant explained that she had been moved to issue the Appeal to soothe the spirits of the College boys who were seething with resentment at their treatment by officers of the CID when they had searched the College for seditious literature. She had also wanted to forestall a possibly violent outburst by members of Benares City Council whom the Commissioner had insulted. She had chosen the example of the train as the least provocative in the circumstances.[41] Her letters to Minto about the affair are a peculiarly Besantine mixture of innocence, cunning, and bravado. They suggest that not only did she know precisely what she was about—one of her objects was to restore her credit with the Extremists—but was hugely enjoying herself. 'Naughty Annie', as the Governor of Bombay remarked.

'The higher officials do not know how much the Indians suffer from the brutality of the lower types of Englishmen', Besant told the Viceroy; 'I have seen it and I know it from intimate friends, and the stories run through India and breed hatred.'[42] Dunlop Smith agreed. The Viceroy and he had often discussed and argued about whether any public appeal to the community would have any effect.' We invariably agreed that any such appeal would do more harm than good in the quarter in which it was meant',[43] he told her. While Minto acknowledged that Besant's intentions were honest, he was disturbed by her over-enthusiasm. 'I do not want to drive her into the enemy's camp', he remarked apropos of her request for a visit, 'but at the same time I can't possibly support her administration [of the Central Hindu College] as it seems to exist.'[44]

Members of that administration were summoned to hear the Commissioner of Benares read a letter from Sir John Hewitt, the current Lieutenant-Governor of the United Provinces. He condemned the Appeal in the strongest terms and took special exception to the fact that it had been circulated with the magazine.[45] The maharaja of Benares was also very dubious about the timing, and, with other trustees, questioned why the matter had not been raised in private. Among those present to receive the official rebuke were Bhagavan Das as College Secretary, and George Arundale, Leadbeater's former pupil, who was now the College Principal. Annie was away.

When she returned to Benares she caused another sensation by going over Hewitt's head to protest to the Viceroy that the College was being punished for something that concerned only her; that is, 'because Mr Leadbeater is my friend'. If Minto's visit to the College could not go ahead for that reason, she would resign her connection with the establishment forthwith.[46]

If, as Hewitt described it, the Appeal was an act of incredible folly, how much more so was Besant's reinstatement of Leadbeater, whose offence was public knowledge? He had been living at Adyar since the beginning of the previous year. It was not that she was ignorant of the distaste he inspired; she expressed regret to Lord Minto that Lady Minto had been obliged to hear his name mentioned in her presence.[47] It might be argued that Annie believed CWL not guilty of anything worse than indiscreet advice, and that her 'warrior and martyr spirit' rejoiced at the prospect of defending him, as, for example, she had defended herself and Charles Bradlaugh against the imputation that they indulged in the kind of sexual licence described in the *Elements of Social Science*.[48] The fact that no further charges were ever proved against Leadbeater lends credence to this theory. But how then to explain Annie's violent reaction to the revelations of the cipher note which she characterized as inimical to family life? Loyalty

must be the first explanation; whatever her friends did to her—
and many of them failed her—she was steadfast in her regard for
them; not least for Leadbeater.

The evidence available, fragmentary though some of it is, points
to a further explanation: that Leadbeater was essential to the
success of her grand scheme. Other writers have suggested that
she held on to him for his psychic power, which she could only
pretend to share.[49] That does not appear to be in character. All
kinds of people—politicians, lawyers, teachers, clerics—testified
to her absolute sincerity. Her behaviour was throughout consistent with a belief in her power as seer and prophet.

Long before 1907 she began to reshape Theosophy with the
help, rather than under the influence, of Leadbeater.[50] As in
every millennarian sect, the expectation of her followers had to
be kept alive. Many faiths, including Christianity and Hinduism,
are familiar with the concept of the avatar, the teacher who brings
rewards for the faithful, and hope for the rest: Blavatsky herself
endorsed it. While Olcott was still alive Besant did no more in
public than lecture on the subject; how far she went in private,
what promises she made to members of the ES, are not known.
But it is this writer's belief that Leadbeater was to be tutor of
the, as yet, unknown child on whom the mantle was to fall, who
would be discovered when the time was right—rather as the
successor to the Dalai Lama is discovered; and that when Leadbeater's public disgrace threatened to upset her intention, she
brushed it aside as, for example, she overcame Bradlaugh's strong
objection to something else she had set her mind and heart on
doing: the publication of the Knowlton pamphlet.[51]

As soon as she became President of the Theosophical Society,
Besant took steps to have Leadbeater reinstated, whereby she
caused many members, among them Blavatsky's former secretary,
G. R. S. Mead and her old friend Herbert Burrows, to depart
in sorrow and disgust.[52] For his part CWL seems to have
promised support for her activities in exchange for her protection,
a bargain which left him considerable room for manœuvre, not
least in his privately expressed opinion of her, which was sometimes highly critical—apart from anything else, Leadbeater did
not like women.[53]

Shortly after he arrived at Adyar at the beginning of 1909,
there occurred the famous encounter on the sea-shore with two

brahmin boys who spoke only Telegu. They were Jiddu Krishnamurti, aged 14, and his younger brother Nityananda. Their mother had died recently, and their father Narianiah was struggling to look after them and his other children in a house just outside the Adyar estate, where he was an employee. CWL took the two boys under his wing; soon they went to live with him in a bungalow on the estate. At the time he was engaged on research into the past lives of members of the Society. His prestige as an occultist was so great that a favourable account in the book he compiled with Besant, *Man: Whence, How and Whither*, conferred standing on the subject, while to be left out produced proportionate dismay. On looking into the past lives of his new protégés, Leadbeater perceived that the elder boy's destiny was nothing less than to be the Vehicle on earth of the coming avatar, the World Teacher, whom he identified as the Lord Maitreya. The boy was to undergo training for this messianic mission; meanwhile his identity was to be protected by referring to him as Alcyone.[54]

When Annie Besant returned from Europe in November 1909, she warned members of the ES to observe the greatest secrecy with regard to Alcyone.[55] In March 1910 she obtained a letter from Narianiah appointing her guardian of both his sons. Though she was absent more often than not while CWL was 'training' Krishnamurti, there was no doubt who was in control. It was by her authority, for example, that the 15-year-old boy was locked with Leadbeater into her rooms for two days and nights, without food, to undergo his 'initiation', at the end of which Leadbeater pronounced him to have been accepted by the Masters. One result of this experience was that the boy himself now looked into the past and recounted what he saw in a book—in perfect English. *The Lives of Alcyone* was also supposed to be a secret.

As one of her most intimate and trusted colleagues, Bhagavan Das knew about the appearance of the boy. His alarm at the way things were going turned to consternation when, contrary to her own strict orders, Annie made a public reference to the *Lives of Alcyone* in the autumn of 1910. His attempts to reason with her produced 'reckless, incoherent, contradictory statements' which seemed to him characteristic of the change that had come over her since she had become President. Bhagavan Das decided that, of the two different qualities in Annie's remarkable nature, higher and lower—altruism and egotism—the former, the wish to serve mankind, had prevailed from 1894, when he first knew her, to 1907. After she moved to Adyar and became President (under very peculiar circumstances, he thought), she fell victim to 'her wish to be regarded as a Saviour'.[56]

CHAPTER 23

'BANDE MATARAM'

'INDIANS who know her tell me that she is imperious and even unrestrained in temper, vain, restless, and ambitious, some of which qualities are not altogether surprising in a woman of nearly 70, of commanding intellectual powers and wonderful bodily vigour.' Thus Lord Pentland, Governor of Madras, who was to be a victim of Besant's political vigour. She was a most capable organizer, he told the Secretary of State, thoroughly versed in all methods of Western political agitation, and a forceful speaker. 'She is an Irishwoman', he added as a further matter of concern.[1] In the years leading up to, and during, the First World War events in Ireland had a marked effect on Annie Besant's attitude to India. While progress was being made in England towards a parliamentary settlement of the Irish problem she was encouraged to believe that India's turn would shortly come. When Home Rule foundered on the rock of Ulster, and Sinn Fein appeared upon the scene, her antagonism to the Raj grew correspondingly fierce.

While peace lasted, the fact that she divided her time between England and India gave credence to her claim to interpret one country to the other. Her Indian audiences heard of her important English contacts, and vice versa. In 1910 Theosophy brought her a new disciple, the 36-year-old Lady Emily Lutyens, whose special interest for Besant was that her father the earl of Lytton had been Viceroy of India. In 1912 her husband Edwin Lutyens was appointed architect to the new imperial capital at Delhi. Lady Emily conceived an unusually intense affection for Krishnamurti, whom she met when Annie brought him and his brother to England for the first time in 1911.[2] Two of her friends, whom she recruited into Theosophy, gave the boys, and Annie as their guardian, much valuable support, financial and otherwise. They were an American, Miss Mary Dodge, and Muriel, Countess de la Warr. The latter's father Lord Brassey, who had been a Liberal junior Minister, was now Lord Warden of the Cinque Ports. That this ceremonial figure took the chair at lectures by Besant raised a good many eyebrows.[3]

She did good service in the cause of Indian nationalism at these lectures by her insistence that the people of the subcontinent were already capable of governing themselves; it was a time when even the most liberally minded among her audience took it for granted that years must pass before they could cope with political responsibility. Another of her arrows that shortly found its mark was the threat of a cataclysmic war in Europe, which had haunted Bradlaugh thirty years before.[4] When natural justice demanded that Indians should be free, how much more pressing would the case be when—not if—her sons were called to arms to defend the Empire, Besant demanded.

Where India was concerned, in spite of its vastness, the number of people engaged in politics was so small she was known to virtually all of them and her influence was disproportionately great. Before she came upon the scene, one of her Indian colleagues remarked, the history of politics was of debating societies in whose chambers small groups of people deliberated over resolutions which were admirably drawn up, but so suave, so reasonable, that no one bothered to pursue them. 'To none of us had come the vision of going to the villages, of speaking to the people at large, of making them realize what they could do, and what it was their duty to do.'[5] Besant taught methods she had learned from Bradlaugh and, in so doing, prepared the way for Gandhi.

It was in 1913 that she launched a determined bid for power, giving two separate reasons for the sudden raising of her sights. One was Theosophical. A member of the Hierarchy, the Rishi Agastya, had directed her to work for Indian self-government, in order to protect the Empire. Since all her energy would be devoted to this cause, her occult powers would be suspended, and she would depend on Leadbeater (who did not approve of her political activities) to transmit messages from the astral plane.[6]

The layman's explanation of her motive, which she gave to the Viceroy, Lord Chelmsford, and the Secretary of State for India, Edwin Montagu, when she met them in 1917, was that, at the time of the anti-partition unrest in Bengal, her work among the boys of the Central Hindu College was made exceedingly difficult by the 'doubts, suspicions, and questionings' of the Government of India. At the same time she was seeing something of the Moderate Congress leader G. N. Gokhale and knew many other members so that her mind was turning in the direction of joining it. 'At last I became convinced that nothing but a change in the Government ... was of any use, and no educational work for the uplift of the people was possible as things were; and that drove me to consider the whole question of the Government of India and to finally resolve to take up definitely political work.'[7] She joined the Indian National Congress in 1913.

Some observers were convinced that her timing was adjusted to distract attention from the snub administered to her by the trustees of her beloved Central Hindu College. In July 1913 they refused a request from the Principal, George Arundale, and over a hundred of her supporters on the teaching staff, for the dismissal of the College Secretary, her old friend Bhagavan Das. His offence was openly to criticize the cult of personality which, for the last three years, had been urged upon the very boys about whose welfare Besant expressed concern to Montagu and Chelmsford. In 1910 students and teachers at the CHC, who were already encouraged to venerate AB, were pressed to commit themselves to exaggerated expressions of esteem for the boy Alcyone. According to Bhagavan Das, 'fuss of the most absurd and mischievous kind' attended the introduction of the Order of the Rising Sun, whose members were taught to expect the Coming of the World Teacher in the person of Krishnamurti/Alcyone. The effect of what he called perverted and gushing language, of emotionalism run riot, was to promote friction between Besant's supporters and those professors and teachers who, endeavouring to keep aloof, saw their authority undermined. When, in 1912, the trustees finally resolved that Hindu boys ought not to be exposed to Krishnamurti worship, Besant disbanded the Order of the Rising Sun with the excuse that, in her absence from Benares, George Arundale had been over-zealous in promoting it.

However, she lost no time in introducing the Order of the Star in the East, whose head was Krishnamurti; she was its Protector. When Theosophists at large pressed her to explain, she annoyed many of them into leaving by declaring that the OSE was nothing to do with Theosophy, but was the embryo of an entirely new religion. Rudolf Steiner, the philosopher and mystic, who had built up the German Section of Theosophy, was one of those who departed at this time, taking his followers with him into the Anthroposophical Society.[9] Another was Nehru's former tutor F. T. Brooks, who published virulent attacks on what he called neo-Theosophy and its props—orders, sashes, jewels, the ritual of co-masonry—everything Besant and Leadbeater toyed with over the years. To Bhagavan Das neo-Theosophy was a sad anticlimax to a noble belief in universal brotherhood.[10] When the trustees retained him as College Secretary, Arundale and many Besant supporters on the staff resigned.

Their departure from Benares opened the way to a compromise over the vexed question of the future of the Central Hindu College. At the same time as Besant had petitioned the king for a Charter for a National University of India, a leading member of the Indian National Congress, Pandit Mohun Malaviya, produced a scheme for a purely Hindu institution of higher education. Government sanction for this seemed beyond dispute in the light of its recent approval of a Muslim College at Aligarh. The desire to avoid intercommunal jealousy, joined to the need for support from every quarter arising from the outbreak of the War in Europe, made the Government very anxious not to refuse Hindus. However, for reasons of finance, any scheme had to be based on the existing College at Benares. The British officials responsible for the decision argued that any establishment connected with Mrs Besant was in danger of becoming a cheap degree factory, in which education would take second place to political propaganda. Her defeat over the dismissal of Bhagavan Das opened the way to a solution. Malaviya's plan was preferred to hers and, with apparent good grace, she accepted an invitation to join the Court of the new Hindu University of Benares. The firm hope was that her appearances would be infrequent.[11]

Meanwhile in Madras 1913 was the year in which Besant's association with the man the Head of the Criminal Investigation Department called 'the disgusting Mr Leadbeater'[12] was constantly in the newspapers, especially the *Hindu*. It printed long extracts from the two hearings—in the High Court, and on appeal—of an action by Narianiah against Besant for the return of his two sons, her wards Krishnamurti and Nityananda.[13]

Since 1911, when Annie first took them to England for the Coronation summer, the boys had spent more time abroad than at home in India. She saw to it that their whereabouts were unknown during the Narianiah case, having convinced herself that he, or those she accused of manipulating him, would kidnap them. Narianiah's complaint was of the undesirability of a man with Leadbeater's reputation having charge of his sons; and of his ignorance, at the time he had given them into Besant's care, of her intention regarding the Coming of the World Teacher. She chose to represent the case as an attempt to destroy her by various enemies, among whom she noted: orthodox Hindus scheming to get control of the new University at Benares; Tilak and his more extreme associates, to whom she was an impediment in their effort for outright independence for India; Mrs Katharine Tingley, the leader after W. Q. Judge of a large and hostile faction of American Theosophists.[14] She lost the case in the High Court and on appeal, and was ordered to restore the boys to their father no later than the summer of 1913. She claimed as victory the fact that Narianiah was not awarded costs in the High Court, the reason being that he, or his advisers, had dragged in allegations about Leadbeater's indecent conduct with Krishnamurti. While they took up a good deal of the court's time—and even more space in the *Hindu*—these could not be substantiated.

The order to give up the boys was a painful reminder of the humiliation visited upon Annie in the case of her own daughter Mabel by the Master of the Rolls.[15] Nevertheless she was not deterred from returning to the English legal system, this time at its most august. As things stood, after she lost her appeal in Madras, if she did not bring the boys back from England and surrender them to their father, she risked imprisonment for contempt each time she stayed at Adyar. She applied for, and was given, leave to appeal to the Judicial Committee of the Privy Council.

The Lord Chancellor who presided was an old acquaintance and sparring partner from Fabian Society days, Lord Haldane. He and the other judges of the Committee decided that the case turned upon the whereabouts of the boys. Since they were in England and of sufficient age to be consulted (Krishnamurti was a few weeks short of 18, Nityananda was 16), Besant could not be expected to comply with the order to return them against their will. Their Lordships expressed no opinion as to what the interests of the 'children' were. They intimated that it remained open to Narianiah to apply for custody in the High Court in England. He did not. Besant had won, though the victory was by a technicality and not, as she liked to claim, a vindication. Although the Judicial Committee did not explore the nature of her relationship with the boys, the judges were not left in ignorance about the most important aspect: counsel for Narianiah complained that the boys were not properly represented: they and everyone about them were under the very powerful influence of Mrs Besant, whose word was law unto Theosophists. 'Anything the minors would say would be really what Mrs Besant says through another mouth.' The Lord Chancellor assured him that, if there were further proceedings, the court in question would make a point of hearing them.[16]

They were never heard on their own account. There are few details of what Krishna and Nitya (as they were called) thought about their difficult and lonely adolescence, over which Annie presided as a loving, but authoritarian, and mostly absent figure. She, who made so much of her delight in the Hindu way of life, chose to have the boys transformed into English gentlemen. From their arrival to stay permanently in England, all through the War, they were kept to their books in a series of often remote, rented houses. As tutor for some of the time, Arundale received his instructions from 'Brother Charles' (Leadbeater), but showed a growing inclination to explore, on his own account, the astral plane. The aim of what turned out to be a test of everyone's endurance was to get the boys into Oxford. Annie's authority had often to be invoked against Krishna's failure to come up to expectation. He was described as vague, vacant, inattentive, dreamy. His letters to Annie were full of assurances that he would try to do better; apparently his attention was never concentrated for long enough to permit him to pass examinations. Hers to him were full of motherly endearments and spiritual exhortation, but, as time went by, she began to wonder if he was, in fact, the Vehicle.[17]

It seems not to have occurred to her that there was anything remotely cruel in putting up a half-grown boy to draw the crowds for the most spectacular performance of her career, nor in translating him to an alien country where the contrast of Norfolk jacket with long black glossy hair, worn christ-like, parted in the middle, made him an object of derision. As yet the great majority of Englishmen were unaccustomed to the sight of persons of his race, and untutored in their prejudice against his colour. No doubt Krishna's remarkable detachment helped him to cope with the constant surveillance; the bickering; the jealous rivalry among his attendants for occult favours. He said later that he believed in what Besant and Leadbeater told him about his spiritual mission for some years; then, as will be seen, he shook himself free. Beyond that there is little to be learned, for he always insisted that he retained no memory of his tutelage. Nor did Nityananda leave a record. His life was one of promise unfulfilled: alert, intelligent, and amiable, he died from tuberculosis in 1925.

The cost of supporting the boys, who learned to recognize excellence whenever they encountered it, was largely met by Miss Dodge. Lady de la Warr helped to supervise and entertain them, and Lady Emily Lutyens gave them, unreservedly, her love. Annie had to send hers by post from India, where she was obliged to stay for the duration of the War, in the knowledge that her political behaviour so distressed the Government of India that, once she departed, it would not allow her to return.[18]

The first step in her decision to take a very active part in Indian politics was to found a new weekly paper, whose role she underlined by giving it the name of the journal of the Socialist League which had devoted much space to her activities in the late 1880s. The new *Commonweal* was barely three months old when she departed for her last annual visit to England before the War, leaving B. P. Wadia in editorial charge. He was one of her most able and trusted lieutenants, Manager of the Theosophical Publishing House at Adyar, a wealthy young Parsee. Like *Our Corner* the *Commonweal* was intended to instruct; Besant told Edwin Montagu its aim was to direct young men away from 'violent and showy methods' by which she meant, principally, bombs. It gave favourable publicity to Gandhi's campaign of passive resistance in South Africa, and reprinted articles by noted English pacifists, among whom were Henry Brailsford, Henry Nevinson, and Philip Snowden.[19] A different kind of contributor was the self-styled revolutionary, the Bengali nationalist leader Bepin Chandra Pal. At the height of the violence against Curzon's partition of Bengal, Pal had edited a newspaper called *New*

India.[20] When Besant returned to Madras in July 1914—much earlier than usual—she found, apparently to her surprise, that Wadia had bought a daily paper, the Madras *Standard*, whose circulation was moribund but which had the necessary licence to publish. Within a month it was on sale in the city and the *mofussil*,[21] renamed *New India*, with Besant as editor in full cry after the British for their refusal to concede political reform for India during the War. The slogan she adopted, 'England's need is India's opportunity', was in sharp contrast to the truce observed by politicians, and to the attitude of thousands of ordinary people who were proud to offer their services to the Empire in its hour of need.

New India was hypercritical of the Raj, often strident: in the opinion of many Anglo-Indians downright poisonous. Besant took full advantage of the means of publicity at her disposal. The Government of Madras was particularly concerned by the circulation of articles reprinted from *New India* in the form of cheap political pamphlets with titles like *Self Government for India*, *The Separation of Judicial and Political Functions*, *The India Council under the Congress Flag*, couched in simple terms which lent themselves to copying. As Governor, Pentland was alarmed by Besant's constant attempts to influence college and high school boys by opening the columns of *New India* to criticism of their teachers and of the hopelessness of their prospects. Besant was invariably, sometimes blindly, the students' advocate, he complained. In a letter to the Secretary of State for India he condemned her use of the newspaper to recruit members to the various organizations she now founded: organizations like the Young Men's Indian Association which she set up in opposition to the Young Men's Christian Association, which she attacked in *New India* as anti-Hindu, therefore anti-India. But what worried him more than anything was her importation of the Irish nationalist poet James Cousins, to be literary sub-editor of *New India*.[22]

He was a 32-year-old Belfast man who was more at home in Dublin where, as poet, he ranked next to Yeats and AE. He joined the Theosophical Society in 1902, excited by the idea that Ireland was to play a leading role in Europe's spiritual regeneration. The Government of Madras suspected him of links to the Irish trade union activist James Larkin. Almost equally disturbing to those in charge of public order was Cousins's wife Margaret, who accompanied him to Madras in 1915. She was a militant suffragette, a veteran of the hunger strike, who had spent time in Holloway Gaol for throwing bits of flowerpot at the windows of 10 Downing Street, and in Mountjoy Prison for a similar assault on Dublin Castle.[23]

In October 1915 Pentland asked the Viceroy, Lord Hardinge, for permission to deport Mrs Besant. The strenuous campaign she was conducting through the *Commonweal* and *New India*, for what she now referred to as Home Rule, made her an All India problem, he ventured optimistically. Hardinge disagreed; her nuisance factor was confined to Madras, whose Government had the necessary means of restraining her; he recommended Pentland to use the Press Act against *New India*. Under this a newspaper could be required to deposit large sums of money as a guarantee of good behaviour. Pentland hesitated, fearing that it would only enhance Besant's standing at the forthcoming Indian National Congress.[24] He tried to remonstrate with her instead. She was delighted at the opportunity of expressing defiance to his face. 'I told Lord Pentland that I would work with him so far as the War was concerned, but in those matters where I disagreed with him I should have to speak out, and I left him saying ... whether he stopped the paper or interned me, I should keep on along the lines I had marked out because I thought that necessary for the well-being of the Empire.'[25]

She took care to send copies of *New India* to the Lord Chancellor, with whom she had kept in touch by the very bold method of writing a letter thanking him nicely for the verdict in the Narianiah case. '*New India* stands for Indian self government within the Empire as I worked with Mr Gokhale', she told Haldane; 'it is too outspoken to please our bureaucracy here and I am dubbed "dangerous". But we who know the real danger growing beneath the surface, of the anger against the Government here because of its antagonism to liberty and the resolution to win freedom ... are a safety valve rather than a danger.'[26]

As she did here Besant often used Gokhale's name to lend credence to her actions. His premature death in February 1915 left a vacancy she aspired to fill. At that time, Gandhi had just returned to India after many years away and was engaged in the exercise recommended to him by his mentor, Gokhale, of re-acquainting himself with the country and its people. Tilak had come out of prison in 1914 and was rallying his supporters.[27] Besant, who had so recently publicly identified him as a personal enemy, now lost no opportunity of saluting him in *New India* as a great patriot and leader. Like a busy sheep-dog she was rounding up the various factions with the intention of leading from the front. Officials of the Raj looked on, undecided whether she was using Indian politicians for her purposes, 'or they, playing on her vanity and feelings, are using her'.[28]

In May 1916 an article by James Cousins for *New India* praising the leaders of the Easter Rising in Dublin caused such a storm of rage Besant was obliged to dismiss him. The Government of Madras ordered the paper to deposit sureties, a move Besant promptly challenged in the courts.[29] Their verdict was that this and other articles in *New India* had brought the Government into hatred and contempt. As Secretary of State, Austen Chamberlain was called upon to defend the Madras Government in the House of Commons. He described Mrs Besant as dangerous, and privately recommended Heads of Government in India to keep away from her.[30]

The mood in parts of the subcontinent was beginning to change. 'Tilak's Home Rule policy is the only live policy before the country, and is being swallowed greedily', the Governor of Bombay reported in July.[31] In April Tilak had begun to organize local groups in and around Bombay to work for Home Rule. In so doing he pre-empted Besant's idea, which she had put before the Congress in December 1915. Then the very words Home Rule had alarmed its members, who were most reluctant to take action at a critical period of the War. The idea was relegated to the standing Committee.[32] Besant warned them that if nothing was done by the end of August 1916 she would proceed unilaterally. Shortly before her deadline expired, she attended the first meeting of the Court of the Hindu University at Benares. The Lieutenant-Governor of the United Provinces was there too, taking the opportunity to observe the Hindu leaders. Pandit Mohun Malaviya told him that they were losing control of the younger men, and that a split had developed over who was to be President of the Congress in December. 'The younger men find Mrs Besant spicy, and now in antagonism to the Government, and are working hard for her election', Sir James Meston reported to the Viceroy.

Malaviya and the orthodox [Hindus] utterly resent the idea of an unbeliever at their head, and they resent even more the refusal of the younger men to obey their dictation. Malaviya had all sorts of complaints about the unconstitutional behaviour of Mrs Besant... If Tilak comes [to the Congress] there will be trouble.[33]

Tilak went to the Congress at Lucknow in December. He and his Extremists [so-called] were readmitted, and a pact for co-operation between the Congress and the Muslim League was concluded.[34] Annie Besant had prepared the way for Tilak's return. Since 1907, when he had stormed out, the Moderates had been in control. In 1914 Besant moved an amendment widening the rules of admission in a way that was acceptable to Tilak as the first step to reconciliation. She maintained that this was Gokhale's wish; in fact he tried to stop her. He thought that, while he was in prison, Tilak had come closer to his own idea, which was 'association with Government where possible, and opposition to it if necessary'. He was distressed to find that Tilak was determined to take over the Congress, when he would substitute opposition pure and simple; 'in short a policy of Irish obstruction', Gokhale warned Besant. In spite of this she moved the amendment, telling the Moderate leader, who was within weeks of death, that she could not go back on a pledge she had given Tilak.[35]

She lost her bid for the Presidency at Lucknow. As soon as the Congress was over, she intensified her anti-government propaganda, courting internment in the expectation—which was realized—that to be perceived as a martyr in the cause of Home Rule must propel her into office next time. The Government was in a dilemma. It was fully aware of the pressure for political advance; the terms were under discussion, the clamour for an announcement was well nigh deafening. But there was deadlock in Whitehall, where the moment was considered inopportune. The War was still going badly for the Allies; English public opinion, which was absorbed in winning it, would become impatient and resentful of any distraction from that tremendous effort. These circumstances provoked anxious discussion among Heads of Government in India about the demand for Home Rule. At what point did a legitimate demand for political reform become seditious? When should action be taken against propagandists like Besant to preserve the authority of Government whose hold upon the country was more precarious than it had ever been?[36]

No such doubts worried Annie Besant. She began her Home Rule League on 3 September 1916, as she had told the Congress she would. She was President; George Arundale (whose time as Krishnamurti's tutor had ended in mutual dislike) was Organizing Secretary; B. P. Wadia, Treasurer. Any three persons subscribing one rupee each could form a branch; these were more numerous in those parts of the country where the Theosophical Society already flourished. Besant's favourite rallying cry was that the British took notice of political demands in proportion to the energy with which they were put forward.

By 1917 the language of her newspapers had become so extreme the new Viceroy, Lord Chelmsford, delivered a public rebuke. But his advice to Heads of Government was, for the time being, to deal with Besant as they saw fit.[37] She was already externed from Bombay and the Central Provinces as a result of the furore over *New India*'s praise for the leaders of the Dublin Easter Rising. Pentland had no such easy way out in Madras. He and many others, Indians among them, believed that her conduct was inciting race hatred between Europeans and Indians, something that had been virtually unknown in sleepy Madras. And the Home Rule propaganda was upsetting the non-brahmin majority of its population. Besant's long connection with, and fervently expressed admiration for, a caste that had exploited the Dravidian people in the past caused her political activities to be regarded with suspicion and dislike. A new party, the Justice Party, was formed to oppose Home Rule (one of its leaders, Dr T. M. Nair, was also a vociferous opponent of Besant over the Narianiah case[38]). The issue was worrying; it concerned the British concept of their responsibility as rulers of India. If Indians were given a larger measure of control who would protect the weaker elements?

The decision to intern Besant and her colleagues Arundale and Wadia was apparently precipitated by the death of a policeman in Calcutta, when *New India* came close to depicting the assassins as martyrs in a noble cause.[39] At about the same time it was learned that Besant had been in touch with terrorists in that city. She maintained she had gone to reason with them; she asked the Governor of Bengal to lift surveillance so they might return to constitutional ways.[40] That, if true, was evidence of such extreme *naïveté* it was a danger in itself. Members of Parliament in England, who were very uneasy at the decision, were told that

Mrs Besant had been engaged in political agitation which might become highly dangerous, and even disastrous, to India. The Secretary of State was very reluctant to enter into detail, but insisted that the internment order was preventive; it was not meant as punishment.[41] It seems the Government feared Besant was about to have recourse to measures they would be hard put to control—boycott, strikes, passive resistance.

Annie was given advance notice of her loss of freedom. She took steps to frustrate its purpose. *New India* was sold for a nominal sum and she made a declaration that she was no longer its printer or publisher. It went on coming out with the aid of Tilak's lieutenant N. C. Kelkar and the English editor of the nationalist *Bombay Chronicle*, Benjamin Horniman. As things turned out, the most effective move was the dispatch of a letter to the President of the United States, Woodrow Wilson. To escape the British censorship it was carried from Adyar to Washington by Olcott's former secretary who was now an American citizen.[42] It commanded attention by virtue of its signatory, who had attained high office under the Raj and received a knighthood for his services. Sir Subramania Aiyar was a retired judge of the High Court in Madras. He was also Recording Secretary of the Theosophical Society, and a devoted colleague of Besant. That she was the real author of the affair may be deduced from its resemblance to the action she took in 1910 with her Appeal. The present document was couched in terms that a member of the legal profession would not normally use; terms so intemperate that, when the letter became public knowledge a year later, the Secretary of State for India described it as disgraceful.

Its message was that Wilson and other leaders had been kept in ignorance of the full measure of misrule and oppression by the British in India where officials of an alien race, speaking a foreign tongue, forced their will on its people. While government officials paid themselves exorbitant salaries, they refused education to Indians and deprived the people of their wealth through crushing taxes. Those who uttered patriotic sentiments were cast into prisons so filthy that they risked death from loathsome diseases. The letter, dated 4 June 1917, was handed to the President who sent it, via the British Ambassador in Washington, to the Foreign Office in London.[43] It is not clear what Wilson made of it; he was already known for his dislike of any form of guardianship by one nation of another. Whether he made representations to Lloyd George or not, there followed a sensational change of direction in policy towards India which, three months later, encompassed Besant's release.

Her internment began on 15 June when she left her 'dear rooms at Adyar', over a path strewn with flowers, for the hill station at Ootacamund, one of six alternatives Pentland offered her. 'Ooty' depressed her from the start. 'Not an Indian name did I see on the gate pillars', she wrote in the diary she began to keep (probably with an eye to publication), 'it was all Brown, Jones and Robinson, a colony of foreigners.' The weather was 'horrible': rain and bitter cold instead of the warmth of the plains, a contrast she did not hesitate to bring to the attention of the authorities (it was the hottest time of year in Madras when she was usually in England). She paced the verandah of the bungalow: twenty-four steps each way; eighty turns to get warm. She described her gaolers as men of the characteristic Teutonic type, given to vulgar boasting over a fallen foe; spiteful, inefficient. As a symbol of defiance she hoisted the red and green flag she had designed for the Congress on a pole in the garden and, at night, a lantern.[44]

News of her internment provoked a huge wave of indignation: 'Everyone is on a soap box gesticulating', Lutyens's assistant wrote to him from Delhi.[45] Before the end of June there were twenty-eight protest meetings in Madras, twelve in Bombay. The Home Rule movement was spreading rapidly, the Chief Commissioner of the Central Provinces warned the Viceroy; he had never known such feeling. He thought the people were not so much concerned about Besant as impatient with the Government for its delay in announcing reform. The Extremists scented victory: something must be done.[46] Indian politicians, who had been less than enthusiastic about Besant, now swung in behind Home Rule. The Nehrus, father and son, went into action in the north, while Jinnah took charge in Bombay.[47] Gandhi, who was not yet the force he would soon become, chose to keep his protest private. 'In my humble opinion the internments are a big blunder', he wrote in confidence to the Viceroy's private secretary. India as a whole had not made common cause with Mrs Besant but now she was in a fair way to commanding India's identity with her methods. 'I myself do not like much in Mrs Besant's method. I have not liked the idea of political propaganda being carried on during the War.' But if she were not let out there would be violence; his life was dedicated to avoiding the spread of it, he told Colonel Maffey.[48]

It was not long before Annie was overcome by her old terror of idleness and solitude. 'Have not had the heart to write all these days', her diary recorded on 30 July, 'my vitality seems to be going.' She had given up her 'quarter deck' walks because she was unable to walk fast enough to get warm. Her visitors were alarmed at her depression. When the authorities learned of her deteriorating health they offered to allow her to resume her Theosophical work as long as she forswore politics. Her refusal, described as 'violently worded', emphasized the indivisibility of her Theosophical and political beliefs.[49] Lord Pentland was terrified she might fall really ill; he wanted to send her home if possible, otherwise remove her to somewhere 'very remote'. He told the Viceroy that she and her companions, Arundale and Wadia, had got in touch with workers at a nearby cordite factory; if they succeeded in causing unrest the factory's war effort would be paralysed. Chelmsford was unmoved. The factory was well enough protected. One did not send an old lady of 70 to sea in a monsoon when one did not send one's own womenfolk: what about German submarines? The most he would do was transfer her to Coimbatore where it was warmer.[50]

On 30 July the Congress, acting jointly with the Muslim League, sent the Viceroy a demand for the implementation of their scheme for political reform together with the release of the internees.[51] Besant's diary welcomed the suggestion that, if she was not free by 30 September, the Congress–League would launch passive resistance in her support. She began to work out a programme for this, and to consider how to adapt passive resistance to Indian conditions. Her diary noted that a rumour was circulating among Anglo-Indians to the effect that the appointment in July of Edwin Montagu to replace Austen Chamberlain as Secretary of State for India was a sign of Lloyd George's disapproval of her internment. She gave no credence to it.[52]

Chamberlain had been obliged to resign following a crisis involving the Indian Army in Mesopotamia. Montagu was well known for his love of India; he had caused a stir by his criticism of the Government of India in Parliament, as recently as 12 July (1917). On 20 August he created a sensation with his statement to the House of Commons setting out a new policy (Ministers were at pains to disclose that Chamberlain had helped draw it up). This envisaged the gradual development of self-governing institutions with a view 'to the progressive realisation of responsible government in India as an integral part of the British Empire'. Montagu proposed to make an extended tour of the subcontinent in the autumn.[53]

The news came as a shock to high officials in India, who were not expecting any announcement before the end of the War. When the impending release of Besant and her colleagues was also announced, Lord Pentland felt betrayed, his authority compromised. The only thing that prevented his resignation was his duty to stay at his post in wartime.[54] The official explanation was that, after the announcement of 20 August, the situation had changed so completely there was no longer any good reason for keeping Besant and her friends shut up.[55] Those who had to implement the decision, against their better judgement, found some consolation when Jinnah privately remarked that Besant was not as popular as it appeared, and that if she were left to her own devices she might well bring about her own downfall.[56]

Public opinion in India and in England was outraged. Invited to lunch at Buckingham Palace on the eve of his departure for India, Montagu was obliged to defend the release to a hostile Queen Mary.[57] He also had to face debates in both Houses of Parliament when he was asked to give assurances that he would show no sympathy to the Home Rule League. When Members pressed him to disclose whether the decision originated in Whitehall or in response to a request from the Government of India, his attitude can only be described as cagey. He refused to give a clear answer and held out against requests in both Houses for papers to be laid. While he let it be known that a prior condition of the release was for assurances to be obtained from Besant for good behaviour, he would not disclose the channels used.[58] In fact it was through Annie's close friend Lady de la Warr, whose sister was married to the Governor of Bombay, Lord Willingdon, who was himself a friend of Montagu.[59]

A great surge of love and enthusiasm greeted her when she reappeared on 16 September. Wherever she went she was lifted high above the chanting happy crowd, garlanded, adored. The two streams of her life had come together; she was high priestess of Theosophy and President Elect of the Indian National Congress (a choice that had not been made without fierce disagreement among the members of the selection committee). As soon as she reached Adyar she offered her services in the cause of reconciliation to the Viceroy,[60] an approach that did not commend itself to Chelmsford's advisers. Willingdon telegraphed to urge him not to receive her. Mrs Besant was not a proper person to be treated as a political leader, or admitted to Government counsels, and must not be allowed to pretend that she had been. 'She has acquired commanding influence over some of the most dangerous elements in Indian politics and is rapidly extending it to those hitherto inclined to the moderate side', Willingdon declared. 'She has used it persistently and with great ability and profound talent to discredit Government and inflame animosities.' She was astute enough to preserve some colour of constitutional form in her own utterances, he went on, but she had permitted, if she had not instigated, a campaign of almost unmeasured vilification, which she could have kept under control.[61]

However, Chelmsford acceded to Besant's request for an interview with himself and Montagu, though he was somewhat defensive about it. The fact that Montagu had brought with him Subramania Aiyar's letter to Woodrow Wilson may have had something to do with his decision.[62] Montagu's diary reveals how frustrated he was by the Government of India's attempt to keep him from contact with Indians. Determined to have a word alone with him, Annie (who had met him in Madras in 1913) turned up at his tent pretending not to know where the interview was to take place. During the short car journey to the viceregal lodge, she urged him to accept the invitation which he had already received, to address the Indian National Congress. Montagu replied that he did not want to make his work, already difficult, impossible. He was trying 'to pull the bureaucracy round' and had to avoid public demonstrations. Privately he lamented in his diary that if Lloyd George were in charge he would dash down to the Congress and make them a tremendous oration. Annie did not press him, but told him that India was very glad to have him.

In her white and gold embroidered Indian clothes, with a voice that was one of the most beautiful he had ever heard, Montagu found Annie very impressive. She gave her account of the history of Home Rule; how she felt it necessary to get hold of the young boys; how if the Home Rule policy could be carried out, she was sure they would be diverted from anarchy; how the League was started to do the propaganda for the 'old fashioned' Congress because she found it went to sleep between the yearly meetings. She and Chelmsford had an amiable skirmish about whether, if she went to England, she would be allowed to return. For his part Montagu decided that her activity and her League had stirred India to such an extent it was no longer possible to say political interest was confined to the educated classes. 'If only the Government had kept this old woman on our side. If only she had been well handled from the beginning.'[63]

CHAPTER 24

BULLETS AND BRICKBATS

HER address as President to the thirty-second annual meeting of the Indian National Congress was an event of tremendous importance to Annie Besant. It took place on 26 December before 9,000 people assembled in a vast *pandal*[1] in the centre of Calcutta. She had ransacked libraries in Madras for books over which she pored for days while composing it. It was not her apotheosis; that was reserved for the day India would achieve self-government within the Empire, a day she believed she would be spared to see however long it was in coming; however old she was.[2] This was her consecration, when she would dedicate herself to the Indian people and they would pledge themselves to her. The speech, which took nearly two hours to deliver, was a huge miscalculation; in failing to hold her audience, she placed herself at the mercy of those Indian politicians who were already very suspicious of her ambition.

The speech was many things combined; a historical account of India's spiritual past, heavily biased towards Hinduism; a review of finance and economics in the manner of the Chancellor of the Exchequer; a Speech from the Throne forecasting detailed political reform; a sermon on the need for moral improvement by, as it were, the archbishop of Canterbury. The single-minded purpose which, years before, Francis Younghusband had divined in her was here revealed as an obsession.[3] Those members of the Esoteric Section who were present knew that she came before them as the Manu, the Lawgiver; the rest could only wonder.

The programme for India's future which she outlined there owed everything to her belief that, like individuals, nations had characters formed over countless generations, to which they must be faithful in order to fulfil their higher destiny. Thus, in working out the structure of a future government, India must have regard for her ancient institutions, like the village council, the panchayat, and be mindful of her ancient rulers who were royal and semi-divine, whose high moral standing raised them above all others and sanctioned their use of power. Besant's long infatuation with the virtuous caste system described in the Book of Manu caused her to reject democracy.[4]

The totalitarian nature of her remarks was underlined when she addressed her own function as President. 'I cannot promise to please you always but I can promise to strive my best to serve the Nation, as I judge of service. I cannot promise to agree with and to follow you always; the duty of a leader is to lead.' The final responsibility before the public must be the leader's; his—or hers—the final decision, she told them. 'Up till now, knowing myself to be of the Nation only by love and service, not by birth, I have claimed no authority of leadership but have only fought in the front of the battle and served as best I might. Now by your election I take the place which you have given.' Her peroration had the cadence of a benediction. She invoked the image of Christ which, if it meant anything to the multitude assembled there, represented an alien authority. 'India the Crucified among Nations now stands on this her Resurrection morning, the Initiated, the Glorious, the Ever Young, and India shall soon be seen, proud and self reliant, strong and free, the radiant Splendour of Asia, as the Light and Blessing of the World.'[5]

Feeling among the other Congress leaders ran high against this speech, and Besant's direction of the meeting as a whole, which they regarded as a failure.[6] They detested the attempt, which she pursued against all advice, to extend her activity as President beyond the three days of the meeting, to the whole of her year of office. They were determined not to give her the mandate she desired. It made no difference. Her actions in the months that followed were such that Montagu thought she was in danger of reneging on the pledges given as a condition of her release. 'She wants to get interned [again] and if she fails to accomplish this she is going downhill so fast that she will disappear... her influence goes day by day', he wrote in April 1918.[7]

Besant had recourse to various devices to preserve that influence. She had neglected to address the Muslim section of the population so far, in the belief that Hinduism, not Islam, came closer to the universal religion that was Theosophy. As soon as she was released she repaired the omission with a strenuous campaign aimed at freeing the Muslim leaders, the brothers Mohammed and Shaukat Ali. Her presence in their homeland, the notoriously volatile Punjab, was not desirable, the Government of India told her[8] (it was nervous of a German attack on India via Afghanistan and the North-West Frontier[9]). She gave it to be understood that, if the terms of the Montagu–Chelmsford Report, due in July, were not to her liking, she would use the Home Rule League to organize hunger strikes against it. She infuriated Chelmsford by publishing the substance of her interview with him and Montagu, in which Montagu was made to appear to agree with her proposition that the British went back on promises. When—no one quite knew how—the Subramania Aiyar letter reached the newspapers in May, pointed questions were put in the House of Commons about Besant's influence on him. The non-brahmin opposition to her in Madras declared the letter treasonable because it sowed discord between the United States and Britain when they were Allies in the War.[10]

The Government of Madras was in a constant state of anxiety about the mood among the textile workers in the city. In the spring of 1918 the mills were disrupted by wage claims, lock-outs, and strikes. Besant's reputation as the victor in the celebrated—and now historic—match girls' strike helped to encourage the cotton workers. But, B. P. Wadia complained years afterwards, her interest was only perfunctory. This was disingenuous of him. At the time, he was about to become chief organizer of the Madras Labour movement. The Bolshevik agent with whom he was in touch warned him on no account to involve Besant.[11]

Her lack of enthusiasm for the workers' cause was consistent with the transformation of her attitude to socialism in general. She described with approval a proletariat in the condition of a child, ready to be governed, ready to admit the superiority of its elders. In her ideal socialist society, members would all be equal at the lowest, local level, that of the panchayat. Thereafter, election to successive tiers of government would depend on qualifications of age, experience, and education, so that the greatest power would reside safely in the hands of the good and the wise. 'A democratic socialism controlled by majority votes, guided by numbers, can never succeed', she wrote as early as 1908; 'a truly aristocratic socialism, controlled by duty, guided by wisdom is the next step upward in civilisation.'[12] This was not just a theory. In 1912, when England was in the throes of severe industrial unrest, she went back to the East End of London. This time her reaction to the misery of the striking workers she saw there, as it appeared in a letter to the earl of Minto, was less than sympathetic. He was in retirement after his term as Viceroy; Besant approached him to join a group she hoped to form of his fellow aristocrats which, under the patronage of King George V, would thwart the revolution she believed was imminent.[13]

This attitude was a serious impediment to understanding between her and members of the Labour Party who were the best hope of support for India's claim to self-government. In 1918 many of Besant's friends in England, including Lady de la Warr, Lady Emily Lutyens, and her solicitor David Graham Pole,[14] campaigned for Labour at the general election. There was some question of Besant herself finding a seat, but this came to nothing. 'Between the attitude of Labour and you there is a divergence of attitude and method', Lady Emily wrote earnestly to Annie. 'Labour has no use for Kings, or the King, or the Prince of Wales, Labour simply does not recognize the King as part of Reconstruction.' Of course Labour had no criticism of Besant's policy with regard to All India affairs, Lady Emily continued (not quite accurately), but that policy ran counter to Labour's ideals as they were recognized in England. 'Those who are "in" with Labour must not be monarchical.'[15]

While the Montagu–Chelmsford Report did make detailed provision for a much greater involvement of Indians in their own affairs, especially at provincial level, the protecting power of the Raj remained inviolate. Besant's first reaction was to denounce the Report as unworthy of those who had produced it. It was a slow eighteenth-century coach lumbering along a road that led eventually to perpetual slavery. India would be a conglomerate of separate and powerless states under a foreign autocracy, *New India* declared.[16] The prospect of reform produced new divisions among Indian politicians. The Moderates, who wanted to make the proposals work, withdrew altogether from the Congress. The Extremist leader Tilak eventually came out in favour of 'responsive cooperation'. The rest, whom *The Times* labelled Super Extremists, would have nothing but complete independence. Suddenly aware that she was in danger of losing what influence she still possessed, Besant softened her attitude towards the reforms in the hope of holding on to the Moderates. Instead it brought her into contumely, and caused her to be howled down in committee.[17]

There were other reasons why her Indian colleagues were out of patience with her, as Tilak explained. 'Though I admire her eloquence, learning, and unfailing energy for work, I cannot bear for a moment the supremacy which she claims for her opinions in matters political under the guise that she is inspired by the Great Souls [the Mahatmas] and that such orders from them as she professes to receive must be unquestionably obeyed', he fumed. 'Autocracy may be, and sometimes is, tolerated in theological and Theosophical Society matters, but in democratic politics we must go by the decisions of the majority ... Congress recognizes no Mahatma to rule over it except the Mahatma of majority.'[18]

For most of 1919 Tilak was in England where his efforts on behalf of the Congress–League scheme earned him the respect of politicians of all parties, though the warmest response came from Labour. His absence from India left the field clear for Gandhi. Though it was Annie Besant who bestowed on him the title of Mahatma, by which he was respectfully and affectionately known for the rest of his career, she was far from being an admirer. From the moment Gandhi arrived home in 1915, they were ill at ease with one another. His dislike of her rampant propaganda has already been remarked; there were more fundamental differences. He was for levelling down in pursuit of equality; she wanted to raise standards through leadership by an élite. His dislike of the caste system prompted him to try to break it down; her preference for the brahmin caste was the cause of serious disagreement in Madras. He gave his life to ending the separation of Hindu from Muslim. While Besant acknowledged the contribution to art and science of the Mogul Empire, she regarded it as no more than an episode in the life of Aryavarta.

Their differences came to a head in early 1919, when Gandhi called for a campaign of passive resistance against the introduction of severe repressive measures. Bills to enable trial without jury and an extension of the power to intern went by the name of Mr Justice Rowlatt, who presided over the committee that recommended their introduction to control unrest. This unrest was partly due to the fact that five million people had died in the influenza epidemic that ravaged the subcontinent in 1918. It was also the effect of the 'seething, boiling political flood' which, according to Montagu, was raging across India.[19] The people's aspirations were greater than ever before, encouraged by the spectacle of their rulers narrowly escaping defeat at the hands of the Germans, and by the Russian Revolution of 1917.

Annie Besant was as inconsistent about passive resistance as she was with regard to the Montagu–Chelmsford reforms. She adopted Bradlaugh's position as it was on the eve of Bloody Sunday: that defiance of the Government was in order as long as he directed it, otherwise it was irresponsible and dangerous. She rationalized her attitude as one of approval for a single act of passive resistance against a specific grievance, but condemned any such action when it was applied to a range of activity, when she accused the organizers of risking the security of the country since they were bound to lose control. In 1919 the declaration by the newly formed Communist International of its intent to foment industrial unrest as a means of destabilizing governments gave point to her concern. Gandhi, on the other hand, argued that passive resistance in the form he called *satyagraha*—soul force—offered a safety-valve for the discontent he also thought the Bolsheviks were poised to exploit.

He initiated the first *satyagraha* against the Rowlatt Bills—the 'black bills'—in February 1919 at a meeting at the Sabarmati ashram near Ahmedabad. Among those present who took the oath of support was one of Besant's Theosophist lieutenants, Jamnadas Dwarkadas, who, with his brother Kanji, had been responsible for directing the Home Rule League in Bombay during the period Besant was externed from that Province. Their desertion to the Gandhian camp was a terrible blow to Besant, who had tried to prevent it. So was the use of the branches of the League she had founded to organize the *satyagraha* nation-wide.[20] She went on arguing against it; passive resistance was either anarchy or futility; Gandhi might be a Mahatma, a great soul, but he was no politician; power was not safe with him. Satyagrahists might be saints but their example could betray the ordinary man into disregard of the law, thereby incurring penalties, she wrote, echoing Bradlaugh.[21]

The public did not want to listen; her audiences turned against her; suddenly garlands gave way to boos and jeers. She was dismayed but not deterred. She was proved right in April when Gandhi called a *hartal* (a cessation of all activity) in Delhi against the Rowlatt measures. When his followers tried to force reluctant shopkeepers to join in, a riot began which soldiers put down by firing into the crowd, killing several. Besant's desire to emphasize her disapproval of Gandhi's methods betrayed her into a too hasty response: 'a Government's first duty is to stop violence', she wrote, reasonably enough, but continued; 'before a riot becomes unmanageable brickbats must inevitably be answered by bullets in every civilized country.' Though she tried to repair the damage by apologizing she was never forgiven and 'bullets for brickbats' became a very effective slogan against her.[22]

Shortly afterwards, on 17 April, she compounded the error by broadly supporting the action of soldiers in the Punjab, before the full extent of what they had done was known.[23] On 10 April four Europeans were murdered by a mob at Amritsar. Three days later a crowd of about 10,000 defied a ban on meetings to assemble in an enclosed space called the Jallianwallah Bagh. Soldiers under the command of a General Dyer fired without warning, leaving the dead lying about in heaps. The report of an enquiry into the incident was not published until months later. Then it was disclosed that 379 people had been killed, and over 1,200 wounded. Dyer, moreover, had inflicted humiliating and savage punishment on the inhabitants of Amritsar.[24] The effect was to make it seem that while it was not all right for Indians to murder Europeans, it was perfectly acceptable for Europeans to shoot Indians. Amritsar became a symbol of all that was most hated about the Raj. It was a tragic irony that Besant, who had been among the first to sound a warning about the dangers of racial discrimination, was here irrevocably identified with what was perceived as tyranny.

In May she acknowledged the growing opposition by resigning as President of the Home Rule League, to be succeeded by Gandhi. 'Forced out', she complained, and immediately set about forming a rival organization to be called—with deliberate intent to confuse—the National Home Rule League.[25] Her need for a group to represent was urgent if she was to be invited to give evidence before the Select Committee of both Houses of Parliament that was to examine the Montagu–Chelmsford proposals in July. It was the time of year when the heat grows daily more intense; Annie was 72, and had never been able to tolerate too much of it. Nevertheless she took to the railway, travelling by night from city to city, in a frantic attempt to bring the new League into being.

Arriving in London in June, she attended every session of the Select Committee hearings. With her short white bristly hair, and her flowing white and gold attire, she was a conspicuous figure. To the annoyance of the other Indian deputations, she told the assembled peers and MPs that many people had asked her to appear because she was unique in touching the life of India at every part. She reiterated her belief that there was danger in delaying the advance to self-government, and gave an assurance, especially poignant in the aftermath of Amritsar, that once Indians were in control racial antagonism would pass away. She made a forceful plea for Indian women to be treated as equals in the new legislation. She ended by repeating her belief that India's permanent government must not reproduce Western forms, but evolve from its ancient institutions.[26] Her desire to monopolize attention was obvious: she was all for herself, Tilak complained, 'evidently intending to show that she it was who alone awakened the sense of political freedom in India'.[27]

Besant's abrupt conversion to the spirit and the letter of the Montagu–Chelmsford reforms made the Congress leaders very suspicious. Some kind of bargain had been offered her in which they were not to share; some undisclosed influence had been brought to bear.[28] As far as one can tell they were deceived in the first instance; they were right as regards the second. In April 1919, Lord Willingdon had succeeded Pentland as Governor of Madras. Willingdon's geniality was matched by that skill in the art of handling people which Montagu discovered to be so sadly lacking in most of the officials of the Raj. Where Besant was concerned, Willingdon had the advantage of his connection, through his wife, with Lady de la Warr. Now that her dangerous influence over Indian politicians had waned he confessed to a sneaking admiration for her combativeness. 'Naughty Annie' was how he described her to Edwin Lutyens.[29] He endeared himself to her by his genuine desire for self-government for India. It was surely he who convinced her that it was vital for English public opinion to be assured that Indians would not succeed in wrecking the Montagu–Chelmsford initiative whether by terrorist activity or passive resistance, and that she could help to turn them from this path. It was a role that appealed to her by reason of her antipathy to Gandhi. She was also encouraged by the hope (which, typically, became anticipation) that the new Viceroy would be Edwin Montagu, in which case she foresaw further possibilities of usefulness and pleasure.[30]

In the summer of 1919, Besant's quarrel with the Congress leaders posed an awkward problem for members of the English auxiliary of her original Home Rule League, more especially for her staunch ally, the editor of the *Daily Herald*, George Lansbury, who interpreted the aims of the League to the Labour movement. When Labour recognized the Indian National Congress, Lansbury, as a supporter of Besant, found himself at odds with the majority. It fell to Lady Emily to tell Annie how things were. Having recognized the Congress, Labour would support its policy, not hers. 'If any in India go to the extreme lengths of demanding separation from England Labour is not going to force them to be within the Empire', she warned her friend. As soon as the India Reform Bill became law, the English auxiliary was quietly wound up.[31]

Chelmsford's successor as Viceroy of India was not Edwin Montagu, but Lord Reading, whose last post had been as Ambassador to Washington (when he had sight of Sir Subramania's letter to the President). He treated Besant with greater consideration than she had received from anyone in his position since Lord Minto. Her privileged access to him was an important factor in the next phase of her political career. By the time Tilak died in 1920, the leadership was already passing to Gandhi. When, in the same year, Congress endorsed his policy of non-co-operation, Besant stormed out. However, in 1922, after serious unrest, Gandhi was sentenced to six years in prison for incitement, and his movement collapsed. Besant stepped in with a proposal that she pushed with great vigour in *New India*, that leading Indian personalities of all shades of opinion should come together to frame a new constitution for the country.

Ever mindful that the opposite pole to India was Ireland, she took her inspiration from the National Convention of Irishmen from all parties (with the exception of Sinn Fein who refused their invitation), which, after deliberating in private for some months, produced a report in April 1918 which formed the basis of the Bill that established the Irish Free State.³² The Indian National Convention, which met several times in Delhi between 1923 and 1925, attracted the support of moderate politicians who had not been able to bring themselves to approve of non-co-operation. Their purpose in attending was not so much to produce a Bill—that was Besant's idea—but to make a demonstration of their effectiveness and, more especially, to make the British Government bring forward the review of policy which, under the India Reform Act, was not due until 1929.³³ Two Liberal leaders, Tej Bahadur Sapru and Srinivasa Sastri, accepted the office of President and Vice-President respectively, while Besant contented herself with being General Secretary.³⁴ A Bill 'to establish within the British Empire a Commonwealth of India' was duly framed. It called for separate tiers of government with, at the centre, an executive to have all powers, excepting only those of defence and foreign relations: these, as in the Irish Free State, were reserved to the Crown. The franchise was to be subject to a test of literacy. This was the closest Besant came to the programme she put forward in her Presidential Address. She must have known that the number of literate Indians was infinitesimal, the problem of teaching them so huge, that it would have been decades before a government elected on this basis could have come anywhere near being representative. One must assume that was her deliberate intention.

Her hopes were raised higher than ever before by the advent in January 1924 of the first Labour Government. One of the first acts of the new Prime Minister, Ramsay MacDonald, was to send an open letter to India which, though it warned that his Government would not be moved by violence of any kind, gave hope of a favourable atmosphere for constitutional advance. When his list of Ministers was published, Besant could be forgiven for thinking she had the Cabinet in her pocket. MacDonald himself was an old acquaintance; Sydney Olivier, the new Secretary of State for India, a friend from Fabian Society times. Philip Snowden, who had briefly served as Vice-Chairman of the auxiliary to the Home Rule League, was Chancellor of the Exchequer; Lord Haldane was Lord Chancellor; one of her closest allies in recent years, Colonel Josiah Wedgwood, was Chancellor of the Duchy of Lancaster. George Lansbury was back in the House as MP for Bow and Bromley.³⁵ A long interview with Lord (as he now was) Olivier convinced her that it was only a matter of time before the Commonwealth of India Bill became law.

Even when MacDonald's government fell in October 1924, to be replaced by a Conservative administration under Stanley Baldwin, Besant remained convinced that Labour would support the Bill, which the Convention directed be placed before Parliament in 1925. She failed in her attempt to broaden its support. Gandhi, Motilal Nehru, C. R. Das all refused to endorse it, while it was said that Jinnah had not even bothered to read it. This, together with the qualified franchise, destroyed its chances at the Labour Party Conference.[36] It was left to the faithful Lansbury to introduce it in December as a Private Member's Bill—there was no second reading.[37]

Besant felt betrayed. In 1927 she joined in the huge wave of anger that greeted the appointment of the Simon Commission to review the progress of the Montagu–Chelmsford reforms two years before the statutory date. By an extraordinary blunder no Indian had been included. Politicians of all parties came together. Besant castigated 'the white seven' in *New India*, and did her share of organizing protest meetings and boycott. In 1928 an All Parties Conference set up a committee under the chairmanship of Motilal Nehru to draft an agreed constitution. That Besant was named as a member gave rise to speculation that she would try to get her own Bill revived. The rumour was without foundation.

It is doubtful whether she was able to exert any influence at all on the Nehru Report, whose appearance marked the end of her political career.[38] Her volatility had been an embarrassment for some time. On the one hand she issued warnings to the English that failure to give India immediate freedom would provoke revolution and invasion by the Bolsheviks; on the other, as an ex-President, she invited the Congress to set up an alternative government on the lines adopted by Sinn Fein. *New India* was in financial difficulties, the *Commonweal* had ceased publication.

The last ten years of her political life were, if anything, more turbulent than the first. In her old age, instead of running down, the motor of her energy began to race, and there was no one to correct it. Some observers thought her ceaseless activity concealed an inner desperation. Her imperiousness kept people at a distance:

'a sort of Church all day feeling', was how Edwin Lutyens described the atmosphere to his wife when Besant stayed with them in Delhi.[39] He was not an impartial witness: he placed all the blame on her for the near breakdown of his marriage. That Lady Emily gave up sleeping with him; that she became deeply involved in occultism which he detested; that she left him unsupported in the years when he was engaged on the exhausting contract to build the imperial capital at Delhi, in order to follow Theosophy and its boy avatar, were solely Annie Besant's fault in his opinion. But even Esther Bright, who was very fond of her,[40] flinched when, like a latter-day Elizabeth I, Besant and her entourage arrived for their annual sojourn at her house.

This activity was extraordinary in a woman in her late seventies; it was amazing in the light of what was happening at the same time in the Theosophical Society. A bitter and debilitating row went on for years over Charles Webster Leadbeater, to whom Besant remained steadfastly loyal. Since 1914 the man himself had been in Australia where, in the company of the usual complement of boys, he pursued his occult studies in a mansion on the shores of Sydney Harbour.[41] As President, Besant came under furious attack for condoning Leadbeater's re-entry into holy orders as a bishop of the Liberal Catholic Church. This came into being after an English Theosophist, James Wedgwood,[42] was received into the Old Catholic Church in England. Historically, it had broken away from the main Roman Church because its members did not accept papal infallibility as defined in the Vatican Council decrees of 1870. Old Catholic bishops claimed apostolic succession and this is what interested Leadbeater when he met Wedgwood in Australia. His thoughts had been returning to Christianity; he now believed that the celebration of the Mass released occult force. When he and Wedgwood became bishops, they were obliged to adopt the name Liberal, because the Old Catholics refused any connection with Theosophy.[43] Though George Arundale found the elaborate ritual of the LCC terribly exciting, other members detested it for being as much of a cuckoo in the nest as co-masonry. Annie was a kind of Visitor; whenever she attended services she was conducted to a seat beside the high altar. This fresh manifestation of what its opponents denounced as neo-Theosophy gave rise to a Back to Blavatsky movement, one of whose leaders was Besant's fellow internee B. P. Wadia.[44]

It was not long before the old accusations with regard to Leadbeater's conduct towards boys created uproar in Australia. There was a wide-ranging police investigation, in the course of which statements were taken from Krishnamurti and Nityananda. The conclusion was that, while the evidence did not justify immediate action, sufficient had been disclosed to warrant keeping Leadbeater under surveillance.[45] Besant's reaction was more of annoyance at having to defend Leadbeater once more, than of censure. What upset her very much, and became a perpetual source of worry in the years that followed, was the tension that developed between Leadbeater and Krishnamurti. When he and his brother accompanied Annie to Australia in 1922 they had not seen Brother Charles for nine years; they were shocked by his new flamboyance as a bishop, and the ritual of the LCC.

Krishnamurti was 26. It was not a matter of regret to him that he had failed to be accepted by any English university; he liked to be quiet and away from people. He spent long periods reading by himself. 'I want to gain everything that the West can give, and then turn my face to India where I am sure I shall work', he told Annie.[46] The instinct to worship was as strong in him as it was in her, but he had an absolute aversion to ceremony. The philosophy he created later on was a form of quietism. He found the peace for which he always longed in the contemplation of nature; his language in describing this was at times too simple for others to understand. The impression of vagueness he gave was belied by the strength he showed in coping with the very peculiar and embarrassing life he was obliged to lead. For many years he neither wholly accepted nor rejected the possibility Besant proclaimed from the roof-tops, that he was the World Teacher. On the rare occasions when, without warning, he spoke in the *voice of the Teacher*, she was in ecstasy at what she believed to be, at last, a manifestation of the Coming.[47]

His attitude to her was one of affection and respect. He was on much closer terms with Lady Emily Lutyens; he was deeply grateful for the warmth she and her children gave him.

In April 1922 Krishnamurti sent Lady Emily an account of a Liberal Catholic Church service in Sydney which Leadbeater had conducted. 'He did it all *very* well but you know I am *not* a ceremonialist and I do not appreciate all the paraphanalia [*sic*] with all those prayers and bobbing up and down, the robes etc:

but I am *not* going to attack it, some people like to so what right have I to attack or disapprove of it.' The service lasted $2\frac{1}{2}$ hours and Krishna was so bored he was nearly fainting, he told Lady Emily. 'I am afraid I rather showed it. I must be careful or else they will misunderstand me and there will be trouble. They are like cats and dogs over this [Liberal Catholic] church affair. They are fools.'[48] Nityananda remarked how much milder Leadbeater had become in nine years, how much less cruel to old ladies. But, just as at Adyar, there was never any question of doubt with him; never any question that anyone else could doubt: 'he is always sure that everything is as real to everybody as it is to him.'[49]

From Australia the young men went on alone to the Ojai Valley in the mountains of California, in the hope that the air would heal Nitya's damaged lungs. There Krishna began to experience regular attacks of agonizing pain which Annie hailed as the raising of the *kundalini*, the spirit force, a sure sign of the Coming. The death of his much loved brother in 1925 was a tremendous shock to Krishnamurti, who had believed the Masters' assertion that Nitya would be spared. His doubts were growing. But when Annie came to Ojai to be with him for some months, he apparently made no objection to the short press release she put out: 'The Divine Spirit has descended once more on a man, Krishnamurti, one who in his lifetime is literally perfect, as those who know him can testify. The World Teacher is here.'[50] She bought land in the Ojai Valley intending it to be the place where the new subrace of more perfect people, which Blavatsky had taught her disciples to expect, would flourish.[51]

Now aged 80, her voice had lost its resonance and she occasionally stumbled over words. Even so, she stuck doggedly to an exhausting schedule of lectures and visits when she was in Europe. These were made easier by the use of a small aeroplane (furnished, according to a photograph with Lloyd Loom chairs). It was a method of transport that delighted her. She was a commanding presence at the annual summer camps for the Order of the Star, which took place in Holland on an estate given by a well-wisher. At these camps George Arundale, who was himself now a bishop in the LCC, directed the thoughts of a small inner group. He 'brought through' exciting messages from the astral plane by which it seemed that the process of Initiation, which had been agonizingly slow when Leadbeater was the medium, had speeded up, so that those involved—Arundale himself, his young Indian wife, Lady Emily, and several of Leadbeater's acolytes—had passed through various stages to reach a level that was divine.[52] For example, Arundale conceived the notion that they were to be among the Twelve Apostles who would attend the Teacher at his Coming. Krishnamurti, who was not present on these occasions, was dismayed when Annie endorsed Arundale's revelations. He described the exercise as, variously, mad, foolish, unnecessary, and absurd. In his opinion, Annie was the only one who was sincere. His distaste for the way he thought she was being used proved greater in the end than his fear of hurting her.

At the Star camp in August 1929 he dissolved the Order. Truth, which was the object of their quest, was a pathless land, he told his audience; it could not be approached by any path, by any religion, by any sect. Truth could not be organized, nor should any organization be formed to lead or coerce people along any particular path. He rejected occultism as a means of progress. 'You can form other organisations and expect someone else', he told them, 'with that I am not concerned, nor with creating new cages, nor with decorations for those cages. My only concern is to set men absolutely, unconditionally free.'[53]

This was ruthlessly straightforward; one might almost infer that Krishnamurti was taking his revenge, so accurately were his remarks aimed at everything Leadbeater and Besant had fashioned over the years. It was her misfortune that, in her extreme old age, Krishnamurti asked her for the one sacrifice she was not inclined to make. Nevertheless she tried. 'Beloved, I have done my best to make a clear field for you', she wrote, on dissolving the Esoteric Section, the centre and symbol of her power. 'You are the only authority.' The effort was too much for her; shortly after, she revived it, causing Krishnamurti to exclaim, 'I must get out of all this rot.'[54]

His opposition devastated Annie. Her strength, her ability to satisfy the demands her abounding talent made on her, endured only so long as her self-assurance remained intact. This was the guiding principle of her life, no less a person than the Prime Minister remarked on reviewing her *Autobiography* after it came out in 1893.[55] As a devout Christian Gladstone regretted her lack of a sense of sin, but he understood what power it gave her to change direction without a qualm. Her self-assurance was badly shaken when the Indian people turned on her; there remained the mission to proclaim the World Teacher. When that was repudiated by the central figure Annie collapsed.

Her last visit to England was in 1930. For three years after that she was a presence in the sanctuary of Adyar, where she was tended with great devotion. Krishnamurti saw her for the last time in May 1933, when he reported her more coherent than usual and very affectionate. 'She said I must not stay long with her as it was bad for me to be in an invalid's room.'[56] He did not see the 'dear rooms' at Adyar again for many years; his presence was not welcome to Arundale, who succeeded Annie as President of the Theosophical Society.

She died on 20 September 1933. They built her funeral pyre on the shore where the River Adyar flows into the vastness of the Indian Ocean.

EPILOGUE

AT her death in 1933 it was forty-two years—half her lifetime—since Annie Besant had put on Madame Blavatsky's ring, the symbol of esoteric power, and twenty-six since she had assumed executive authority as Henry Olcott's successor. Long before she became President, she had begun to bring the Theosophical Society round to the course she was determined it would take, at whose landfall the millennium would begin. Her navigation was a peculiar blend of masterly propaganda—learned in the Hall of Science—political acumen, and religious magic. When this captured the imagination of Hindus it hastened the pace of change at a critical moment in India's history. Who knows what might have happened if Besant had been able to consolidate the influence she possessed for a brief moment in 1917? As it was, she was halted and turned aside by the new generation of Indian politicians. Gandhi, Nehru, and their colleagues were determined that an independent India would be secular and socialist: ideas that Besant had already tried and found wanting.

Her successes were divisive: Indians as well as Europeans blamed her for inciting race hatred and caste hatred. Few were disposed to excuse her conduct during the First World War as prompted by her Irish origins; the great majority of Anglo-Indians considered it treasonable. But the overriding reason for the loss of prestige she suffered, as her obituary in *The Times* did not fail to point out,[1] was her unswerving support for Charles Webster Leadbeater.

He reached Adyar shortly before she died, and played a leading part in her obsequies. He stayed on for the annual convention in December, the occasion for tributes to the late President. Then, although he was very unwell, he insisted on departing: he had always detested India. Shortly after returning to Australia he died, being of an advanced age; 86 according to himself (the same age as Annie), 80 according to his biographer.[2] He was cremated with the ceremony befitting a bishop of the Liberal Catholic Church. Krishnamurti, who was in Sydney at the time on business of his own, was among those who attended.

Krishnamurti himself was destined to become a teacher of world-wide renown, one of the foremost of his kind of the twentieth century. His books have been translated into many languages, and the number of his disciples continues to increase. After Annie died he lived very simply in California until after the Second World War. During these years Aldous Huxley and Gerald Heard were among those who came to know and value his philosophy. Krishnamurti gave expression to something within himself which, to those who understood it, gave enlightenment and healing. He used plain words and uncomplicated images drawn from nature to illustrate his teaching. Its meaning was not always clear: peace through self-knowledge was a part of it. This opacity contributed to the striking impression he made. His disciples founded schools, especially in India.

He did not return to Adyar until after Dr Radha Burnier was elected President of the Theosophical Society in 1979. It was his habit to walk with her and other friends on the beach where Leadbeater first encountered him. He died in 1986, aged 91, having fulfilled the destiny for which Annie Besant and Leadbeater had so rigorously prepared him, though not at all in the manner they had intended.

In the last decade of her career, Besant's obsession with the Coming of the World Teacher again created tension among Theosophists, and was the cause of a serious falling off in the size of the audiences at her public lectures. The Society itself increased its membership during her Presidency. From just under 15,000 in 1907 it rose to a peak of 45,000 in 1928, after which it began to decline. Towards the end of her life Annie's financial situation deteriorated to the point where Digby Besant felt obliged to provide his mother with a regular allowance. He had prospered in his career as an actuary, was a keen amateur photographer, and a member of the Athenaeum. He and his wife welcomed Annie as the grandmother of their children whenever she was in England. Digby became Chairman of Besant & Co., a publishing business, which developed into the Theosophical Bookshop, whose premises, until recently, were in Great Russell Street opposite the British Museum.

The final version of Annie's will gave this business to Digby. He and Mabel, who kept in touch with her mother after the breakdown of her own marriage, were allowed to choose a memento from among Annie's possessions. Among other bequests, Madame Blavatsky's ring with its seal was left to the use of the President while in office, and Annie's masonic jewels to the Museum of Records at Adyar. She directed that the portraits of the Masters which accompanied her whenever she travelled were to be kept at Adyar, where her rooms were to remain as they had been during her lifetime. She asked that suitable financial provision be made for her faithful servants. The last clause made her successor, George Arundale, residuary legatee. He was instructed to do with the rest of her property what he knew she would have intended.

When Annie committed herself to Theosophy in 1891 it was as if a line were drawn under what had gone before: friends she left behind were moved to strike the balance of her career. Those who found the subject incomprehensible, or distasteful, wrote her off entirely: they expressed regret for promise unfulfilled, and astonishment at what, in their opinion, she threw away.

No other woman could rouse and dominate an audience like Annie Besant at her best, H. M. Hyndman wrote; she was the most valuable recruit to the socialist movement in any country. In 1889 her influence was considerable, due in great part to her admirable work on the London School Board.

What more could a woman of great ability want in the way of a career under existing conditions [Hyndman demanded] than that she should be the leading champion not only in the metropolis, but throughout Great Britain, and indeed all over the world, of the physical, intellectual and moral development of children by relieving them of the wretched results of competitive profitmongering and anarchical indifference through direct social organisation for their benefit: holding as she did at the same time a position in the world of thought and letters and [here he differed from Beatrice Webb] an unchallengeable place in the only growing party of the time?

Yet in the prime of her life and vigour she went off into the mystic groves of Hindu philosophy and religion. He hoped she found compensation there for what she gave up at home.[4]

Although this remark placed the practical aspects of Besant's career above the spiritual, Hyndman was himself famous for hailing the millennium at each and every crisis. Mystical ideas, movements tending towards a new world order—socialism, secularism, spiritualism, vegetarianism, to name but a few—proliferated during the ten years of Besant's greatest activity at home; she was involved with the most significant.

'And the world to end shall come in eighteen hundred and eighty one', Edward Carpenter wrote, quoting Mother Shipton, in an attempt to convey the spirit of the moment. He valued Besant highly among the leaders of what he judged to be a reaction to the smug commercialism of the mid-Victorian era, especially after she became a Theosophist. By making the words karma and reincarnation familiar she formed a new channel for religious thought, and sought to bring the Western public in touch with the great age-old ideas and inspirations of the old Indian sages, he wrote. She helped in the creation of the great twentieth-century bridge which, because he was a millennialist himself, Carpenter believed would lead into another world.

He could only fault her in one curious respect.

With all her enthusiasm for the subject, Mrs Besant does not appear to have the intuitive perception, the mystic quality of mind which should enable her to reach the very heart of the old Vedantic teaching. Her intellect, clear and systematic . . . has little of the poetic or inspirational, and it may be doubted whether it has ever quite fathomed the religious writings with which it has been so much occupied.

But, he hastened to add, she had done great work and shown much kindness. And in later years, he thought she had achieved a 'mental calmness', which was in great contrast to the restless bitterness that had marked her when they first met.[5]

Bhagavan Das was another who pondered the apparently irreconcilable halves of Annie's life. Her progress from altruism to the opposite extreme, once she had become President of the Theosophical Society, startled and dismayed him.[6] In fact there was no real conflict: egotism—in the highest meaning of the word—was the mainspring of her character; all that happened in 1907 was that the last necessity of concealing her ambition was removed. The concern for others reflected in her work, her 'mission', was the expression of an absorbing sense of self, the means of release for the extraordinary force within her.

Circumstances—her father's early death; lack of money; her mother's apprehension, which betrayed her into a suffocating marriage—conspired to keep Annie down. She proved strong enough to overcome all these; she made her way in spite of a society that, notoriously, gave no quarter to its victims. When it condemned her for abandoning husband and Church, she turned her back on it; in her mid-twenties she was suddenly accountable to no one but herself, free to exercise her remarkable ability. Probably that is why, though she spoke and wrote much in support of the women's cause, some who were closely involved with it complained of her lack of real commitment.[7] Having struggled free from Frank Besant, Annie ceased to think of herself as an oppressed female.

Pride, unshakeable self-confidence—the gift of Ellen Marryat—and gallantry sustained her progress which, while it was headlong, was by no means as haphazard as some liked to think. Bernard Shaw noted the rapidity with which she changed direction. That, together with the enthusiasm and dramatic flair she brought to every undertaking, gave her a reputation for volatility, for extreme impressionism. But her autobiography and correspondence show that she devoted private study and reflection to each step before announcing it. Each time her position in an organization was sufficiently established, she began to pull in her own direction, as she did the National Secular Society in the case of the Knowlton pamphlet. Charles Bradlaugh's daughter, who understood her method, never forgave the way, in her opinion, Annie used and discarded her father. Annie's passage through Fabian socialism—Shaw's phrase—was a similar attempt to impose her own ideas. When the opposition proved intractable she cut her losses, as she did when the Social Democratic Federation fragmented.

Theosophy was no exception. The removal of Madame Blavatsky and her entourage to 19 Avenue Road gave its chatelaine a great advantage over other members. Soon Annie was not only speaking for the sick woman at Theosophical gatherings, but controlling her visitors. HPB's death merely hastened the process whereby Annie began to build on the existing framework of the Society. As a means of obtaining the new order she so passionately desired, this promised far better than the fiasco of the Ironside Circles.

As far as *The Secret Doctrine* went, in the light of what we know about her, it is likely that the shock she experienced on first reading it was of recognition, rather than of revelation. Here were the same ideas that had enthralled her as a girl; stories, legends, tales of magic that, on her release from Miss Marryat's supervision, she had discovered for herself in Harrow School Library.[8] The doctrine of ancient wisdom, of Superhuman Men, restored the spiritual dimension that was essential to her complete well-being. By embracing Theosophy she ended a long exile. Beyond that it is not possible to say at what point belief gave way to imagination.

Today, in certain quarters, the occult writings of Charles Webster Leadbeater are a source of inspiration, not only to Theosophists. Purists confine themselves to the works of Madame Blavatsky, whose life and personality remain the subject of serious research and speculation. One recent study examined the possibility that she might have been a Sufi. There is no school of Besant. Edward Carpenter thought her philosophical writings, unlike her lectures, dull. She always seemed to be repeating something, corroborating some preconception, never describing something she had perceived, he complained: 'the pages bristle with sanskrit jargon, but no living or creative idea moves among them.'

The exception is her *Autobiography*, which compels admiration, astonishment, and sympathy in equal measure. To use the language of chivalry in which she liked to express the high and noble purpose that inspired her: when most of the other members of her sex were condemned to languish in the bower, Annie Besant rode out to break lances, whereby she gave proof of much valour.

NOTES

Chapter 1
1. Birth certificate.
2. Annie Besant, *Annie Besant: An Autobiography* (Theosophical Publishing House, 2nd ed., 1908), 3 (henceforth *Autobiography*).
3. *DNB*, 'Matthew Wood' (1768–1843); Annie Besant, *Autobiographical Sketches* (Freethought Publishing Co., 1885), 5 (henceforth *Sketches*).
4. Birth certificate; *Sketches*, 7.
5. Besant, *Sketches*, 9.
6. Ibid. 3.
7. Ibid. 8.
8. Ibid.
9. Ibid. 11.
10. Ibid. 28, 29.
11. Ibid. 12.
12. Ibid. 12.
13. Ibid. 19.
14. Ibid. 13.
15. Ibid. 14.
16. Ibid. 22.
17. Ibid. 16.
18. C. T. Davis, *The Family of Marryat* (privately printed, n.d.); Marryat family tree and papers, in possession of Mrs Elizabeth Sewell; C. T. Davis, 'Memoir of Mrs Charlotte Marryat of Wimbledon House' (privately printed, 1900).
19. Besant, *Sketches*, 19.
20. The material in *Sketches* was first published in the magazine *Our Corner*, ed. Annie Besant, in 1884.
21. Besant, *Sketches*, 20.
22. Ibid. 18.
23. Besant, *Autobiography*, 30.
24. Ibid. 32.
25. Besant, *Sketches*, 37.
26. Ibid. 23.
27. Not $16\frac{1}{2}$, as stated in *Autobiography*, 40.
28. Besant, *Sketches*, 25.
29. Esther Bright, *Old Memories and Letters of Annie Besant* (Theosophical Publishing House, 1937), 11.
30. Besant, *Sketches*, 26.
31. Ibid. 42.
32. Ibid. 31.
33. Ibid.; *Autobiography*, 44.
34. *Sketches*, 32, 33.

Chapter 2

1. Besant, *Sketches*, 34.
2. R. C. Welch (ed.) *Harrow School Register* (Longman's, Green & Co., 1894); *DNB*, 'Charles John Vaughan' (1816-97).
3. Besant, *Autobiography*, 51.
4. See Welch, *Harrow School Register*, 'Henry Trueman Wood' (1845-1929).
5. Besant, *Autobiography*, 53.
6. Arthur Digby Besant, *The Besant Pedigree* (Besant & Co., 1930), 189-90.
7. Ibid. 140-1; Walter Besant, *Autobiography* (Hutchinson, 1902), 33-4.
8. W. Besant, *Autobiography*, 55; Besant, *The Besant Pedigree*, 193.
9. W. Besant, *Autobiography*, 67.
10. Though Jeremy Bentham and his associates intended students to be admitted to London University without religious bar, King's College owed its existence to a group which, while supporting the main purpose of the University, desired to maintain Christian teaching in a part of it.
11. W. Besant, *Autobiography*, 67.
12. Ibid. 93.
13. Besant, *The Besant Pedigree*, 195.
14. Ibid.
15. Ibid. 197; Besant, *Sketches*, 36, 37.
16. Besant, *Sketches*, 37.
17. S. and B. Webb, *The History of Trade Unionism* (Longman's, Green, 1929), 182-5; Besant, *Sketches*, 37.
18. G. J. Holyoake, *Sixty Years of an Agitator's Life* (Fisher Unwin, 1893), i. 105.
19. Besant, *Sketches*, 38.
20. Ibid. 39.
21. Ibid.
22. Ibid. 35.
23. *Who Was Who, 1897-1916*.
24. Besant, *Sketches*, 38.
25. Ibid. 40-3.
26. Ibid.
27. Hastings is on the marriage certificate. *Sketches*, 45, puts it at St Leonards.
28. *Review of Reviews* (Oct. 1891), character sketch of Annie Besant by W. T. Stead.

Chapter 3

1. Besant, *Autobiography*, 55.
2. Arthur Digby Besant, *The Besant Pedigree* (Besant & Co., 1930), 100.
3. A. A. Hunter (ed.), *Cheltenham College Register, 1841-1889* (Bell & Sons, 1890).
4. *Cheltenham and Gloucester Directory* (1869).
5. Hunter, *Cheltenham College Register*.
6. Ibid.

7. Besant, *Sketches*, 45.
8. Hunter, *Cheltenham College Register*. I am grateful to the Revd S. Salter for his help regarding the history of Cheltenham College.
9. Ibid.
10. George Lansbury, *My Life* (Constable, 1928), 2.
11. *Cheltenham and Gloucester Directory*.
12. G. J. Holyoake, *Sixty Years of an Agitator's Life* (Fisher Unwin, 1893), i. 146.
13. Besant, *Autobiography*, 66.
14. Besant, *The Besant Pedigree*, 195.
15. *Our Corner* (1884).
16. Besant, *Autobiography*, 64.
17. Ibid.
18. 45 & 46 Victoria, c. 75.
19. Besant, *Sketches*, 46.
20. Besant, *The Besant Pedigree*, 201.
21. Otherwise referred to as 'Besant Parochial Diary', Lincolnshire Archives Office, Sibsey, PAR 23/1.
22. Marriage certificate.
23. Public Record Office, Chancery, J. 4, 507, 5999. And see *National Reformer* for 6 Apr. 1879.
24. PRO, Chancery, J. 4, 507, 5999.
25. Besant, *Autobiography*, 73, 81. The *Autobiography* is far more outspoken about the marriage than the *Sketches*. The reason may be that when the latter was written, in 1884, Digby and Mabel were minors and in their father's care. By 1893, when Annie published her *Autobiography*, the children were grown up and reconciled to her.
26. Ibid.
27. *The Times*, report of Besant case (28 Mar. 1879), and *National Reformer* (6 Apr. 1879).
28. Besant, *Sketches*, 47, 48.
29. Ibid. 49.
30. Ibid. 51.
31. *Crockford's Clerical Directory* (1870); Edward Walker, *Sermons on Old Testament Subjects* (Wishart, 1873).
32. *The Times* (8 Mar. 1870; 13, 17 Feb. 1871); M. D. Conway, *The Voysey Case* (Thomas Scott, 1871).
33. Walter Besant, *Autobiography* (Hutchinson, 1902), 55.
34. Besant, *Sketches*, 53, 54.
35. J. B. Atlay, *The Victorian Chancellors* (Smith Elder, 1908), ii. 350.
36. *Crockford's*.

Chapter 4
1. Graham Swift, *Waterland* (Heinemann, 1983).
2. *Kelly's Directory* (1876).
3. Besant, *Sketches*, 56.
4. Arthur Digby Besant, *The Besant Pedigree* (Besant & Co., 1930), 201.

5. Frank Besant, 'Parochial Diary', Lincolnshire Archives Office, Sibsey, PAR 23/1.
6. Ibid.
7. Ibid.
8. Besant, *Sketches*, 60.
9. Besant, *The Besant Pedigree*, 201. And see J. Stovin (ed.), *Journals of a Methodist Farmer* (Croom Helm, 1982), 48.
10. Besant, *Sketches*, 60.
11. Ibid. 61.
12. Matthew Arnold, 'Literature and Dogma', 1st pub. in *Cornhill Magazine* (1871).
13. Besant, 'Parochial Diary'.
14. PRO, Chancery, J. 4, 507, 5999.
15. Besant *Autobiography*, 89.
16. M. D. Conway, *Autobiography, Memories and Experiences* (Cassell, 1904), ii. 290.
17. M. D. Conway, *The Voysey Case* (Thomas Scott, 1871).
18. Correspondence of Thomas Allsop, Manchester College, Oxford, MSS Misc. 6, Voysey to Allsop, 5 Mar. 1870.
19. Ibid. 2 Jan. 1869.
20. Conway, *The Voysey Case*.
21. *Crockford's Clerical Directory* (1870).
22. Conway, *The Voysey Case*.
23. *The Times* (13 Feb. 1871).
24. Allsop correspondence MSS. Misc. 6, Voysey to Allsop, 22 Mar. 1870.
25. Ibid. 12 Feb. 1870.
26. *Inquirer*, Scott obituary (4 Jan. 1879).
27. Ibid.; *DNB*, 'Thomas Scott' (1808–78).
28. Scott, Thomas, *A Farewell Address* (T. Scott, 1876).
29. Ibid.
30. *Inquirer* (4 Jan. 1879).
31. Besant, *Sketches*, 68.
32. Edward Walker, *Sermons on Old Testament Subject* (Wishart, 1873), editor's preface.
33. Besant, *Sketches*, 65.
34. Ibid. 66, 67.
35. Ibid.

Chapter 5

1. See e.g. *The Parish Registers of Boston, Copied by the Revd Frank Besant* (Lincoln Record Society, 1914).
2. Besant, *Sketches*, 71.
3. Ibid. 72.
4. PRO, Chancery, J. 4, 497, 3049.
5. Besant, *Sketches*, 73.
6. Ibid. 70.
7. Adyar Archives, B/7, Annie Besant, 'Autobiographical Notes'.
8. Besant, *Sketches*, 72.

9. *The Diary of Beatrice Webb*, ed. N. and J. Mackenzie, iv: *1924–1943*: *The Wheel of Life* (Virago, in association with London School of Economics and Political Science, 1985), 305.
10. PRO, Chancery, J. 4, 497, 3093.
11. Ibid. 3049.
12. Ibid. 3093.
13. Frank Besant, 'Parochial Diary', Lincolnshire Archives Office, Sibsey, PAR 23/1, 19 Mar. 1873.
14. Walter Besant, *Autobiography* (Hutchinson, 1902), 110.
15. *National Reformer* (6 Apr. 1879).
16. Besant, *Sketches*, 71.
17. W. Besant, *Autobiography*, 175.
18. Besant, *Sketches*, 32.
19. Besant, 'Parochial Diary', 20 July 1873.
20. Besant, *Autobiography*, 99.
21. Besant, *Sketches*, 74.
22. *National Reformer* (6 Apr. 1879).
23. PRO, Chancery, J. 4, 497, 2983.
24. 36 & 37 Victoria, c. 12.
25. PRO, Chancery, J. 4, 497, 2983.

Chapter 6
1. *National Reformer* (25 May 1878).
2. Besant, *Sketches*, 75.
3. M. D. Conway, *Autobiography, Memories and Experiences* (Cassell, 1904), ii. 261.
4. Ibid.
5. M. D. Conway, *The Voysey Case* (Thomas Scott, 1871).
6. Holyoake Collection, Co-operative Union Library, Manchester, Letter no. 2,178, T. H. Huxley to G. J. Holyoake, n.d.
7. M. I. Burtis, *Moncure Conway 1832–1907* (Rutgers University Press, 1952), 132; J. d'Entremont, 58th Conway Memorial Lecture (South Place Institute, 1977).
8. D'Entremont, Lecture.
9. Conway, *Autobiography*, i. 87.
10. C. M. Davies, *Heterodox London* (Tinsley Bros., 1871), 60.
11. Besant, *Sketches*, 76.
12. PRO, Chancery, J. 4, 496, 2642.
13. Besant, *Sketches*, 77.
14. I. Beeton, *The Book of Household Management* (Ward, Lock & Co., rev. edn., 1888), 1574.
15. Besant, *Sketches*, 79.
16. Ibid. 80.
17. Ibid. 80, 81.
18. Annie did not reveal her mother's long friendship with Stanley's sister Catherine Vaughan, nor the dean's close acquaintance with her cousin Lord Hatherley, effectively underlining the general significance of her situation at the expense of the particular. See J. B. Atlay, *The Victorian*

Chancellors (Smith Elder, 1908), ii. 350, for Hatherley's friendship with Stanley.
19. R. Prothero, *The Life of Arthur Penrhyn Stanley* (John Murray, 1893), ii. 451.
20. Diary of Henry Liddon, Liddon House, London, 8 Apr. 1889. Stead's office was above Temple Station on the Embankment.
21. Liddon Diary, 16 Nov. 1873.
22. Besant, *Sketches*, 81.
23. Death certificate.
24. Besant, *Sketches*, 87.
25. Besant, *Autobiography*, 108.
26. PRO, Chancery, J. 4, 497, 2974; *National Reformer* (25 May 1878).
27. Ibid. (2 June 1878).
28. Besant, *Sketches*, 88.
29. Holyoake Collection, Letter no. 1,658, Scott to Holyoake, 27 Mar. 1866.
30. G. J. Holyoake, 'The Warpath of Opinion' (1896) (Bradlaugh Papers, no. 2,651).
31. Besant, *Sketches*, 86.
32. *National Reformer* (16 Aug. 1874).
33. Besant, *Sketches*, 89, 90.
34. Bradlaugh Bonner Family Papers, Hypatia Bradlaugh Bonner's memorandum. And see H. B. Bonner, *Charles Bradlaugh: A Record of His Life and Work by His Daughter* (Fisher Unwin, 1895), i. 300-1.
35. *National Reformer* (25 Aug. 1874).
36. *The Times* (28 Mar. 1879).
37. Holyoake, 'The Warpath of Opinion'.
38. Holyoake Collection, Letter no. 2,274, Scott to Holyoake, 12 Sept. 1874.

Chapter 7

1. *National Reformer* (30 Aug. 1874).
2. Bradlaugh Papers, National Secular Society, 388, Bradlaugh to Alice and Hypatia Bradlaugh, n.d.
3. E. Royle, *Radicals, Secularists and Republicans: Popular Freethought in Britain 1866-1915* (Manchester University Press, 1980), 10-11.
4. W. E. Adams, *Memoirs of a Social Atom* (Hutchinson, 1903), ii. 413.
5. D. Tribe, *President Charles Bradlaugh MP* (Elek, 1971), 149.
6. T. P. O'Connor, 'Charles Bradlaugh', *T.P.'s Weekly* (21 Aug. 1903), repr. in J. P. Gilmour (ed.), *Champion of Liberty: Charles Bradlaugh Centenary Volume* (C. A. Watts, 1933), 43-6.
7. Ibid.
8. G. J. Holyoake, 'The Warpath of Opinion' (1896) (Bradlaugh Papers, no. 2,651).
9. Besant, *Autobiography*, 157.
10. J. M. Robertson, *Charles Bradlaugh* (Watts & Co., 1920), 59.
11. H. B. Bonner, *Charles Bradlaugh* (Fisher Unwin, 1895).
12. Robertson, *Charles Bradlaugh*, 9.
13. 17 & 18 Victoria, c. 125: Common Law Procedure Act.
14. C. Mackay, *Charles Bradlaugh* (D. J. Gunn, 1888), 420.

15. O'Connor, 'Charles Bradlaugh'.
16. *Northern Echo* (26 May 1880).
17. Tribe, *President Charles Bradlaugh*, 110.
18. Mackay, *Charles Bradlaugh*, 80.
19. Besant, *Autobiography*, 117.
20. Bradlaugh Bonner Family Papers, Hypatia Bradlaugh Bonner's memorandum.
21. *Bernard Shaw: Collected Letters 1874–1897*, ed. D. H. Laurence (Reinhardt, 1965), 67.
22. Bonner, *Charles Bradlaugh*, ii. 15; H. Hyndman, *Further Reminiscences* (Macmillan, 1912), 3.
23. *National Reformer* (30 Aug. 1874).

Chapter 8

1. Bradlaugh Bonner Family Papers, Hypatia Bradlaugh Bonner's memorandum. And see Ch. 6n. 39.
2. *National Reformer* (4 Sept. 1874).
3. D. Tribe, *President Charles Bradlaugh MP* (Elek, 1971), 160.
4. Ibid.; H. P. Bonner, *Charles Bradlaugh* (Fisher Unwin, 1895), esp. vol. i.
5. *National Reformer* (11 Oct. 1874).
6. Ibid.
7. Ibid.
8. PRO, Chancery, J. 4, 496, 2642.
9. Bradlaugh Papers, National Secular Society, 392, Charles Watts to Bradlaugh, 19 Dec. 1874.
10. *National Reformer* (30 Aug. 1874).
11. Ibid. (6 Sept. 1874).
12. Quoted in A. Nethercot, *The First Five Lives of Annie Besant* (Hart-Davis, 1961), 83.
13. *DNB*, 'C. M. Davies' (1828–1910).
14. C. M. Davies, *Unorthodox London* (Tinsley Bros., 1873), 137.
15. Besant, *Autobiography*, 165.
16. PRO, Chancery, J. 4, 497, 5999.
17. *National Reformer* (6 Sept. 1874).
18. British Library Add. 44448, fo. 304, Scott to Gladstone, 16 Dec. 1875.
19. Besant, *Autobiography*, 170.
20. Bradlaugh Papers, 184.
21. C. M. Davies, *Orthodox London* (Tinsley Bros., 1875), 50.
22. *Weekly Dispatch* (8 June 1879).
23. *T. P.'s Weekly* (21 Aug. 1903).
24. M. Quin, *Memoirs of a Positivist* (Allen & Unwin, 1924) 53.
25. [George Drysdale], *Elements of Social Science; or, Physical, Sexual and Natural Religion* (E. Truelove, 1861), 15, 162, 338, 347.
26. Nethercot, *The First Five Lives*, 108.
27. Besant, *Autobiography*, 53.
28. Annie Besant, *Auguste Comte: His Philosophy, His Religion and His Sociology* (C. Watts, 1875).

29. Moncure Daniel Conway Papers, Rare Book and Manuscript Library, Columbia University, New York, Annie Besant to Ellen Conway, 10 Mar. 1875.
30. G. W. Foote, *Reminiscences of Charles Bradlaugh* (Progressive Publishing Company, 1891).
31. Bradlaugh Papers, correspondence with Prince Napoleon, 1872, 1873, 1876.
32. Conway Papers, Annie Besant to Ellen Conway, 4 May 1875.
33. *National Reformer* (22 Aug. 1875).
34. *DNB*, 'George Henry Lewis' (1833–1911).
35. PRO, Chancery, J. 4, 497, 2974 and 5999.
36. Ibid.
37. Ibid.
38. Ibid.

Chapter 9

1. Conway Papers, Rare Book and Manuscript Library, Columbia University, New York, Annie Besant to Ellen Conway, 2 June 1888.
2. Besant, *Autobiography*, 44.
3. M. D. Conway, *Autobiography, Memories and Experiences* (Cassell, 1904), ii. 361.
4. Conway Papers, Bradlaugh to Moncure Conway, 15 Dec. 1875.
5. Bradlaugh Papers, National Secular Society, 448.
6. Besant, *Sketches*, 109.
7. Besant, *Autobiography*, 152.
8. M. I. Burtis, *Moncure Conway 1832–1907* (Rutgers University Press, 1952), 156.
9. *National Reformer* (17 Sept. 1876). Printed in E. Royle, *The Infidel Tradition from Paine to Bradlaugh* (Macmillan, 1976), 215–18.
10. Besant, *Sketches*, 103.
11. Besant, *Autobiography*, 173.
12. Conway Papers, Annie Besant to Ellen Conway, 4 May 1876.
13. Ibid., Annie Besant to Moncure Conway, 8 Apr. 1876.
14. Bradlaugh Bonner Family Papers, Hypatia Bradlaugh Bonner's memorandum.
15. Ibid.
16. Ibid.
17. Ibid., Annie Besant to Hypatia Bradlaugh Bonner, 4 Mar. 1891, 'my possible legacy under the Wood will'. Kitty O'Shea was descended from Annie's great uncle Alderman Matthew Wood.
18. John Galsworthy, *The Man of Property* (Penguin Books 1951), 84.
19. PRO, Chancery, J. 4, 496, 2641.
20. *National Reformer* (18 and 25 June 1876).
21. British Library Add. 45452, fo. 157, Gladstone to Annie Besant, 23 Nov. 1876.
22. National Secular Society, *Almanack* for 1877.
23. Bradlaugh Papers, 478, Charles Watts to Henry Cook, 11 Dec. 1876.
24. *National Reformer* (11 Feb. 1877).

25. Kate Watts, *Mrs Watts Reply to Mr Bradlaugh's Misrepresentations* (Feb. 1877).
26. Ibid.
27. *National Reformer* (11 Feb. 1877).
28. Ibid.
29. For Watts's character and career, see E. Royle, *Radicals, Secularists and Republicans* (Manchester University Press, 1980), 97.
30. Watts, *Mrs Watts Reply*.
31. Bradlaugh Papers, 493 and 506; *National Reformer* (11 Feb. 1877).
32. Bradlaugh Papers, 488.

Chapter 10

1. 20 & 21 Victoria, c. 83.
2. J. MacCabe, *Life and Letters of G. J. Holyoake* (Watts & Co., 1908), ii. 82.
3. J. C. Smith and B. Hogan, *Criminal Law* (Butterworth, 1988), 729.
4. Charles Knowlton, *Fruits of Philosophy: An Essay on the Population Question* (Freethought Publishing Co., 1877), preface.
5. See, among others, N. Himes, *Medical History of Contraception* (repr. Gamut Press, 1963); R. A. Soloway, *Birth Control and the Population Question in England, 1877–1930* (University of Carolina Press, 1982); J. A. Banks, *Prosperity and Parenthood: A Study of Family Planning among the Victorian Middle Classes* (Routledge, 1954).
6. British Library Add. 37949, Robert Owen to Francis Place, 17 Mar. 1818; Add. 35153, Francis Place to T. Hodgskin, 8 Sept. 1819; *Republican*, 11 (1825) (article by Thomas Carlile).
7. R. Leopold, *Robert Dale Owen* (Octagon Books, 1969), 60, 61.
8. Ibid.
9. Kate Watts, *Mrs Watts Reply to Mr Bradlaugh's Misrepresentations* (Feb. 1877); Bradlaugh Bonner Family Papers, Hypatia Bradlaugh Bonner's memorandum.
10. MacCabe, *Life and Letters*, ii. 80.
11. Ibid. ii. 82.
12. H. B. Bonner, *Charles Bradlaugh* (Fisher Unwin, 1895), ii. 23.
13. Bradlaugh Bonner Family Papers, Hypatia's memorandum.
14. B. and P. Russell (eds.), *Amberley Papers* (Allen & Unwin, 1937), ii. 167.
15. Ibid.
16. Bonner, *Charles Bradlaugh*, ii. 18.
17. *National Reformer* (15 Apr. 1877).
18. Ibid. (27 May 1877).
19. M. D. Conway, *Autobiography, Memories and Experiences* (Cassell, 1904), ii. 261.
20. Bonner, *Charles Bradlaugh*, ii. 21.
21. Ibid. ii. 23.
22. Russell and Russell, *Amberley Papers*, ii. 247; but see MacCabe, *Life and Letters*, ii. 63.

23. Quoted in J. and O. Banks, *Feminism and Family Planning in Victorian England* (Liverpool University Press, 1964), 93.
24. Ibid. 92; Soloway, *Birth Control*, 134.
25. J. and O. Banks, 'The Bradlaugh/Besant Trial and the English Newspapers', *Population Studies*, 8 (1954-5), 22-34.
26. Ibid.
27. *National Reformer* (10 June 1877).
28. Later Lord Halsbury.
29. Bonner, *Charles Bradlaugh*, ii. 22.
30. *The Queen* v. *Charles Bradlaugh and Annie Besant*, printed in full in the *National Reformer*, special issue (June 1877).
31. E. Royle, *Radicals, Secularists and Republicans* (Manchester University, Press, 1980), 251.
32. *The Queen* v. *CB and AB*.
33. See Charles Darwin, *The Descent of Man*: 'Our natural rate of increase, though leading to many and obvious evils, must not be greatly diminished by any means. There should be open competition for all men; and the most able should not be prevented by laws or customs from succeeding best and rearing the largest number of offspring.' Quoted in Ronald W. Clark, *The Survival of Charles Darwin* (Weidenfeld & Nicolson, 1984), 207.
34. Bonner, *Charles Bradlaugh*, ii. 24.
35. *The Queen* v. *CB and AB*.
36. Banks, *Prosperity and Parenthood*, 155.
37. *The Queen* v. *CB and AB*.
38. Bradlaugh was 'polite, insinuating and deferential', before judges; 'he...had a...superstitious veneration' for them. G. W. Foote, *Reminiscences of Charles Bradlaugh* (Progressive Publishing Co., 1891).
39. Besant, *Autobiography*, 187.

Chapter 11

1. M. D. Conway, *Autobiography, Memories and Experiences* (Cassell, 1904), ii. 265.
2. In his autobiography *The Record of an Adventurous Life* (Macmillan, 1911), 337, Henry Hyndman called secularism 'the fanaticism of negation', and Bradlaugh its prophet, priest, and king.
3. J. and O. Banks, *Feminism and Family Planning in Victorian England* (Liverpool University Press, 1964), 90.
4. Annie Besant, *Law of Population*, preface to 1882 edition. Some people thought the steep decline in the birth rate that occurred after 1877 was a direct result of the Bradlaugh/Besant trial, e.g. J. M. Robertson, *Charles Bradlaugh* (Watts & Co., 1920), 63. And see R. K. Ensor, *England 1870-1914* (Oxford University Press, repr. Readers Union, 1964), 104 n. A more modern view is that it was the result of many different causes, one of which was the spread of knowledge as a result of the trial.
5. *Annual Register* (1877), 96.
6. 27 & 28 Victoria, c. 85; 29 & 30 Victoria, c. 35; 32 & 33 Victoria, c. 96.

7. Besant, *Law of Population*.
8. H. B. Bonner, *Charles Bradlaugh* (Fisher Unwin, 1895), ii. 32.
9. Besant, *Autobiography*, 156.
10. Bonner, Charles Bradlaugh, ii. 13.
11. Bradlaugh Bonner Family Papers, Hypatia Bradlaugh Bonner's memorandum.
12. Ibid.
13. See p. 124 above.
14. Bodleian Library Eng. Misc. 182, W. T. Stead to Olga Novikoff, 26 June 1880.
15. J. P. Gilmour (ed.), *Champion of Liberty: Charles Bradlaugh Centenary Volume* (C. A. Watts, 1933), 39.
16. e.g. Sri Prakasa in *Annie Besant as Woman and as Leader* (Theosophical Publishing House, 1941), 37. But in an interview with the author, Mary Lutyens expressed her belief that they were lovers.
17. Bonner, *Charles Bradlaugh*, ii. 35.
18. Besant, *Autobiography*, 157.
19. Bradlaugh Bonner Family Papers, Hypatia Bradlaugh Bonner's memorandum.
20. Besant, *Autobiography*, 180.
21. British Library Add. 44111, fo. 75, Bradlaugh to Gladstone, 9 June 1869 and 22 June 1881.
22. *National Reformer* (17 Mar. 1878).
23. Besant, *Sketches*, 159. This was to ignore the considerable criticism voiced by, among others, John Bright, Joseph Cowen, Joseph Chamberlain, and A. J. Mundella.
24. D. Tribe, *President Charles Bradlaugh MP* (Elek, 1971), 181–2.
25. J. R. Bulwer (ed.), *Law Reports: Queen's Bench Division*, iii: 1877–78 (incorporated Council for Law Reporting, 1878).
26. Besant, *Sketches*, 161.
27. PRO, Chancery, J. 4, 495, 2265.

Chapter 12
1. PRO, Chancery, J. 4, 496, 2642.
2. Besant, *Sketches*, 159.
3. PRO, Chancery, J. 4, 495, 2265.
4. *The Times* (17 Apr. 1878).
5. *National Reformer* (4 May 1878).
6. *DNB*, 'Sir George Jessel' (1824–83).
7. Besant, *Autobiography*, 190.
8. G. W. Hemming (ed.), *Law Reports, Chancery, 1879* (Incorporated Council for Law Reporting, 1879), ii. 508–9.
9. M. D. Conway, *Autobiography, Memories and Experiences* (Cassell, 1904), ii. 261; *Dictionary of American Biography*, 'Thomas Wentworth Higginson' (1823–1911).
10. Hemming, *Law Reports*; *National Reformer*, special issues (26 May, 2 June 1878).

11. Quoted in S. Maccoby, *English Radicalism 1853-86* (Allen & Unwin, 1938), 255 n.
12. PRO, Chancery, J. 4, 497, 2643.
13. Ibid. 3048.
14. Hemming, *Law Reports*; *National Reformer* (2 June 1878).
15. Conway, *Autobiography*, ii. 265.
16. *National Reformer* (26 May 1878).
17. H. B. Bonner, *Charles Bradlaugh* (Fisher Unwin, 1895), ii. 37.
18. Besant, *Autobiography*, 193.
19. *National Reformer* (9 June 1878).
20. Ibid. (4 Aug. 1878).
21. Ibid. (26 Jan. 1879).
22. Ibid. (13 Apr. 1879).
23. For a similar case see J. Galsworthy, *In Chancery* (Penguin, 1962), 36: Soames Forsyte's advice to Winifred Dartie.
24. Besant, *Autobiography*, 194; Bradlaugh Bonner Family Papers, Hypatia Bradlaugh Bonner's memorandum.
25. Ibid.
26. *National Reformer* (10 Nov.-22 Dec. 1878).
    It was probably more than a coincidence that Annie's pamphlet followed hard on the sensation created by H. B. Hyndman's article in the October issue of the *Nineteenth Century*, 'The Bankruptcy of India'.
27. Presidential address to 32nd Indian National Congress, Calcutta, Dec. 1917.
28. *The Times* (5 Apr. 1879).
29. *National Reformer* (13 Apr. 1879).
30. Ibid. (6 Sept. 1879).
31. Ibid.
32. Ibid.
33. Arthur Digby Besant, *The Besant Pedigree* (Besant & Co., 1930), 211-12.
34. Ibid.
35. *National Reformer* (12 Oct. 1879).
36. Ibid.

Chapter 13

1. Archives of the Theosophical Society at Adyar, Madras, India (afterwards referred to as Adyar Archives).
2. *National Reformer* (4 Apr. 1879).
3. *Secular Review* (5 Apr. 1879). Under pressure from AB's supporters Thurlow renounced office and she was reinstated.
4. *National Reformer* (19 Sept. 1880).
5. Bradlaugh Bonner Family Papers, Hypatia Bradlaugh Bonner's memorandum.
6. Women were not admitted to the English Bar until 1921.
7. 'Thorough' was a sobriquet of Bradlaugh's.
8. A. Nethercot, *The First Five Lives of Annie Besant* (Hart-Davis, 1961), i. 153.
9. *National Reformer* (9 Feb. 1879).

10. Announcements of his lectures stated 'Ladies specially invited'.
11. Bradlaugh Bonner Family Papers, Hypatia Bradlaugh Bonner's memorandum.
12. *National Reformer* (19 Jan. 1879).
13. See p. 20 above.
14. *Republican* (Dec. 1881).
15. In 1884 Aveling entered into a common law marriage with Eleanor, youngest daughter of Karl Marx.
16. Besant, *Autobiography*, 220.
17. *National Reformer* (7 June 1879).
18. Havelock Ellis in *Adelphi*, NS 10, (Apr.–Sept. 1935).
19. Bradlaugh Bonner Family Papers, Hypatia Bradlaugh Bonner's memorandum. Kathleen, reprieved, was sold to the Army.
20. Ibid.
21. Hesketh Pearson, *Bernard Shaw: His Life and Personality* (Reprint Society, 1948), 126.
22. Friedrich Engels to R. Sorge, 8 Aug. 1887. Quoted in Thompson, *William Morris: Romantic to Revolutionary* (Lawrence & Wishart, 1955), 435 n.
23. Bradlaugh Bonner Family Papers, Hypatia's memorandum.
24. Ibid.
25. Bradlaugh Papers, National Secular Society, 559.
26. *National Reformer* (14 Dec. 1879).
27. Ibid. (1 Feb. 1880).
28. Ibid. (10 Aug. 1879).
29. Thompson, *William Morris*, 431.
30. H. Salt, *Seventy Years among Savages* (Allen & Unwin, 1921), 80.
31. Bradlaugh Papers, 1122, 1131.
32. F. G. Bettany, *Stuart Headlam: A Biography* (John Murray, 1926).
33. *Republican* (Dec. 1881).
34. *National Reformer* (19 Jan. 1879).
35. *The Times* (11 Feb. 1880).
36. *Annual Register* (1879).
37. *National Reformer*, various dates in first three months of 1880.
38. Ibid. (29 Mar. 1878).
39. H. B. Bonner, *Charles Bradlaugh* (Fisher Unwin, 1895), ii. 192. This was the same Kelly whose arrest precipitated the execution of the Manchester martyrs.
40. J. M. Robertson, *Charles Bradlaugh* (Watts & Co., 1920), 37.
41. Besant, *Autobiography*, 63.
42. Elizabeth Longford, *A Pilgrimage of Passion: The Life of Wilfrid Scawen Blunt*, (Weidenfeld & Nicolson, 1979), 402.
43. *The Times* (11 Feb. 1880).
44. Ibid., Burrows's obituary (20 Dec. 1922).
45. Besant *Autobiography*, 227.

Chapter 14
1. Besant, *Autobiography*, 160.

2. The best account of this long and complicated affair is in W. L. Arnstein, *The Bradlaugh Case: A Study in Late Victorian Opinion and Politics* (Clarendon Press, 1965).
3. See e.g. Hansard, 3rd Series, (1880), e.g. cclii, col. 20; ccliii, cols. 622–31, 647–9; ccliv, col. 463.
4. *The Times* (24 June 1880).
5. F. and R. Black (eds.), *The Harney Papers* (Internationaal Instituut vor Sociale Geschiedenis, 1969), 185.
6. *The Times* (25 June 1880).
7. Arnstein, *The Bradlaugh Case*, 200.
8. *National Reformer* (4 July 1880).
9. Ibid.
10. Besant, *Autobiography*, 232. 'Tory lordlings' was a reference to Lord Randolph Churchill, who made the question of the oath an issue in his attempt to launch the so-called Fourth Party. See Arnstein, *The Bradlaugh Case*, 44–7, 195–6.
11. Arnstein, *The Bradlaugh Case*, 80–1.
12. *National Reformer* (5 Sept. 1880).
13. Ibid. (26 Dec. 1880).
14. H. Varley, *An Address to the Electors in the Borough of Northampton* (John F. Shaw & Co., 1881).
15. BL Add. 44111, fo. 93.
16. Arnstein, *The Bradlaugh Case*, 119 and n.
17. Besant, *Autobiography*, 237–8.
18. T. P. O'Connor, in J. P. Gilmour (ed.), *Champion of Liberty: Charles Bradlaugh Centenary Volume* (C. A. Watts, 1933), 45.
19. G. Lansbury, *Looking Backward and Forward* (Constable, 1935), 169.
20. *The Times* (4 Aug. 1881). And see D. Tribe, *President Charles Bradlaugh MP* (Elek, 1971), 211.
21. *National Reformer* (13 Aug. 1881).
22. Hansard, 3rd Series, cclxiv, cols. 1385–6.
23. So described by, among others, Gladstone's daughter Mary.
24. D. Bahlman (ed.), *The Diary of Sir Edward Hamilton* (Clarendon Press, 1972), ii. 437. The picture was on sale in a shop in St James's.
25. Quoted in Tribe, *President Charles Bradlaugh*, 227.
26. University College London, Council Papers, AB to Ely, 7 May 1883, 2 May 1883; R. Morison to Ely, 3 May 1883.
27. Ibid., petitions for an EGM signed by Huxley and, among others, John Maynard Keynes.
28. Tribe, *President Charles Bradlaugh*, 227. It was Huxley who first used the term 'agnostic'.
29. Bradlaugh Papers, National Secular Society, 2934, F. Pollock to Hypatia Bradlaugh Bonner, 25 May 1912.
30. *National Reformer* (19 Sept. 1880).
31. Royal Botanic Gardens, Kew, Archives, English Letters, 386, AB to Sir J. Hooker, 23 Oct. 1882.
32. Arnstein, *The Bradlaugh Case*, 166.

Chapter 15

1. Besant, *Autobiography*, 285. See e.g. N. and J. Mackenzie, *The First Fabians* (Weidenfeld & Nicolson, 1977), 45.
2. In August 1880, AB and the Bradlaugh girls were on holiday in Jersey. When Aveling joined them, Hypatia thought it 'most extraordinary'. Bradlaugh Bonner Family Papers, Hypatia Bradlaugh Bonner's memorandum.
3. Thompson, *William Morris: Romantic to Revolutionary* (Lawrence & Wishart, 1955), 328.
4. British Library of Political and Economic Science, Coll. Misc. 706. Beesly was a Positivist and trade union adviser. Taylor (J. S. Mill's stepdaughter) had been elected to the London School Board in 1876, and was a staunch supporter of the Irish Land League. Cowen was a former Chartist, MP for Newcastle. Gladstone's Egypt policy ended in the disaster of Gordon's murder at Khartoum.
5. H. Hyndman, *Further Reminiscences* (Macmillan, 1912), 4. And see Ch. 11 n. 2.
6. T. Burt, in *Review of Reviews* (July 1891).
7. C. Bradlaugh, *Labour and Law* (Forder, 1891), 183.
8. *National Reformer* (11 May 1884).
9. Cambridge University Library, MSS division 6257/66, AB to Taylor, 30 Nov. 1883; *The Times* (29 Nov. 1883).
10. *National Reformer* (3 Feb. 1884).
11. Ibid. (10 Feb. 1884).
12. H. Hyndman, *The Text Book of Democracy: England for All* (E. W. Allen, 1881). This presented the Marxist doctrine without, in Marx's view, due acknowledgement.
13. C. Bradlaugh, *Will Socialism Benefit the English People?* (Freethought Publishing Co., 1884).
14. H. Hyndman, *The Record of an Adventurous Life* (Macmillan, 1911), 337.
15. *National Reformer* (27 Apr. 1884).
16. BLPES, Hyndman Papers, 522/1/6.
17. *National Reformer* (11 May 1884).
18. Ibid.
19. *George Bernard Shaw: An Autobiography, 1856–1898: Selected from His Writings*, ed. S. Weintraub (Reinhardt, 1969), 114. The first English translation of *Das Kapital* was by Edward Aveling and Samuel Moore, in 1887.
20. *National Reformer* (25 May 1884).
21. Ibid. (4 May 1884).
22. Bradlaugh Bonner Family Papers, Hypatia's memorandum.
23. *National Reformer* (28 Nov. 1883).
24. Beatrice Webb, *Diary*, ed. N. and J. Mackenzie (Virago/LSE, 1980–5), i. 88. And see Y. Kapp, *Eleanor Marx* (Virago, 1979), i. 205–7, ii. 519.
25. Shaw, *Autobiography*, 114.

26. The author was the Revd Andrew Mearns. See A. S. Wohl, 'The Bitter Cry of Outcast London', *International Review for Social History*, 13 (1968), 189–245.
27. Hyndman, *The Record*, 51.
28. *The Correspondence of Friedrich Engels and Paul and Laura Lafargue*, trans. Y. Kapp (Lawrence & Wishart, 1960), i. 168.
29. Marx called this 'a calumny, ridiculous as it is infamous', D. Tribe, *President Charles Bradlaugh MP* (Elek, 1971), 124–7.
30. *National Reformer* (13 Jan. 1884).
31. Bradlaugh Papers, National Secular Society, 1128.
32. *National Reformer* (23 Dec. 1883).
33. Bradlaugh Papers, 939.
34. Ibid. 1156.
35. Ibid. 1036.
36. *National Reformer* (8 June and 27 July 1884).
37. Bradlaugh Papers, 1127.
38. Ibid. 1128.
39. Ibid. 1130. Havelock Ellis tells a story, said to have originated with Eleanor Marx, of Bradlaugh coming round to the lodging she shared with Aveling to repossess letters written to the latter by AB. *Adelphi*, NS 10, (1935), Ellis on 'Eleanor Marx'.
40. Bradlaugh Papers, 585.
41. Ibid. 1163.
42. BLPES, Hyndman Papers, 522/14a, Burrows to Hyndman, 11 Aug. 1884.
43. Ibid. 522/1/59, Burrows to Champion, 15 Sept. 1884.
44. *Justice* (27 Sept. 1884). William Morris told a colleague that while the Aveling matter did not come to 'extremity', it made enmity with Hyndman. BL Add. 45345, fo. 84, Morris to Joynes, 25 Dec. 1884.
45. Bradlaugh Papers, 1150.
46. Ibid. 1158.
47. Ibid. 1163.
48. See e.g. Thompson, *William Morris*, 453; E. Bernstein, *My Years of Exile: Reminiscences of a Socialist* (Parsons, 1921), 162.
49. *National Reformer* (12 Oct. 1884).
50. Bradlaugh Bonner Family Papers, Note by? (probably Arthur Bonner).

Chapter 16

1. *Justice* (21 June 1884).
2. Ibid. (12 July 1884).
3. *National Reformer* (26 Oct. 1884).
4. Ibid. (21 Dec. 1884). Besant, *Autobiography*, 275.
5. *National Reformer* (25 Jan. 1885).
6. Besant, *Autobiography*, 273.
7. *National Reformer* (5 Apr. 1885); Besant, *Autobiography*, 281.
8. Besant, *Sketches*, 37.
9. *National Reformer* (17 Aug. 1884).

10. National Library of Scotland, Geddes Papers, 10523, fo. 229, Robertson to Geddes, 1 Nov. 1884. For Geddes (1854–1932), see *DNB*. Robertson describes him here as 'botanist, morphologist, sociologist, psychologist, economist ... and reformer of things in general'.
11. 'Mrs Besant as a Fabian Socialist', *Theosophist* (Oct. 1917); G. B. Shaw, *An Autobiography, 1856–1898*, ed. S. Weintraub (Reinhardt, 1969), 101.
12. M. Holroyd, *Bernard Shaw: The Search for Love* (Chatto, 1988), i. 83.
13. *Bernard Shaw: Collected Letters, 1874–1897*, ed. D. H. Laurence (Reinhardt, 1965), 113.
14. *Bernard Shaw: The Diaries, 1885–1897*, ed. S. Weintraub, (Pennsylvania State University Press, 1986), entry for 21 Jan. 1885.
15. Shaw, *An Autobiography*, 141.
16. Shaw, *Diary*, 31 Jan. 1885.
17. Holroyd, *Bernard Shaw*, 168.
18. Besant, *Autobiography*, 276–7.
19. Conway Papers, Rare Book and Manuscript Library, Columbia University, New York, AB to Ellen Conway, 3 Apr. 1885.
20. *National Reformer* (22 and 29 Mar. 1885).
21. See e.g. Holroyd, *Bernard Shaw*, i. 167; G. M. Williams, *The Passionate Pilgrim* (A. Knopf, 1946), 133.
22. Bradlaugh Papers, National Secular Society, 1093, 30 Jan. 1884.
23. Bradlaugh Bonner Family Papers, Hypatia Bradlaugh Bonner's memorandum.
24. Conway Papers, AB to Moncure D. Conway, 28 Dec. 1886.
25. Annie Besant, *Why I am a Socialist* (printed by Annie Besant and Charles Bradlaugh, 63 Fleet Street, 1886), price 1d.
26. Holroyd, *Bernard Shaw*, i. 131.
27. E. R. Pease, *The History of the Fabian Society* (Fifield, 1916), 31.
28. Holroyd, *Bernard Shaw*, i. 168.
29. Fabian Society Papers, C. 36, Minutes of 19 June 1885.
30. BL Add. 45, 345, fo. 107, AB to W. Morris, 9 Mar. 1886.
31. Annie Besant, *Modern Socialism* (Freethought Publishing Co., 1886).
32. Besant, *Autobiography*, 258.
33. Quoted in Nethercot, *The First Five Lives of Annie Besant* (Hart-Davis, 1961), 236.
34. Shaw, *Diary*, 7 Aug. 1885.
35. Besant, *Autobiography*, 282.
36. Besant, *Why I am a Socialist*.
37. Shaw, *Diary*, 23 Sept. 1885.
38. Besant, *Autobiography*, 283.
39. Shaw, *An Autobiography*, 140.
40. Conway Papers, AB to Ellen Conway, 3 Dec. 1885.
41. *Annual Register* (10 Feb. 1886).
42. See e.g. *The Times* (11 Feb. 1886 and following issues).
43. Fabian Society Papers, Executive Committee Minutes, C. 1, 17 Feb. 1886.
44. Ibid.
45. Conway Papers, AB to Ellen Conway, 3 Dec. 1885.

46. *Our Corner* (July 1886); Shaw, *Diary*, Note, i. 175.
47. Fabian Society Papers, Nuffield College, Oxford, C. 36; Pease, *History of the Fabian Society*, 67.
48. Fabian Society Papers, 53/2, Manifesto of the Fabian Parliamentary League.
49. Nethercot, *The First Five Lives*, i. 252; Shaw, *Diary*, 8 July 1887. In the 'Socialist' government, Champion was Prime Minister; Bland, Foreign Secretary; Webb, Chancellor; Besant, Home Secretary; Headlam, Irish Secretary; Shaw, President of the Local Government Board.
50. Shaw, *An Autobiography*, 139.
51. Pease, *History of the Fabian Society*, 98.
52. Ibid. 64.
53. Ibid. 60.
54. G. B. Shaw, 'Mrs Besant as a Fabian Socialist', *Theosophist* (Oct. 1917).
55. Pease, *History of the Fabian Society*, 61.
56. Besant, *Autobiography*, 281.
57. E. Carpenter, *My Days and Dreams* (Allen & Unwin, 1916), 221.
58. Bradlaugh Bonner Family Papers, Hypatia's memorandum.
59. Conway Papers, AB to Ellen Conway, 3 Dec. 1885.
60. Bradlaugh Bonner Family Papers, Hypatia's memorandum.
61. It was held in the Adelphi.
62. Shaw, *Diary*: various entries.
63. BL Add. 50,529, fo. 62, AB to GBS, 11 June 1887.
64. See e.g. Shaw, *Diary*, 19 Mar. 1887.
65. S. Winsten, *Days with Bernard Shaw* (Hutchinson, 1951), 111; H. Pearson, *Bernard Shaw: His Life and Personality* (Reprint Society, 1948), 115.
66. Shaw, *An Autobiography*, 139.
67. Winsten, *Days with Bernard Shaw*, 111. And see Shaw, *Diary*, 20 May 1887.
68. Shaw, *An Autobiography*, 140.
69. Besant, *Autobiography*, 274.
70. Bradlaugh Bonner Family Papers, Hypatia's memorandum.
71. Ibid.
72. Shaw, *Letters*, i. 210.
73. Pearson, *Bernard Shaw*, 115.
74. Shaw, *Diary*, i. 366; Holroyd, *Bernard Shaw*, i. 221. (The woman was Geraldine Spooner.)
75. Shaw, *Letters*, i. 107.
76. Shaw, *Diary*, 25 Dec. 1887, and i. 34.
77. Pearson, *Bernard Shaw*, 116.
78. Shaw, *Diary*, i. 326.
79. Shaw, *An Autobiography*, 170.
80. Stead Papers, Churchill College, Cambridge, AB to WTS, 17 Apr. 1888.
81. NLS, Geddes Papers, 10524, fo. 115, Robertson to Geddes, 9 Mar. 1887.
82. Shaw, *Diary*, 19 Feb. 1886.
83. Ibid. i. 34.

84. NLS, Geddes Papers, 10524, fo. 137, Robertson to Geddes, 8 Nov. 1887.
85. Stead Papers, Dec. 1887–Apr. 1888.

Chapter 17
 1. *National Reformer* (23 Oct. 1887); Besant, *Autobiography*, 290–1.
 2. Besant, *Autobiography*, 292.
 3. *Our Corner* (Mar. 1887). The Coercion Bill was known as the Crimes Act.
 4. Ibid.
 5. *Commonweal* (22 Oct. 1887).
 6. *Our Corner* (Nov. 1887); *Justice* (19 Nov. 1887).
 7. *Commonweal* (26 Nov. 1887), 'The Truth about the Unemployed', by one of Them [James Allmann].
 8. Ibid.
 9. *Bernard Shaw: The Diaries, 1885–1897*, ed. S. Weintraub (Pennsylvania State University Press, 1986), 7 Nov. 1887, *The Times* (8 Nov. 1887); *Commonweal* (26 Nov. 1887).
10. *Commonweal* (26 Nov. 1887).
11. Ibid.; *Our Corner* (Dec. 1887).
12. BL Add. 46288, fo. 85, AB to Mrs Burns, 28 Jan. 1888.
13. *The Times* (12 Nov. 1887).
14. Ibid. (10 Nov. 1887).
15. *National Reformer* (20 Nov. 1887).
16. Ibid.
17. Ibid.; *Pall Mall Gazette* (9 Nov. 1887).
18. *The Times* (14 Nov. 1887).
19. Ibid.; *Pall Mall Gazette* (14 Nov. 1887).
20. Shaw, *Diary*, 13 Nov. 1887.
21. *Pall Mall Gazette* (15. Nov. 1887).
22. Ibid. (14. Nov. 1887).
23. *The Times* (14 Nov. 1887). The magistrate was prepared to read it, but was not called upon to do so.
24. Text printed by, among others, *The Times* (15 Nov. 1887).
25. Bradlaugh Papers, National Secular Society, 1439, Gladstone to CB, 2 Dec. 1887.
26. *The Times* (17 Nov. 1887); Shaw, *Diary*, 16 Nov. 1887.
27. Shaw, *Diary*, 16 Nov. 1887.
28. BL Add. 45345, fo. 138, GBS to Morris, 22 Nov. 1887.
29. Besant, *Autobiography*, 296; *Commonweal* (26 Nov. 1887).
30. *Pall Mall Gazette* (15 Nov. 1887).
31. Besant, *Autobiography*, 296; *National Reformer* (18 and 25 Dec. 1887).
32. *Commonweal* (24 Dec. 1887).
33. BL Add. 46288, fo. 37, AB to Mrs Burns, 21 Jan. 1888.
34. Stead Papers, Churchill College, Cambridge, AB to WTS, 22 Jan. 1888.
35. *The Times* (15 Nov. 1887).
36. *Pall Mall Gazette* (17 Nov. 1887).
37. *National Reformer* (20 Nov. 1887).

38. Letter by Bradlaugh in the *Link* (31 Mar. 1888).
39. 'I find the work harder for tongue and body than it was thirteen years ago. I feel I shall be all the better when I can afford a little rest.' Bradlaugh in Northumberland, *National Reformer* (16 Nov. 1887).
40. Stead Papers, AB to WTS, 20 Dec. 1887.
41. Ibid. 19 Jan. 1888.
42. Quoted in A. S. Wohl, 'The Bitter Cry of Outcast London', *Int. Review for Social History*, 13, (1968).
43. Frederic Whyte, *The Life of W. T. Stead* (Cape, 1925).
44. H. Pearson, *Bernard Shaw: His Life and Personality* (Reprint Society, 1948), 108. And see Whyte, *Life of Stead*, ii. 341.
45. Wohl, 'Bitter Cry of Outcast London'. And see J. Robertson Scott, *The Life and Death of a Newspaper (Pall Mall Gazette)* (Methuen, 1952), 85.
46. Whyte, *Life of Stead*, i. 159–86.
47. Ibid. i. 258.
48. Stead Papers, AB to WTS, 20 Dec. 1887.
49. Ibid. 1 Jan. 1888.
50. Conway Papers, Rare Book and Manuscript Library, Columbia University, New York, AB to Ellen Conway, 2 Jan. 1888.
51. Stead Papers, AB to WTS, 1 Jan. 1888.
52. Bodleian Library, Eng. Misc. d. 182, Stead to O. Novikoff, various dates.
53. *Link* (4 Feb. 1888).
54. *Justice* (14 Jan. 1888); A. Nethercot, *The First Five Lives of Annie Besant* (Hart-Davis, 1961), i. 265.
55. BLPES, Wallas Collection, AB to Graham Wallas, 9 Jan. 1888.
56. Stead Papers, AB to WTS, 27 Mar. 1888.
57. Ibid. 4 Mar. 1888.
58. Ibid. 8 Mar. 1888.
59. See p. 24 above.
60. Stead Papers, AB to WTS, 20 Apr. 1888.
61. W. T. Stead, *The MP for Russia: Memories and Correspondence of Mde Olga Novikoff* (Melrose, 1909), ii. 238.
62. Whyte, *Life of Stead*, ii. 313.
63. Stead Papers, AB to WTS, 3 July 1888.

Chapter 18

1. *Link* (11 Feb. 1888).
2. Ibid. (4 Feb. 1888). Davitt also called for a petition to Parliament asking for an enquiry into the conduct of the police on 13 November 1887 and, in the longer term, for the Metropolitan force to be under the control of a body elected by the citizens of London.
3. *National Reformer* (23 Dec. 1887); *Justice* (14 Jan. 1888).
4. *Link* (25 Feb. 1888); *Justice* (18 and 25 Feb. 1888).
5. *Commonweal* (28 Jan. 1888); H. B. Bonner, *Charles Bradlaugh* (Fisher Unwin, 1895), ii. 387.
6. *Link* (21 Apr. 1888).

7. Ibid. (16 June 1888).
8. Ibid. And see Hansard, 3rd Series, cccxxvi (1888), cols. 437–8.
9. *Link* (19 May, 14 July 1888).
10. Parliamentary Papers, 'First Report of the Lords' Committee on the Sweating System, together with the Proceedings of the Committee, Minutes of Evidence, and Appendices' (1888), xx. 1–2.
11. *Link* (16 June 1888).
12. *Justice* (3 Mar. 1888). 'The inspectors come with naive delight to visit any person who has information to give about the people over whose welfare they are supposed to watch.' Margaret Harkness, surveying female labour in the City.
13. Bradlaugh's first successful attempt at legislation.
14. Truck Amendment Act 1887.
15. *Link* (16 June 1888).
16. Stead Papers, Churchill College, Cambridge, AB to WTS, 4 Mar. 1888.
17. Hansard, 3rd Series, cccxxvi (1888), cols. 1014–15.
18. 'Lords' Committee on the Sweating System', xx. 248.
19. Bryant & May Papers, Hackney Archives Dept., B/B/BRY, A. C. Bryant to Lord Rothschild, 4 June 1888.
20. *Justice* (26 May 1888).
21. *Link* (23 June 1888). Shaw was sufficiently impressed by the fact that the resolution was put by non-members of the Fabian Society to record it in his diary, 15 June 1888.
22. *Link* (16 June 1888).
23. *Justice* (7 July 1888).
24. *Link* (23 June 1888).
25. *Link* (14 July 1888).
26. Bryant & May Papers, W. W. Bryant to G. Dutton & Sons, 24 July 1888.
27. Ibid., G. P. Bartholomew to Clarke, Nicholls & Coombs, 27 July 1888; id. to Bristow & Carpmael, 27 July 1888; id. to Morris, 2 Aug. 1888. And see *Daily Telegraph* (1 Aug. 1888).
28. *Link* (30 June 1888).
29. *Justice* (7 July 1888).
30. Bryant & May Papers, 'G. P. Bartholomew's account of the strike'.
31. Hansard, 3rd series, cccxxviii (1888), cols. 563–4, 1095.
32. *Justice* (14. July 1888).
33. *Link* (14 July 1888).
34. Printed in full in *Justice* (11 Aug. 1888).
35. Ibid. (14 July 1888).
36. *Star* (14 July 1888).
37. *Link, Justice* (14 July 1888). And see Fabian Society Papers, Nuffield College, Oxford, C. 36/84, 6 July 1888.
38. W. J. Fishman, *East End 1888* (Duckworth, 1988), 259–65.
39. Ben Cooper of the Cigar Makers' Union and Bill Steadman of the Barge Builders. *Link* (21 July 1888).
40. Ibid. For the difference in wages paid before and after the strike see C. Booth (ed.), *Life and Labour of the People in London*, 1st series,

Poverty (Macmillan, 1902), 287. In May 1888, 21.59% of the women employed earned 4–6s. per week; 29.73% earned 6–8s.; 29.63%, 8–10s.; 14.86%, 10–12s.; 3.96%, 12–15s.; 0.23%, over 15s. In September 11.48% earned 4–6s.; 17.97%, 6–8s.; 27.16%, 8–10s.; 30%, 10–12s.; 12.03%, 12–15s.; 1.36%, over 15s.

41. Bryant & May Papers, Bartholomew's account.
42. *Link* (21 July 1888).
43. Besant, *Autobiography*, 306.
44. *Justice* (21 Dec. 1889).
45. S. and B. Webb, *The History of Trade Unionism, 1680–1920* (TU edn., 1919), 402.
46. H. A. Clegg, A. Fox, and A. F. Thompson, *A History of British Trade Unions since 1889, 1889–1910* (Clarendon Press, 1964), i: 55–6, 61, 292.
47. *Justice* (10 Nov. 1888).
48. *Bernard Shaw: The Diaries, 1885–1897*, ed. S. Weintraub (Pennsylvania State University Press, 1986), 29 Aug. 1888.
49. Fabian Society Papers, 63/2.
50. See Ch. 16 n. 54.
51. B. Webb, *Diary*, ed. N. and J. Mackenzie (Virago/LSE, 1980–5), i. 222–3.
52. D. Rubinstein, 'Annie Besant and Stuart Headlam: The London School Board Elections of 1888', *East London Papers*, 13, (1970), 3–24.
53. A. Nethercot, *The First Five Lives of Annie Besant* (Hart-Davis, 1961), appendix to vol. i.
54. *Justice* (3 Nov. 1888).
55. Besant continued the campaign, begun by Helen Taylor, to have the money in City charities redirected to the schools. Florence Fenwick Miller (1854–1935) defied the electors of Hackney to demonstrate her concern for free speech; she did not approve of neo-Malthusianism. See *National Reformer* (7 Dec. 1879).
56. Ibid. (1 Nov. 1885).
57. *Final Report of the School Board for London, 1870–1904* (2nd edn., King, 1904), 202.
58. Quoted in B. Simon, *Education and the Labour Movement 1870–1920* (Lawrence & Wishart, 1965), 119.
59. *Justice* (1 Sept. 1888).
60. Ibid. (17 Nov. 1888).
61. Fishman, *East End 1888*, 292.
62. Stead Papers, AB to WTS, 8 Mar. 1888.
63. Fishman, *East End 1888*, 33–6.
64. *National Reformer* (8 Nov. 1885).
65. Ibid. (9 Dec. 1888).
66. P. Hollis, *Ladies Elect: Women in English Local Government, 1865–1914* (Clarendon Press, 1987). For Henrietta Müller (who later became a Theosophist) see Karl Pearson Papers, University College London, 793, M. Sharpe to Pearson, 12 July 1886, 1 July 1888.
67. Stead Papers, AB to WTS, 24 Dec. 1887.
68. Hollis, *Ladies Elect*, 118.

69. Ibid. 113.
70. Ibid. 114–5; *Justice* (25 Mar. 1889).
71. *Justice* (16 Mar. 1889).
72. *Final Report of the LSB*, 202.
73. Ibid. 207.
74. *Justice* (20 July, 10 Aug. 1889).
75. Carpenter Papers, City of Sheffield Central Library, MSS 386, AB to EC, 7 July 1889. Shaw offered to let AB have *Love among the Artists* for nothing, 'as the magazine is entailing a loss on her', Shaw, *Diary*, 27 May 1888.
76. Bradlaugh Bonner Family Papers, AB to HBB, 4 Mar. 1891.
77. Ibid., Hypatia's memo.
78. Webb, *Diary*, ii: *All the Good Things of Life, 1892–1905*, 322.

Chapter 19

1. H. B. Bonner, *Charles Bradlaugh* (Fisher Unwin, 1895), i 248, 343.
2. Besant, *Autobiography*, 309.
3. *Bernard Shaw: Collected Letters, 1874–1897*, ed. D. H. Laurence (Reinhardt, 1965), ii. 497.
4. Bradlaugh Bonner Family Papers, Hypatia Bradlaugh Bonner's memorandum.
5. B. Webb, *Diary*, ed. N. and J. Mackenzie (Virago/LSE, 1980–5), i. 223.
6. Besant, *Autobiography*, 308.
7. *Justice* (5 May, 9 June 1888).
8. Besant, *Autobiography*, 309.
9. For Besant's doubts see e.g. *National Reformer* (18 June 1888). For the Fox sisters, see *Pall Mall Gazette* (8 Nov. 1888).
10. *DNB*, 'D. D. Home' (1833–86); E. Jenkins, *The Shadow and the Light: A Defence of Daniel Dunglass Home, the Medium* (Hamish Hamilton, 1982).
11. Besant, *Autobiography*, 309.
12. Quoted in A. Nethercot, *The First Five Lives of Annie Besant* (Hart-Davis, 1961), appendix to vol. i.
13. A. P. Sinnett, *The Occult World* (Trübner, 1881); *Esoteric Buddhism* (Trübner, 1883); *Incidents in the Life of Madame Blavatsky* (Redway, 1886).
14. J. Ransom, *A Short History of the Theosophical Society, 1875–1937* (Theosophical Publishing House, 1938).
15. Besant, *Autobiography*, and *Sketches*; B. Webb, *My Apprenticeship* (Longman's, Green & Co., 1926).
16. Webb, *My Apprenticeship*, 83.
17. Webb, *Diary*, ii. 321.
18. Ransom, *Short History*, 1.
19. *DNB*, 'William Winwood Reade' (1838–75).
20. Webb, *My Apprenticeship*, 132.
21. BL Add. 46439, fo. 72, H. P. Blavatsky to A. R. Wallace, 7 Nov. 1875; Ransom, *Short History*, 94.

22. H. P. Blavatsky, *Isis Unveiled* (W. J. Bouton, 1877); H. S. Olcott, *Old Diary Leaves* (Putnam's, 1895), i. 205.
23. BL Add. 46439, fo. 72. The address was 302 W. 47th Street.
24. C. E. Bechofer-Roberts *The Mysterious Madame* (J. Lane/Bodley Head, 1931), 80; Ransom, *Short History*, 40.
25. Annie Besant in the *Review of Reviews* (July 1891).
26. Sinnett, *Incidents*.
27. Ransom, *Short History*, 30–1.
28. Sinnett, *Incidents*, 150–1.
29. e.g. Bechofer-Roberts, *The Mysterious Madame*; V. S. Soloviev, *A Modern Priestess of Isis*, trans. W. Leaf (Longman's, Green & Co., 1895).
30. R. Leopold, *Robert Dale Owen* (Octagon Books, 1969), 400–7; Olcott, *Old Diary Leaves*, i. 34; D. D. Home, *Lights and Shadows of Spiritualism* (Carleton, 1877); Ransom, *Short History*, 63, 76.
31. Ransom, *Short History*, 78.
32. Home, *Lights and Shadows*, 324, 327. Home objected to reports that the Society had sent a member to Africa in search of a magician, and to the publicity given to its support for the new practice of cremation.
33. Professor A. Aksakoff, Prince E. Wittgenstein; see Ransom, *Short History*, 16, 17. D. D. Home's second wife was related to Aksakoff.
34. Sinnett, *Incidents*, 239.
35. *The Collected Letters of W. B. Yeats*, ed. J. Kelly, 1865–1895 (Clarendon Press, 1986), 164, Yeats to J. O'Leary, 7 May 1889.
36. M. D. Conway, *My Pilgrimage to the Wise Men of the East* (A. Constable, 1906), 200; J. Farquhar, *Modern Religious Movements in India* (Macmillan, 1915), 226.
37. Ransom, *Short History*, 85–6; Bechofer-Roberts, *The Mysterious Madame*, 84.
38. Bechofer-Roberts, *The Mysterious Madame*, 76.
39. Bodleian Library, Dep. d. 71, fo. 139, Olcott to Max Müller, 25 Apr. 1893.
40. Yeats, *Collected Letters* 164.
41. Adyar Archives K 2/8.
42. Ransom, *Short History*, 35–6; *Dictionary of American Biography*, 'H. S. Olcott' (1832–1907); H. Murphet, *The Hammer on the Mountain: The Life of Henry Steele Olcott* (Theosophical Publishing House, Wheaton, 1972).
43. C. Sandberg, *Abraham Lincoln* (Harcourt Brace, 1939), iii. 460, iv. 127.
44. Olcott, *Old Diary Leaves*, i. 382–3; Ransom, *Short History*, 98; Adyar Archives, A. 21.
45. *National Reformer* (20 July 1879).
46. Ransom, *Short History*, 128.
47. Ibid.
48. J. D. Smith, *Servant of India: The Diary of James Dunlop Smith*, ed. M. Gilbert (Longman's, Green & Co., 1966), 175 n.
49. Olcott, *Old Diary Leaves*, ii. 25; Ransom, *Short History*, 50–1.

50. *DNB*, 'A. O. Hume' (1829–1912). Hume's father Joseph, the radical MP, was one of Bradlaugh's heroes. And see Penderel Moon, *The British Conquest and Dominion of India* (Duckworth, 1989), 885.
51. Bradlaugh Papers, National Secular Society, 1166, Hume to—Knight, Sept. 1884.
52. Moon, *The British Conquest*, 889. Bradlaugh gave much coverage to Indian affairs in the *National Reformer*; see e.g. his support for the Ilbert Bill in the latter (Sept.–Oct. 1883).
53. Adyar Archives, A. 21.
54. Olcott, *Old Diary Leaves*, i. 245–7.
55. See e.g. *National Reformer* (22 Dec. 1878), 25 Apr. 1880, 20 Jan. 1884, and esp. 17 Feb. 1884).
56. Ibid. (7 Nov. 1880).
57. Ibid. (18 June 1882).
58. *Theosophist* (Aug. 1882).
59. Advance information and secret service cash. See B. B. Misra, *The Indian Political Parties* (Oxford University Press, 1976), 88.
60. BL Add. 45287 (presented in 1939).
61. Ransom, *Short History*, 209–10.
62. W. T. Stead, *The MP for Russia: Memories and Correspondence of Mde Olga Novikoff* (Melrose, 1909), i. 132.
63. *National Reformer* (2 Nov. 1884).
64. Bradlaugh Bonner Family Papers, Hypatia Bradlaugh Bonner's memorandum.
65. H. P. Blavatsky, *The Secret Doctrine*, 2 vols. (Theosophical Publishing Co., 1888).
66. Ransom, *Short History*, 215.
67. Stead, *The MP for Russia*, i. 130.
68. Besant, *Autobiography*, 311.
69. Bradlaugh Bonner Family Papers, Letter by? [Arthur Bonner]. And see G. W. Foote, *An Open Letter to Madame Blavatsky* (Progressive Publishing Co., 1889).

Chapter 20

1. Constance Wachtmeister, *HPB and the Present Crisis in the Theosophical Society* (London, 1895).
2. A. Nethercot, *The First Five Lives of Annie Besant* (Hart-Davis, 1961), appendix, AB to Ashman, 22 Mar. 1889.
3. E. Carpenter, *My Days and Dreams* (Allen & Unwin, 1916), 244.
4. Besant, *Autobiography*, 310.
5. Ibid. 318; 'The Truth of Reincarnation', *Bibby's Annual* (1912). And see Annie Besant, *The Ancient Wisdom: An Outline of Theosophical Teachings* (London, 1897), 339; V. Chirol, 'India in Travail', *Edinburgh Review*, 228 (1918).
6. *Pall Mall Gazette* (6 Jan. 1887).
7. *Theosophist* (Oct. 1917).
8. Ibid.

9. *Times of Ceylon* (14 Mar. 1947). And see *Bernard Shaw: The Diaries, 1885–1897*, ed. S. Weintraub (Pennsylvania State University Press, 1986), 19 Mar. 1889.
10. Besant, *Autobiography*, 312–13.
11. *Pall Mall Gazette* (6 Jan. 1887, 3 Jan. 1889).
12. M. D. Conway, *My Pilgrimage to the Wise Men of the East* (A. Constable, 1906), 205.
13. Ibid.
14. Annie Besant, *Why I Became a Theosophist* (Freethought Publishing Co., 1889).
15. Bradlaugh Bonner Family Papers, Hypatia Bradlaugh Banner's memorandom.
16. *National Reformer* (30 June 1889), printed in Besant, *Autobiography*, 319–21.
17. M. Holroyd, *Bernard Shaw: The Search for Love* (Chatto, 1988), 188.
18. BLPES, Coll. Misc., 98, fo. 360.
19. Annie Besant, 'Industry under Socialism', in *Fabian Essays in Socialism* (Fabian Society, 1889).
20. S. Webb, preface to *Fabian Essays in Socialism*, 1920.
21. BLPES, Hyndman Papers, 522/3/1, Burrows's report. And see e.g. Y. Kapp, *Eleanor Marx* (Virago, 1979), ii. 293–4, 296–8.
22. Besant, *Autobiography*, 302–3.
    And see G. W. Foote, *Mrs Besant's Theosophy* (Progressive Publishing Co., 1889).
23. Foote, *Mrs Besant's Theosophy*, 321–2.
24. J. Ransom, *A Short History of the Theosophical Society, 1875–1937* (Theosophical Publishing House, 1938), 52, 255.
25. *Lucifer* (15 June 1891); *Westminster Gazette* (24 Apr. 1895).
26. Ransom, *Short History*, 42–56.
27. Ibid. Conway's statement that the name Koot Humi derived from an amalgam of Olcott and A. O. Hume is well known. Less attention has been paid to the fact that, when Conway saw a portrait at Adyar that Blavatsky claimed was of Koot Humi, he recognized it as a likeness of the Hindu religious leader Ram Mohun Roy (1772–1833) painted by James Phelp whose brother gave him, Conway, a copy. Conway, *My Pilgrimage*, 201.
28. T. Besterman, *Mrs Annie Besant* (Kegan Paul, Trench, Trübner, 1934), 148–54. Quoted in Nethercot, *The First Five Lives*, i. 322.
29. E. Bright, *Old Memories and Letters of Annie Besant* (Theosophical Publishing House, 1937), 20.
30. For references to action initiated or supported by Besant see *Final Report of the School Board for London, 1870–1904* (2nd edn., King, 1904), 202, 217, 322, 340.
31. D. Tribe, *President Charles Bradlaugh MP* (Elek, 1971), 278; G. West, *The Life of Annie Besant* (Gerald Howe, 1929), 130.
32. H. B. Bonner, *Charles Bradlaugh* (Fisher Unwin, 1895), ii. 406; BL Add. 46284, fo. 20, AB to Graham, 29 July 1889; fo. 18, Graham to Burns, 31 July 1889.

33. Bonner, *Charles Bradlaugh*, ii. 395. Besant, *Autobiography*, 327.
34. Karl Pearson Papers, University College London, 10/25, AB to Pearson, 12 Jan. 1887.
35. AB's desire to compensate for this 'error' prompted her to take a keen interest in the science of eugenics.
36. Besant, *Autobiography*, 214–18.
37. Shaw, *Diary*, introduction, 1890; 28 May, 29 June, 15 July 1890.
38. A. D. Besant, *The Besant Pedigree* (Besant & Co., 1930), 221, 224. Digby Besant prospered, becoming Secretary of the Society of Actuaries, and a member of the Athenaeum.
39. Ibid. Nethercot, *The First Five Lives*, i. 351–3. Professor Nethercot interviewed Digby Besant in London in 1954.
40. Ransom, *Short History*, 251.
41. F. Tuohy, *William Butler Yeats* (Macmillan, 1976), 31–2; P. Kuch, *Yeats and A. E. [G. W. Russell]* (Smythe & Barnes & Noble, 1986), 56.
42. Besant, *Autobiography*, 4.
43. Tuohy, *Yeats*, 32; Kuch, *Yeats and A.E.*, 106; *The Collected Letters of W. B. Yeats*, ed. J. Kelly, i: *1865–1895* (Clarendon Press, 1986). 'They [Theosophists] all looked to Ireland to produce some great spiritual teaching. The ark of the covenant is at Tara.'
44. Kuch, *Yeats and A.E.*, 9, 12.
45. Yeats, *Collected Letters*, Yeats to O'Leary, 7 May 1889; to Rhys, Aug. 1889; to K. Tynan, 27 Apr. 1890.
46. Besant, *Autobiography*, 330.
47. Bradlaugh Bonner Family Papers, Hypatia's memorandum.
48. *The Letters of Sidney and Beatrice Webb*, ed. N. Mackenzie, i: *1873–1892* (Cambridge, 1978), 222.
49. E. R. Pease, *The History of the Fabian Society* (Fifield, 1919), 98.
50. Besant, *Autobiography*, 4.
51. *Labour World* (11 Oct. 1890).
52. *Bernard Shaw: Collected Letters, 1874–1897*, ed. D. H. Laurence (Reinhardt, 1965), i. 210.
53. Fabian Society Papers, Nuffield College, Oxford, A. 6, AB to Pease, 17 Mar., 27 July 1890; C. 3, 21 Oct., 4 Nov. 1890.
54. Shaw, *Letters*, i. 273.
55. Fabian Society Papers, C. 55/56, Lists of Members.
56. Bonner, *Charles Bradlaugh*, ii. 407.
57. Ibid. ii. 408–21.
58. Ibid. ii. 420.
59. G. B. Shaw, *The Quintessence of Ibsenism* (Walter Scott, 1891), 'The Moral of the Plays', 127. AB took the chair at the first delivery of the lecture with this title, on 18 July 1890. 'She must have felt Ibsenism was not her faith', Shaw wrote, 'He handled idealism pitilessly; she still believed.' *Theosophist* (Oct. 1917).
60. J. P. Gilmour (ed.), *Champion of Liberty: Charles Bradlaugh Centenary Volume* (C. A. Watts, 1933), 49–50.
61. *Review of Reviews* (Mar. 1891).
62. Bradlaugh Bonner Family Papers, Hypatia's memorandum.

63. M. Pyrelal, *Mahatma Gandhi: The Early Phase* (Navajivan, 1965), i. 259, 260; J. D. Hunt, *Gandhi in London* (Promilla, 1978), 35.
64. Ransom, *Short History*, 252–3, 287.
65. J. Farquhar, *Modern Religious Movements in India* (Macmillan, 1915), 268.

Chapter 21

1. Theosophical Society Scrapbook for 1891.
2. G. W. Foote, *Mrs Besant's Theosophy* (Progressive Publishing Co., 1889).
3. E. Bright, *Old Memories and Letters of Annie Besant* (Theosophical Publishing House, 1937), 1 Sept. 1891.
4. Ibid.; Annie Besant, 'A Fragment of Autobiography: 1847–1891', *Lucifer*, 9, (1891–2).
5. A. Nethercot, *The Last Four Lives of Annie Besant* (Hart-Davis, 1963), ii. 28.
6. *Daily Chronicle* (5 Oct. 1891).
7. *The Letters of Sidney and Beatrice Webb*, ed. N. Mackenzie (Cambridge University Press, 1978), i. 222.
8. *Theosophist* (Oct. 1917).
9. *Lucifer*, 6 (1890).
10. H. Murphet, *Hammer on the Mountain* (Theosophical Publishing House, 1972), 259.
11. *Lucifer*, 12 (1893).
12. Theosophical Society Scrapbook for 1891.
13. J. Farquhar, *Modern Religious Movements in India* (Macmillan, 1915), 194; S. N. Dhar, *A Comprehensive Biography of Swami Vivekananda* (Vivekenanda Prakashan Kendra, 1975), ii. 800, 875.
14. J. H. Barrows (ed.), *The World Parliament of Religions* (Parliament Publishing Co. (Chicago), 1893), ii. 886–8.
15. *The Theosophical Movement: A History and a Survey* (E. P. Dutton, 1925), 443, 453; H. S. Olcott, *Old Diary Leaves* (Putnam's, 1895), iii. 192; W. T. Stead in *Borderland*, 2: 170.
16. *Borderland*, 2: 170.
17. Wood performed the same office at the 1893 Chicago World's Fair.
18. *Dictionary of American Biography*, 'Susan B. Anthony' (1820–1906); 'Julia Ward Howe' (1819–1910). Both were prominent in the women's cause. Howe was author of the 'Battle Hymn of the Republic'.
19. Dhar, *Swami Vivekananda*, i. 484 n.
20. *Lucifer*, 12 (1893).
21. Ibid. 9 (1892–3); Bodleian Library, Dep. d. 171, fo. 133, Olcott to Max Müller, 15 Dec. 1892.
22. M. D. Conway, *My Pilgrimage to the Wise Men of the East* (A. Constable, 1906), 177.
23. Edward Binfield Havell (1861–1934).
24. Bodleian Library, Dep. d. 171, fos. 127, 129, 145; *Nineteenth Century* (May 1893); *Lucifer*, 10 (1893).
25. Adyar Archives, Annie Besant's Diary.
26. Annie Besant, 'India: Her Past and Her Future', *Lucifer*, 13 (1893–4).

27. C. Wachtmeister, *HPB and the Present Crisis in the Theosophical Society* (London, 1895).
28. *Lucifer*, 13 (1893–4).
29. Adyar Archives, Annie Besant's Diary, various dates.
30. *Lucifer*, 13 (1893–4).
31. H. M. Hyndman, *Further Reminiscences* (Macmillan, 1912), 5–6.
32. *Lucifer*, 13 (1893–4).
33. V. Chirol, 'India in Travail', *Edinburgh Review*, 228 (1918). And see V. Chirol, *Indian Unrest* (Macmillan, 1910), 28–9.
34. They counted from 1875.
35. Annie Besant, *India; a Nation. A plea for Indian self government*. T. C. and E. C. Sack, 1915. *How India wrought for freedom*, Theosophical Publishing House, 1915.
36. See e.g. B. P. Sitaramayya, *The History of the Indian National Congress* (Chand, 1969), i. 11.
37. *Lucifer*, 13 (1893–4).
38. *Indian Mirror* (18 Jan. 1894). The editor, Norendrath Sen, was a brother of Keshub Chunder Sen. India Office Library, Eur. D. 767/9, Ghose to Sir W. Digby, 9 Aug. 1892.
39. *The Times* (5 Feb. 1894).
40. Ibid.
41. India Office Library, Eur. E. 243, p. 69. These were prophetic words where Besant was concerned. See Ch. 23 above.
42. Ibid. Eur. D. 558/6, p. 52, Lansdowne to Kimberley, 23 Aug. 1893.
43. R. P. Masani, *Naoroji, Dadabhai, 1825–1917* (Allen & Unwin, 1939), esp. 531.
44. India Office Library, Eur. D. 558/6, p. 52.
45. G. P. Gokhale (1866–1915). See B. R. Nanda, *The Indian Moderates and the British Raj*. (Oxford University Press, 1977); S. A. Wolpert, *Tilak and Gokhale* (University of California Press, 1962).
46. Bal Gangadhur Tilak (1856–1920); Wolpert, *Tilak and Gokhale*.
47. *The Times* (30 Mar. 1894).
48. Sri Prakasa, *Annie Besant as Woman and as Leader* (Theosophical Publishing House, 1941), 1.
49. Ben Tillett in J. Cousins, *The Annie Besant Centenary Book* (Besant Centenary Committee, Adyar, 1947), 163.
50. Olcott, *Old Diary Leaves*, vi. 136.
51. Stead (ed.), *Borderland*, i. 354; H. T. Edge *The Plot against the Theosophical Society* (C. H. Collings, 1895), Olcott, *Old Diary Leaves*, v. 257, 307.
52. See e.g. Ch. 20 n. 48.
53. Besant, *The Neutrality of the Theosophical Society* (A Besant, 1894).
54. Stead (ed.), *Borderland*, ii. 170–1; Edge, *Plot against the Theosophical Society*, 69–70; Olcott, *Old Diary Leaves*, vi. 330–1.
55. Farquhar, *Modern Religious Movements in India*, 242.
56. Besant, *Neutrality of the Theosophical Society*, 13.
57. *Westminster Gazette* (29 Oct.–8 Nov. 1894). The papers were given to Garrett by a Theosophist who disagreed with the decision not to

proceed against Judge. A. Nethercot, *The Last Four Lives of Annie Besant* (Hart-Davis, 1963), ii. 36.
58. Stead (ed.), *Borderland*, ii. 33. And see *DNB*, 'F. Edmund Garrett' (1865-1907).
59. Stead (ed.), *Borderland*, ii. 207.
60. Ibid. ii. 344.
61. Edge, *Plot against the Theosophical Society*, 59.
62. Stead (ed.), *Borderland*, ii. 206.
63. E. Nesbit, *The Story of the Amulet* (Fisher Unwin, 1906), 191.

Chapter 22

1. See e.g. Annie Besant, *India, Bond or Free?* (Putnam's, 1926), 161.
2. E. Bright, *Old Memories and Letters of Annie Besant* (Theosophical Publishing House, 1937), various dates in 1898.
3. Kor—the ancient city in Rider Haggard's *She*.
4. E. B. Havell, *Benares: The Sacred City* (Blackie, 1905).
5. *The Bhagavad Gita*, trans. A. Besant (A. Besant, 1895).
6. *Central Hindu College Magazine: A Journal for Hindu Boys* (1901), i. 28-9.
7. S. N. Dhar, *A Comprehensive Biography of Swami Vivekananda* (Vivekenanda Prakashan Kendra, 1975), i. 914. Sri Prakasa, *Annie Besant as Woman and as Leader* (Theosophical Publishing House, 1941), 111; Bepin Chandra Pal, *Mrs Annie Besant: A Psychological Study* (Ganesh, 1917).
8. *Central Hindu College Magazine*, i. 32.
9. Brooks broke with Besant in 1913 (see Ch. 23). He became a member of the Arya Somaj and earned a precarious living lecturing and writing about Hinduism.
10. Pal, *Mrs Annie Besant*, 361.
11. Aparna Basu, *The Growth of Education and Political Development in India, 1898-1920* (Oxford University Press, 1974), 10.
12. Prakasa, *Annie Besant*, 4.
13. Bright, *Old Memories and Letters*, 19 Mar. 1898.
14. Bhagavan Das, *The Central Hindu College and Mrs Besant* (Divine Life Press, 1913).
15. See e.g. G. Tillett, *The Elder Brother: A Biography of Charles Webster Leadbeater* (Routledge & Kegan Paul, 1982).
16. Annie Besant and C. W. Leadbeater, *Occult Chemistry: Clairvoyant Observations on the Chemical Elements* (2nd ed., Theosophical Publishing House, Adyar, 1919).
17. *National Reformer* (19 Sept. 1880).
18. Müller took Besant's place on the lecture tour of India with Olcott in 1892. After some years in the Society she transferred her support to the Ramakrishna Mission.
19. Bright, *Old Memories and Letters*, 17 Aug. 1913.
20. University of Chicago Library, Helen Dennis Papers, Besant letters, various dates, esp. 21 Jan. 1905.

21. G. N. Chakravarti, 'Spirituality and Psychism', Speech to TS Convention, Adyar, 1900.
22. Among the large amount of material concerning this affair see: Dennis Papers; Tillett, *The Elder Brother*, 77–8; H. Burrows and G. R. S. Mead, *The Leadbeater Case: The Suppressed Speeches Herbert Burrows and G. R. S. Mead made at the Annual Convention of the British Section of the Theosophical Society* (London, 1908); Eugene Lévy, *Mrs Besant and the Present Crisis in the Theosophical Society* (H. J. Heywood-Smith, 1913); J. Ransom, *A Short History of the Theosophical Society, 1875–1937* (Theosophical Publishing House, 1938) 356–7.
23. Dennis Papers, AB to her 'dear friends and fellow workers', 9 June 1906.
24. Ibid., Laura Cooper to AB, 4 May 1906.
25. Edward Carpenter wrote: 'There *is* an organic connexion between the homosexual temperament and unusual psychic or divinatory powers.' It is not known whether Besant, who greatly admired Carpenter's work in the 1880s, followed him in this. See T. Chushichi, *Edward Carpenter, 1844–1929* (Cambridge, 1980), 146.
26. Dennis Papers, Helen Dennis to AB, 20 May 1906, 9 Jan. 1907.
27. Ibid., enclosure.
28. Ibid. 9 June 1906. And see Burrows and Mead, *The Leadbeater Case*.
29. Dennis Papers, Fullerton to AB, 11 Feb. 1907; Olcott to AB, 7 Jan. 1907.
30. Adyar Archives, B/6/47, Sinnett to AB, 16 July 1907.
31. J. D. Smith, *Servant of India: The Diary of James Dunlop Smith*, ed. M. Gilbert (Longman's, Green, 1966), 58.
32. B. B. Misra, *The Indian Political Parties* (Oxford University Press, 1976), 118.
33. Bright, *Old Memories and Letters*, 8 Oct. 1908.
34. Smith, *Servant of India*, 142.
35. Ibid. 59.
36. See e.g. T. Morison, *Imperial Rule in India* (A. Constable, 1899).
37. Smith, *Servant of India*, 123.
38. National Library of Scotland, Minto Papers, 12762, fos. 71, 20, 212.
39. Adyar Archives, B6/23c, Gokhale to AB, 10 Mar. 1910.
40. Ibid. SN6, Sir A. Lawley to AB, 20 Feb. 1910.
41. NLS, Minto Papers, 12762, fo. 247.
42. Ibid.
43. Ibid., Dunlop Smith to AB, 7 Oct. 1910.
44. Ibid. fos. 347, 348.
45. Adyar Archives, B6/24, Bhagavan Das to AB, 19 Feb. 1910.
46. NLS, Minto Papers; 12762, fo. 297.
47. Ibid. fo. 346.
48. See Ch. 8.
49. Lévy, *Mrs Besant and the Present Crisis*.
50. See e.g. Dennis Papers; Bright, *Old Memories and Letters*; F. T. Brooks, *Neo Theosophy Exposed* (Vyasashrama Bookshop, 1914).
51. See Ch. 10.

52. Burrows and Mead, *The Leadbeater Case*.
53. Tillett, *The Elder Brother*, 199.
54. Ibid. 103–4; P. Jayakar, *Krishnamurti: A Biography* (Harper & Row, 1986).
55. Adyar Archives, B6/24.
56. Das, *The Central Hindu College*.

Chapter 23

1. India Office Library, Eur. E. 264/17, Pentland to A. Chamberlain, 7 July 1916.
2. Lady E. Lutyens, *Candles in the Sun* (Hart-Davis, 1957); P. Jayakar, *Krishnamurti: A Biography* (Harper & Row, 1986); Mary Lutyens, *Krishnamurti: The Years of Awakening* (Rider, 1984).
3. Lutyens, *Krishnamurti*; IOL, Eur. E. 264/17, Pentland to Chamberlain, 7 July 1916.
4. See Ch. 15.
5. The politician in question was Ramaswami Aiyar. He was counsel for the plaintiff in the Narianiah case. Besant was so impressed with him she asked him to work with her, which he did until the end of her life. J. Cousins, *The Annie Besant Centenary Book* (Besant Centenary Committee, Adyar, 1947), 118–22.
6. G. Tillett, *The Elder Brother* (Routledge & Kegan Paul, 1982), 159.
7. Adyar Archives, 'Interview with Mr Montagu', typescript with emendations by Besant.
8. Bhagavan Das, *The Central Hindu College and Mrs Besant* (Divine Life Press, 1913).
9. E. Lévy, *Mrs Besant and the Present Crisis in the Theosophical Society* (Heywood-Smith, 1913). And see R. Steiner, *The Way of Initiation* (Theosophical Publishing House, 1908), esp. Besant's introduction.
10. Das, *The Central Hindu College*; F. T. Brooks, *Neo Theosophy Exposed* (Vyasashrama Bookshop, 1914), and *The Theosophical Society and its Esoteric Bogeydom* (Vyasashrama Bookshop, 1914).
11. A. Basu, *The Growth of Education and Political Development in India, 1898–1920* (Oxford University Press, 1974).
12. IOL, Eur. E. 264/19, Cleveland to Maffey, 18 Oct. 1917.
13. Tillett, *The Elder Brother*, 149–50.
14. Das, *The Central Hindu College*.
15. See Ch. 12.
16. 'The Besant Privy Council Appeal', *Law Weekly*, Madras (1914).
17. Lady E. Lutyens, *Candles in the Sun*; Mary Lutyens, *Krishnamurti*; Krishnamurti Foundation of America, Ojai, Calif., Krishnamurti/Besant correspondence.
18. 'Interview with Mr Montagu'.
19. *DNB*, 'Nevinson' (1856–1941), 'Brailsford' (1873–1958), 'Snowden' (1864–1937).
20. R. K. Raj, *Social Conflict and Political Unrest in Bengal, 1875–1927* (Oxford University Press, 1984), 105, 141.
21. *Mofussil*—the surrounding district.

22. IOL, Eur. E. 264/17, Pentland to Chanberlain, 7 July 1916.
23. Ibid.; J. and M. Cousins, *We Two Together* (Ganesh, 1950).
24. IOL, Eur. E. 264/17, Pentland to Chamberlain, 7 July 1916.
25. Baroness Pentland, *Lord Pentland: A Memoir* (Methuen, 1928), 250; 'Interview with Mr Montagu'.
26. National Library of Scotland, Haldane Papers, 5903, fo. 228; 5911, fo. 29.
27. They called him Lokamanya, saviour of the people.
28. IOL, Eur. E. 264/17, Pentland to Chamberlain, 7 July 1916.
29. One of Bradlaugh's favourite tactics.
30. IOL, Eur. E. 264/17, Chelmsford to Meston, 16 Sept. 1916.
31. Ibid., Willingdon to Chelmsford, 18 July 1916.
32. The All India Congress Committee.
33. IOL, Eur. E. 264/17, Meston to Chelmsford, 16 Aug. 1916.
34. The Congress–League scheme for self-government, drawn up at this meeting, was aimed at subordinating the executive to the legislature. Separate electoral rolls for Muslims would, it was hoped, safeguard their minority position.
35. Adyar Archives, B6/23c, Gokhale to Besant, 9 Jan. 1915; Besant to Gokhale, 23 Jan. 1915.
36. P. Robb, *The Government of India and Reform, 1916–1921* (Oxford University Press, 1976).
37. H. F. Owen, 'Towards Nationwide Agitation and Organisation: The Home Rule Leagues, 1915–1918', in D. Low (ed.), *Soundings in Modern South Asian History* (University of California, 1968).
38. Nair revived the accusations of 1906 against Leadbeater in *The Antiseptic*, published in Madras. In 1913 Besant sued for libel, and lost. And see E. F. Irschick, *Politics and Social Conflict in South India* (University of California, 1969).
39. Hansard, 5th Series, HC xcv (1917) col. 198.
40. 'Interview with Mr Montagu'.
41. Hansard, 5th Series, HC xcv (1917), col. 198.
42. Mrs H. Hotchener (Marie Russak).
43. Hansard, 5th Series, HL xxx (1918), cols. 231–5.
44. Adyar Archives, B6/37A, Internment Diary. She was made to take the flag down.
45. Quoted in C. Hussey, *The Life of Edwin Lutyens* (Country Life, 1953), 378.
46. IOL, Eur. E. 264/19, Robertson to Chelmsford, 15 July 1917.
47. R. Kumar (ed.), *Essays on Gandhian Politics: The Rowlatt Satyagraha of 1919* (Clarendon Press, 1971), 153.
48. *The Collected Works of Mahatma Gandhi* (Government of India, 1915–17), xii. 464–5.
49. Hansard, 5th Series, HC xcv (1917), col. 1880.
50. IOL, Eur. E. 264/19, Pentland to Chelmsford, 23 July 1917; Chelmsford to Pentland, July 1917.
51. Hansard, 5th Series, HC xxvi (1917), cols. 744–92.
52. Internment Diary.

53. Hansard, 5th Series, HC xcvl (1917), cols. 863–4.
54. Pentland, *Lord Pentland*, 253.
55. House of Lords Record Office, Lloyd George Papers, F. 39/3/3.
56. Robb, *Government of India and Reform*, 125.
57. S. D. Waley, *Edwin Montagu: A Memoir* (Asia Publishing House, 1964), 138.
58. Hansard, 5th Series, HC xcviii (1917), cols. 55–74; HL xxvi (1917), cols. 746–92.
59. IOL, Eur. E. 264/19, Besant to Chelmsford, 10 Sept. 1917.
60. Ibid., Besant to Maffey, 22 Sept. 1917.
61. Ibid., Willingdon to Chelmsford, 28 Oct. 1917.
62. Hansard, 5th Series, HL xxx (1918), col. 232.
63. V. Montagu (ed.), *An Indian Diary* (Heinemann, 1930), 58–9; 'Interview with Mr Montagu'.

Chapter 24

1. A temporary canopy for large gatherings. Besant was the first woman, but not the first European, to hold office as President.
2. J. Cousins (ed.), *The Annie Besant Centenary Book* (Besant Centenary Committee, Adyar, 1947), 88–93.
3. See Ch. 22.
4. See Ch. 21.
5. Adyar Archives, B. 291. There is a copy among the Lloyd George Papers, House of Lords Record Office, F. 172/1/5.
6. B. P. Sitaramayya, *The History of the Indian National Congress* (Chand, 1969), 150; V. Montagu (ed.), *An Indian Diary* (Heinemann, 1930), 137, 234.
7. Montagu, *An Indian Diary*, 368.
8. India Office Library, Eur. E. 264/19, AB to Maffey, 27 Sept., 7 Oct. 1917; Willingdon to Chelmsford, 16 Oct. 1917.
9. See e.g. A. Link (ed.), *The Papers of Woodrow Wilson* (Princeton University Press, 1966), xlvi. 590; xlvii. 356, 366.
10. *The Times* (1 July 1918); Hansard, 5th Series, HC cvi (1918), col. 1218. There were rumours of German money backing the Home Rule League; Besant would not have sought it, believing that the Kaiser's Government was under evil influence.
11. B. B. Misra, *The Indian Political Parties* (Oxford University Press, 1976), A. Nethercot, *The Last Four Lives of Annie Besant* (Hart-Davis, 1963), ii. 285–6.
12. Annie Besant, 'The Future of Socialism', *Bibby's Annual*, (summer 1908).
13. NLS, Minto Papers, 12406, fos. 183, 189 AB to Minto, 3 Aug. 1912.
14. D. G. Pole (1877–1952), Theosophist. Solicitor to the Privy Council, 1913, when AB persuaded him to work for her. Labour MP 1929.
15. Adyar Archives, Lady E. Lutyens to AB, fragment, 1918.
16. *The Times* (15 Aug. 1918).
17. Misra, *The Indian Political Parties*, 177.
18. Adyar Archives, Indian National Congress Papers, 585.

NOTES

19. Montagu, *An Indian Diary*, 58.
20. Kanji Dwarkadas, *Dr Annie Besant* (Popular Prakashan, 1966); S. A. Wolpert, *Tilak and Gokhale* (University of California Press, 1962), 70.
21. Dwarkadas, *Dr Annie Besant*, 74; *New India* (21 Mar. 1919); H. F. Owen, 'Towards Nationwide Agitation and Organisation: The Home Rule Leagues 1915–1918', in D. A. Low (ed.), *Soundings in Modern South Asian History* (University of California Press, 1968), 159–95.
22. *New India* (31 Mar. 1919).
23. *New India* (17 Apr. 1919).
24. People living on the street where an Englishwoman was attacked were forced to crawl. Amritsar's water and electricity were cut off. No third-class tickets were allowed to be sold on trains, effectively stopping travel for most people.
25. Dwarkadas, *Dr Annie Besant*, 100–1.
26. Parliamentary Papers, 'Joint Select Committee on the Government of India Bill' (1919), iv. 172. And see appendix B.
27. Adyar Archives, INC papers, 585; Sitaramayya, *Indian National Congress*, 176.
28. Sitaramayya, *Indian National Congress*, 176.
29. Adyar Archives, B/6 23c, Willingdon letters to AB, 1919–24; *The Letters of Sir Edwin Lutyens to His Wife*, ed. C. Percy and J. Ridley (Collins, 1985), 386.
30. Adyar Archives, Willingdon/Besant letters, 1919.
31. Ibid., Lady E. Lutyens to AB, n.d.; Home Rule League papers, 'Report of the Principal Officers of the H.R. League for India, 1918.
32. See e.g. *Annual Register* (1917, 1918).
33. S. R. Mehrotra, *India and the Commonwealth 1885–1929* (Allen & Unwin, 1965), 165–6.
34. Sir Tej Bahadur Sapru (1875–1949), one of the principal architects of the Nehru Report, 1928. Rt. Hon. VSPC Srinivasa Shastri (1869–1949) succeeded G. K. Gokhale as head of the Servants of India Society.
35. George Lansbury (1859–1940), twice MP for Bow and Bromley (1910–12 and 1922–40). During his period as editor of the *Daily Herald* (1913–22) the paper received financial support from Lady de la Warr and Miss Dodge. His acquaintance with AB dated from the early 1890s when his branch of the Social Democratic Federation met in the club she founded for the match girls. His allegiance to Theosophy did not go beyond the doctrine of universal brotherhood. Colonel Josiah Wedgwood (1873–1943), MP for Newcastle under Lyme, Vice-Chairman of the Labour Party (1921–4).
36. *The Times* (1 Oct. 1925).
37. Hansard, 5th Series, HC clxxxix (1924–5), col. 1633.
38. She was not invited to the discussions that followed the publication of the Report. Nethercot, *The Last Four Lives*, ii. 421.
39. Mary Lutyens, *Edwin Lutyens by His Daughter* (John Murray, 1950), 227.
40. Ibid. 127, 281.
41. G. Tillett, *The Elder Brother* (Routledge & Kegan Paul, 1982), 171–2.

42. He was Colonel Wedgwood's cousin; both were members of the china manufacturing family.
43. S. Morison, *Some Fruits of Theosophy* (Hardinge 1919).
44. Tillett, *The Elder Brother*, 195, 260.
45. Ibid. 185-6.
46. Krishnamurti Foundation of America, Ojai, Calif., Krishnamurti to AB, 4 Feb., 1 July 1915; 10 Sept. 1920.
47. Mary Lutyens, *Krishnamurti: The Years of Awakening* (Rider, 1984), 54-5, 233-4.
48. Ibid. 142.
49. Ibid. 140.
50. Ibid. 241.
51. According to *The Secret Doctrine*, mankind would evolve towards perfection through a sequence of seven root races; each race was subdivided. The fifth root race was the Aryan race. The sixth subrace was expected to develop on the west coast of the United States.
52. Mary Lutyens, *Krishnamurti*, 213.
53. Ibid. 272-4.
54. Krishnamurti Foundation of America, AB to Krishnamurti, 31 Oct. 1928; Krishnamurti to Lady E. Lutyens, 26 Dec. 1929.
55. *Nineteenth Century* (Sept. 1894), 'True and False Conceptions of the Atonement'.
56. Krishnamurti Foundation of America, Krishnamurti to Lady Emily Lutyens, 17 May 1933.

Epilogue
1. *The Times* (22 Sept. 1933).
2. G. Tillett, *The Elder Brother* (Routledge & Kegan Paul, 1982), 16.
3. Copy in Somerset House, London.
4. H. M. Hyndman, *Further Reminiscences* (Macmillan, 1912), 4-5. And see p. 213 above.
5. E. Carpenter, *My Days and Dreams* (Allen & Unwin, 1916), 221-2.
6. See p. 291 above.
7. See e.g. J. Cousins (ed.), *The Annie Besant Centenary Book* (Besant Centenary Committee, Adyar, 1947), 331.
8. See p. 15 above.
9. P. Johnson, *Madame Blavatsky: The 'Veiled' Years* (Theosophical History Centre, 1987).
10. Carpenter, *My Days and Dreams*, 222.

SELECT BIBLIOGRAPHY

Principal Unpublished Sources
Archives of the Theosophical Society, Adyar, Madras, India.
Archives of the Theosophical Society, London.
Bodleian Library, Oxford: letters of W. T. Stead to Madame O. Novikoff; Max Müller correspondence.
Bradlaugh Bonner Family Papers (in private hands).
British Library: Blavatsky letters; Burns Papers; Chelmsford Papers (Oriental and India Office Collections); letters of Annie Besant to Bernard Shaw.
British Library of Political and Economic Science: Hyndman Papers.
Churchill College, Cambridge: letters of Annie Besant to W. T. Stead.
Columbia University, New York, Rare Book and Manuscript Library: Moncure D. Conway Papers.
Co-operative Union Library, Manchester: Holyoake Collection.
Hackney Archives Department: Bryant & May Papers.
Joseph Regenstein Library, University of Chicago: Helen I. Dennis Papers.
Lincolnshire Archives: Parochial Diary of the Revd Frank Besant.
National Library of Scotland: Geddes Papers; Haldane Papers; Minto Papers.
National Secular Society: Bradlaugh Papers.
Nuffield College, Oxford: Fabian Society Papers.
Public Record Office: *Besant* v. *Wood*.
University College London: Council Papers.

Newspapers, Periodicals
Bibby's Annual.
Borderland.
Central Hindu College Magazine.
Commonweal, London.
Commonweal, Madras.
Justice.
Labour Elector.
Labour World.
Link.
Lucifer.
National Reformer.
National Secular Society Almanacks.

New India.
Nineteenth Century.
Our Corner.
Pall Mall Gazette.
Pioneer.
Review of Reviews.
Star.
The Times.
Times of India.
Theosophist.
Westminster Gazette.

Published Works
For a complete list of Annie Besant's works see T. Besterman, *A Bibliography of Annie Besant* (Theosophical Society in England, 1924). For a complete list of Charles Bradlaugh's works see D. Tribe, *President Charles Bradlaugh MP* (Elek, 1971).

ADAMS, W. E., *Memoirs of a Social Atom* (Hutchinson, 1903).

SELECT BIBLIOGRAPHY

ANONYMOUS, *The Theosophical Movement: A History and a Survey* (E. P. Dutton, 1925).
ARNSTEIN, W. L., *The Bradlaugh Case: A Study in Late Victorian Opinion and Politics* (Clarendon Press, 1965).
BANKS, J. and O., *Feminism and Family Planning in Victorian England* (Liverpool University Press, 1964).
—— 'The Bradlaugh/Besant Trial and the English Newspapers', *Population Studies*, 8 (1954–5), 22–34.
BECHOFER-ROBERTS, C. E., *The Mysterious Madame* (John Lane/Bodley Head, 1931).
BESANT, ANNIE, *On the Deity of Jesus of Nazareth* (Thomas Scott, 1872).
—— *Auguste Comte: His Philosophy, His Religion, and His Sociology* (C. Watts, 1875).
—— *Law of Population* (Freethought Publishing Co., 1878).
—— *Why I am a Socialist* (A. Besant and C. Bradlaugh, 1886).
—— *Modern Socialism* (Freethought Publishing Co., 1886).
—— *Autobiographical Sketches* (Freethought Publishing Co., 1885).
—— *An Autobiography* (1893; 2nd edn., Theosophical Publishing House, 1908).
—— *Why I Became a Theosophist* (Freethought Publishing Co., 1889).
—— *Wake up India: A Plea for Social Reform* (A. Besant, Adyar, Madras, 1913).
—— *India, Bond or Free?* (Putnam's, 1926).
—— *The Coming of the World Teacher* (Theosophical Publishing House, 1925).
—— (ed.), *The Secret Doctrine*, iii (Theosophical Publishing Society, 1897).
—— and BRADLAUGH, CHARLES, introduction to Charles Knowlton, *Fruits of Philosophy* (Freethought Publishing Co., 1877).
—— and LEADBEATER, C. W., *Occult Chemistry* (2nd edn., Theosophical Publishing House, 1919).
—— —— *Man: Whence, How and Whither* (Theosophical Publishing House, 1913).
BESANT, ARTHUR DIGBY, *The Besant Pedigree* (Besant & Co., 1930).
BESANT, WALTER, *Autobiography* (Hutchinson, 1902).
BESTERMAN, T., *Mrs Annie Besant: A Modern Prophet* (Kegan Paul, Trench, Trübner, 1934).
BLAVATSKY, H. P., *Isis Unveiled* (W. J. Bouton, 1877).
—— *The Secret Doctrine*, 2 vols. (Theosophical Publishing Co., 1888).
—— *The Voice of the Silence* (London, 1889).
BONNER, H. B., *Charles Bradlaugh: A Record of His Life and Work by His Daughter* (Fisher Unwin, 1895).
BRADLAUGH, CHARLES, *Jesus, Shelley and Malthus; or, Pious Poverty and Heterodox Happiness* (1861).
—— *Poverty and its Effect on the Political Condition of the People* (1863).
—— *The Irish Question* (1868).
—— *The Land Question* (1870).
—— *The Impeachment of the House of Brunswick* (1871).
—— *Letter to the Prince of Wales on His Indian Visit* (1875).

SELECT BIBLIOGRAPHY

—— *Hints to Emigrants to the United States of America* (1879).
—— *Some Objections to Socialism* (1884).
—— *The Channel Tunnel: Ought the Democracy to Oppose or Support it?* (1887).
—— *The Eight Hours Movement* (1889).
—— *Labour and Law* (1891).
BRIGHT, E., *Old Memories and Letters of Annie Besant* (Theosophical Publishing House, 1937).
BROOKS, F. T., *Neo Theosophy Exposed* (Vyasashrama Bookshop, 1914).
BURROWS, H., and MEAD, G. R. S., *The Leadbeater Case* (London, 1908).
BURTIS, M. I., *Moncure Conway 1832–1907* (Rutgers University Press, 1952).
CARPENTER, E., *My Days and Dreams* (Allen & Unwin, 1916).
CHIROL, V., 'India in Travail', *Edinburgh Review*, 228, (1918).
—— *Indian Unrest* (Macmillan, 1910).
CONWAY, M. D., *Autobiography, Memories and Experiences* (Cassell, 1904).
—— *The Voysey Case* (Thomas Scott, 1871).
—— *My Pilgrimage to the Wise Men of the East* (A. Constable, 1906).
COUSINS, J. (ed.), *The Annie Besant Centenary Book* (Besant Centenary Committee, Adyar, 1947).
—— and COUSINS, M., *We Two Together* (Ganesh, 1950).
DAS, BHAGAVAN, *The Central Hindu College and Mrs Besant* (Divine Life Press, 1913).
DAVIES, C. M., *Heterodox London* (Tinsley Bros., 1871).
DAVIS, C. T., *The Family of Marryat* (privately printed, n.d.).
—— 'Memoir of Mrs Charlotte Marryat of Wimbledon House' (privately printed, 1900).
DHAR, S. N. *A Comprehensive Biography of Swami Vivekananda* (Vivekenanda Prakashan Kendra, 1975).
[DRYSDALE, GEORGE], *Elements of Social Science; or, Physical, Sexual and Natural Religion* (E. Truelove, 1861).
DWARKADAS, K., *Dr Annie Besant* (Popular Prakashan, 1966).
EDGE, H. T., *The Plot against the Theosophical Society* (C. H. Collings, 1895).
FARQUHAR, J., *Modern Religious Movements in India* (Macmillan, 1915).
FISHMAN, W. J., *East End 1888* (Duckworth, 1988).
FOOTE, G. W., *Reminiscences of Charles Bradlaugh* (Progressive Publishing Co., 1891).
—— *An Open Letter to Madame Blavatsky* (Progressive Publishing Co., 1889).
—— *Mrs Besant's Theosophy* (Progressive Publishing Co., 1889).
HAVELL, E. B., *Benares: The Sacred City* (Blackie, 1905).
HOLLIS, P., *Ladies Elect: Women in English Local Government, 1865–1914* (Clarendon Press, 1987).
HOLROYD, M., *Bernard Shaw: The Search for Love* (Chatto, 1988).
HOLYOAKE, G. J., *Sixty Years of an Agitator's Life* (Fisher Unwin, 1893).
HOME, D. D., *Lights and Shadows of Spiritualism* (Carleton, 1877).
HUSSEY, C., *The Life of Edwin Lutyens* (Country Life, 1953).

HYNDMAN, H. M., *The Record of an Adventurous Life* (Macmillan, 1911).
—— *Further Reminiscences* (Macmillan, 1912).
IRSCHICK, E. F., *Politics and Social Conflict in South India* (University of California, 1969).
JAYAKAR, P., *Krishnamurti: A Biography* (Harper & Row, 1986).
KAPP, Y., *Eleanor Marx*, i: *Family Life*; ii: *The Crowded Years* (Virago, 1979).
KRISHNAMURTI, J., *Notebooks* (Gollancz, 1976).
—— *Commentaries on Living* (Gollancz, 1977).
KUCH, P., *Yeats and A.E.* (Smythe & Barnes & Noble, 1986).
KUMAR, R., *Annie Besant's Rise to Power in Indian Politics, 1914–1917* (Concept Publishing Co., 1981).
—— (ed.), *Essays on Gandhian Politics: The Rowlatt Satyagraha of 1919* (Clarendon Press, 1971).
LANSBURY, G., *My Life* (Constable, 1928).
LÉVY, E., *Mrs Besant and the Present Crisis in the Theosophical Society* (Heywood-Smith, 1913).
LUTYENS, E., *The Letters of Sir E. Lutyens to his Wife*, ed. C. Percy and J. Ridley (Collins, 1985).
LUTYENS, Lady Emily, *Candles in the Sun* (Hart-Davis, 1957).
LUTYENS, MARY, *Edwin Lutyens by His Daughter* (John Murray, 1950).
—— *Krishnamurti: The Years of Awakening* (Rider, 1984).
MACCABE, J., *The Life and Letters of G. J. Holyoake* (Watts & Co., 1908).
MACCOBY, S., *English Radicalism 1853–86* (Allen & Unwin, 1938).
MACKAY, C., *Charles Bradlaugh* (D. J. Gunn, 1888).
MACKENZIE, N. and J., *The First Fabians* (Weidenfeld & Nicolson, 1977).
MISRA, B. B., *The Indian Political Parties* (Oxford University Press, 1976).
MONTAGU, V. (ed.), *An Indian Diary* (Heinemann, 1930).
MORISON, S., *Some Fruits of Theosophy* (Hardinge, 1919).
MORISON, T., *Imperial Rule in India* (A. Constable, 1899).
MURPHET, H., *Hammer on the Mountain: The Life of Henry Steele Olcott* (Theosophical Publishing House, Wheaton, 1972).
NETHERCOT, A., *The First Five Lives of Annie Besant* (Hart-Davis, 1961).
—— *The Last Four Lives of Annie Besant* (Hart-Davis, 1963).
OLCOTT, H. S., *Old Diary Leaves* (Putnam's, 1895).
OWEN, H. F., 'Towards Nationwide Agitation and Organisation: The Home Rule Leagues, 1915–1918', in D. Low (ed.), *Soundings in Modern South Asian History* (University of California, 1968).
PAL, BEPIN CHANDRA, *Mrs Annie Besant: A Psychological Study* (Ganesh, 1917).
PEASE, E. R., *The History of the Fabian Society* (Fifield, 1919).
PENTLAND, Lady, *Lord Pentland: A Memoir* (Methuen, 1928).
PRAKASA, SRI, *Annie Besant as Woman and as Leader* (Theosophical Publishing House, 1941).
PROTHERO, R., *The Life of Arthur Penrhyn Stanley* (John Murray, 1893).
PYRELAL, M., *Mahatma Gandhi: The Early Phase* (Navajivan, 1965).
QUIN, M., *Memoirs of a Positivist* (Allen & Unwin, 1924).

SELECT BIBLIOGRAPHY 373

RANSOM, J., *A Short History of the Theosophical Society, 1875–1937* (Theosophical Publishing House, 1938).
ROBB, P., *The Government of India and Reform, 1916–1921* (Oxford University Press, 1976).
ROBERTSON, J. M., *Charles Bradlaugh* (Watts, & Co., 1920).
ROYLE, E., *Radicals, Secularists and Republicans: Popular Freethought in Britain, 1866–1915* (Manchester University Press, 1980).
RUBINSTEIN, D., 'Annie Besant and Stuart Headlam: The London School Board Elections of 1888', *East London Papers*, 13 (1970), 3–24.
SCOTT, THOMAS, *A Farewell Address* (T. Scott, 1876).
SHAW, BERNARD, *Collected Letters 1874–1897*, ed. D. H. Laurence (Reinhardt, 1965).
—— *George Bernard Shaw: An Autobiography, 1856–1898: Selected from His Writings*, ed. S. Weintraub (Reinhardt, 1969).
—— *Bernard Shaw: The Diaries 1885–1897*, ed. S. Weintraub (Pennsylvania State University Press, 1986).
SINNETT, A. P., *The Occult World* (Trübner, 1881).
—— *Esoteric Buddhism* (Trübner, 1883).
—— *Incidents in the Life of Madame Blavatsky* (Redway, 1886).
SITARAMAYYA, B. P., *The History of the Indian National Congress* (Chand, 1969).
SMITH, J. D., *Servant of India: The Diary of James Dunlop Smith*, ed. M. Gilbert (Longman's, Green, 1966).
SOLOVIEV, V. S. *A Modern Priestess of Isis*, trans. W. Leaf (Longman's, Green, 1895).
STEINER, R., *The Way of Initiation* (Theosophical Publishing House, 1908).
TILLETT, G., *The Elder Brother: A Biography of Charles Webster Leadbeater* (Routledge & Kegan Paul, 1982) (contains a comprehensive list of Leadbeater's works).
TRIBE, D., *President Charles Bradlaugh MP* (Elek, 1971).
TUOHY, F., *William Butler Yeats* (Macmillan, 1976).
WACHTMEISTER, C., *HPB and the Present Crisis in the Theosophical Society* (London, 1895).
WALEY, S. D., *Edwin Montagu: A Memoir* (Asia Publishing House, 1964).
WEBB, B., *My Apprenticeship* (Penguin, 1971).
—— *The Diary of Beatrice Webb*, ed. N. and J. Mackenzie, i–iv (Virago, in association with LSE, 1980–5).
WEBB, S. and B., *The History of Trade Unionism* (1919; Longman's, Green, 1924; 1929).
—— —— *The Letters of Sidney and Beatrice Webb 1873–1892*, ed. N. Mackenzie (Cambridge University Press, 1978).
WEST, G., *The Life of Annie Besant* (Gerald Howe, 1929).
WHYTE, F., *The Life of W. T. Stead* (Cape, 1925).
WILLIAMS, G. M., *The Passionate Pilgrim: A Life of Annie Besant* (A. Knopf, 1946).
WOLPERT, S. A., *Tilak and Gokhale* (University of California Press, 1962).

INDEX

According to St John; On the Deity of Jesus of Nazareth, Part II (Besant) 57
Adamson, Revd 206
Adyar, India 235–6, 238, 239, 263, 265, 269, 277, 280, 285, 290–1, 299, 326, 327, 329
Agastya, Rishi 294
Aiyar, Sir Subramania 305, 309
Ali, Mohammad 313
Ali, Shaukat 313
Allen (Manchester martyr) 26–7, 190
Allman, James 191
Amberley, Lord 111
Amrita Bazaar Patrika 270, 272
Amritsar, India 317, 367
Annie Besant; An Autobiography 133, 146–7, 176, 247, 250, 326, 332
Anthony, Susan B. 264, 360
Arch, Joseph 43, 85, 144–5, 147
Arnold, Edwin 264
Arnold, Matthew 44
Arundale, Francesca 281
Arundale, George 295, 296, 298, 304, 307, 322, 324–5, 326, 329
Arya Somaj 232–3, 235, 362
Ashman, Revd William 224
Autobiographical Sketches (Besant) 2, 11, 16, 29, 35, 58, 173
Aveling, Edward Bibbins 147, 148, 177, 192, 247, 345, 348; friendship with Annie, 140–1, 189, 199, 214, 347; Hall of Science, 142–4; as teacher, 152, 158, 160, 178, 226; as socialist, 161, 163, 165, 166, 167–71, 173
Aveling, Eleanor, née Marx 166, 167, 168–9, 170, 177, 186, 189, 247, 345, 348
Avenue Road, St John's Wood, London 165–6, 171, 173–4, 179, 183, 220, 252–4, 260–1, 264, 275, 282

Bagally, Lord Justice 135
Baines, Mrs 137
Baldwin, Stanley 321
Ball, W. P. 172
Bardswell, Mr 130, 136

Barker, Ambrose 161
Barnett, Revd Samuel 211
Barnum, Phineas Taylor 224
Barry, Dr 29
Bartholomew, G. P. 209, 211
Beesly, Professor 162, 164, 347
Beeton, Mrs Isabella 65, 174
Bennett, D. M. 236
Bentham, Jeremy 108, 334
Besant, Albert 95
Besant, Annie, née Wood (1847–1933) 165; Irish descent, 1, 185; and father, 2; relationship with mother, 3, 5–7, 14–15, 24, 34, 44, 52, 59, 61, 64, 65, 66–7, 69; education, 6–12, 13, 359; and reading, 11, 14, 31–2, 82; confirmation, 12–13; Evangelicanism, 9–11, 12, 16–17, 24; Roman Catholicism, 12, 15; Frank as suitor, 18–19, 21; and Roberts, 22–3; engagement, 24; marriage, 27, 28–31, 34, 57, 58–9; relationship with Frank, 31–5, 52, 53, 55, 90, 123, 125, 135–6; writing, 31–3, 76, 81–4, 86, 99, 188; and Digby Besant, 33, 335; and Mabel Besant, 33, 36, 54, 94–5, 128, 135, 138–9, 335; religious doubt, 36–9, 43, 50, 52, 54, 66–7; illness, 39, 57, 84, 132–4, 153, 175–6; Theism, 44–5, 47; as orator, 55, 70, 73, 74, 85, 88–9, 97–8, 125, 149, 178, 189, 223, 262, 293–4, 311–12; independence, 60, 68; poverty, 61–2, 64; on love, 89–91; relationship with Bradlaugh, 91–3, 95, 96–7, 104–5, 123–5, 129, 144, 150–2, 154–6, 160, 173, 175, 176, 185, 205–6, 244, 256, 257, 259, 272; land reform, 85, 145, 147; as free thinker, 86, 97, 126, 138, 140, 160, 172; National Secular Society, 87–8, 127; and Hypatia Bonner, 99–100; 'Monster Petition', 101; Knowlton trial, 109–13, 114–15, 117–19; on jail, 112, 120; and contraception, 121–2, 251–2; Besant v. Wood, 128–33, 136; and India, 122, 134; Ireland,

INDEX

Besant, Annie (*cont.*):
146; university degree, 139–40, 143, 158–9, 183; and Aveling, 141–2, 143, 168–9, 170; as socialist, 161–3, 164, 167, 172, 177–80, 181–2, 188, 190–3, 195–7, 213, 215, 313–14; and Shaw, 165, 185–6; and Eleanor Aveling, 166; *Our Corner*, 173–4; and Fabian Society, 182–3; on Shaw, 184; and Stead, 198–202; influence of, 203–5, 212–13, 318–21, 327; and Bryant and May strike, 207–12; School Board, 214, 215–20, 249, 254, 262–3; occultism, 222, 224; as Theosophist, 223, 234, 236, 239, 240, 243, 244–5, 252–4, 255, 260–1, 266, 273, 290, 331–2; and Blavatsky, 228, 241, 247–9, 259–60, 275, 327; on reincarnation, 242; and politics, 245–7, 254, 350; and dockers' strike, 250; Freethought Publishing Company, 255–6; in USA, 258, 262; and Chakravarti, 263–4, 283; and caste system, 266–7, 312; in India, 267–70, 271, 277, 287–8; on renunciation, 268–9; Shanti Kunj, Benares, 277; Central Hindu College, 278–80, 286–7, 288–9; and Leadbeater, 282–5, 289–90, 322–3; and Esther Bright, 282; and Indian politics, 294–5, 299–300, 301–5, 308–10; neo-Theosophy, 295–6, 322–3, 354; and Krishnamurti, 296–9, 323, 325, 328; internment at Ootacamund, 306–7; and Gandhi, 315; passive resistance, 316–17; Order of the Star, 324–5; death, 326; will, 328–9; Hyndman on, 329; Carpenter on, 330, 332; and Lansbury, 367
Besant, Arthur Digby; baptism, 33–4; illness, 36; childhood, 41–2; and father, 60, 137; relationship with mother, 66, 68, 94, 132–3, 136, 172, 252, 328, 335; and Mabel, 131
Besant, Revd Frank (1840–1917), 137; as suitor, 18–19, 21; religious belief, 19, 44; education, 20, 140; as schoolmaster, 20, 25, 28–9, 30; engagement, 24; marriage, 27, 28, 31, 34, 52, 57; relationship with Annie, 31, 34–5, 44, 53–5, 59, 85, 90, 109, 123, 125, 135, 331; and Annie's writing, 32–3; as minister, 33–4, 36, 39, 41–2, 56–7, 58; separation, 60; and children, 68, 94, 95, 99, 127, 128, 137, 252; Besant *v.* Wood, 129–33, 136, 141
Besant, Mabel Emily; birth, 34–5; illness, 36, 54; relationship with mother, 60, 61, 68, 84, 94–5, 138–9, 172, 176, 252, 259, 328, 335; in Folkestone, 64–5; relationship with father, 99, 127, 128; education, 100; Custody of Infants Act, 130–3, 135, 136, 137, 251, 297
Besant, Sarah 19
Besant, Walter 19, 20, 56–7, 62, 86, 140, 252
Besant, William, father of Frank 19, 28
Besant, William, brother of Frank 20
Bhagavadgita 278
Bingham, Captain 175
Black, Clementina 207, 210, 212
Blake, Sophia Jex 29
Blake, Thomas Jex 29
Bland, Hubert 181, 245, 276, 350
Bland, Mrs Hubert 186
Blavatsky, Helena Petrovna 227–9; Theosophical Society, 224, 226, 273; Home on, 230–1; and Olcott, 232–3, 267; and India, 234, 235, 262; *Theosophist*, 236–7; reputation, 237–40; *Secret Doctrine*, 239, 241–2, 243, 245, 253, 259, 331–2, 368; relationship with Annie, 247–9, 250, 251–2, 256, 258, 260, 327, 331; influence of, 265, 275, 280–1, 290, 324, 332, 358
Blavatsky, Nicephore 228
Bloody Sunday 187, 194, 196, 316
Boehme, Jakob 225
Bonner, Arthur 183
Bonner, Hypatia, née Bradlaugh 72; relationship with father, 74, 112, 148, 155, 257; relationship with Annie, 82, 99, 100, 110, 132–3, 183, 185, 197–9, 220, 222–3, 244, 254, 256, 347; and Mme de Brimont, 93; relationship between father and Annie, 124–5, 176; education, 139–40, 144; and Aveling, 141, 168
Booth, Catherine 261
Bradlaugh, Alice 72; relationship with father, 74, 148, 155; and Mme de Brimont, 93; relationship with Annie, 99, 100, 110, 132–3, 197–8, 347; relationship between father and Annie, 124–5; and Aveling, 141;

education, 144, 157–8; death, 220
Bradlaugh, Charles (1833–91) 72, 75–7, 123, 185, 342; as editor of *National Reformer*, 47, 74, 81, 121–2, 140, 171, 188, 222, 357; relationship with Annie, 70, 91–2, 124–5, 139, 144, 172–3, 175–7, 181, 186–7, 220, 250, 259, 272, 287, 289; as secularist, 71, 79, 86; and Holyoake, 73; as Parliamentary candidate, 75, 80–1, 83–4, 126, 148; religious doubt, 77–8; as solicitor, 79; as Iconoclast, 79, 134, 192, 352; and land reform, 85, 145–7; and Hall of Science, 88, 280; and love, 89–90; and Mme de Brimont, 92–3; and Conway, 96–7; and Hypatia Bonner, 99; *Fruits of Philosophy*, 102–6; and Knowlton trial, 109, 110–14, 115, 117, 119–20, 214, 251; and National Secular Society, 127, 244; Besant *v.* Wood, 94–5, 128, 129, 130, 132–3, 137; and Aveling, 141, 142, 143, 165, 168–9, 170, 348; as MP, 149–57, 176, 205–6, 209–10, 353; and socialism, 162–3, 164, 167, 193, 194, 195, 196–7; Ironside Circles, 200–1; India, 234–5; Theosophy, 238–9, 245; Freethought Publishing Company, 255–6; Shaw on, 256–7; Bloody Sunday, 316
Bradlaugh, Susannah 72, 84, 99, 125
Brailsford, Henry (1873–1958) 299
Bramwell, Lord Justice 135
Brand, Mr Speaker 149, 150, 153, 156
Brassey, Lord 294
Bray, Charles 49
Brett, PC 26
Bright, Esther 260, 282, 286, 322
Bright, Jacob 195, 282
Bright, John 25, 71, 151
Bright, Ursula 277, 282
British Secular Union 127
Broad Church 44–7, 67
Brooke, Stopford 44
Brooks, F. T. 279, 296, 362
Bryant, A. C. 206
Bryant, William Wilberforce 207, 208–9
Bryant and May match girls' strike 206–11, 313
Büchner, Dr Friedrich 152–3, 186
Burnier, Dr Radha 328
Burns, John 195, 196–7, 204, 213, 247, 250

Burrows, Herbert; Land League 147; and socialism, 162, 164, 190, 205, 223, 250, 251; and Aveling, 169–70; and Social Democratic Federation, 177, 207; relationship with Annie, 189, 213–14, 240, 259, 275; Law and Liberty League, 196; Bryant and May strike, 209–12; School Board, 217; Theosophical Society, 247, 258, 290
Burt, Thomas 101, 162
Burton, Richard 264
Butler, Josephine 29, 261

Campbell, McLeod 38
Carlile, Richard 77, 78, 108, 109, 122
Carpenter, Edward 183, 220, 242, 330, 332, 363
Central Hindu College, Benares, India 278–80, 286, 288, 295, 296
Chakravarti, Gyanandra Nath 263–4, 272, 273, 277, 279, 283
Chamberlain, Austen 302, 307–8
Champion, H. H. 169, 207, 247, 250, 350
Charrington, Frederick 211
Cheltenham 28–9, 39
Chelmsford, Viscount 294, 295, 304, 307, 309, 310, 313, 319
Chicago anarchists 189, 192
Chirol, Valentine 268
Christian Evidence Society 111
Churchill, Lord Randolph 346
Clarke, William 245
Cluseret, Gustave 146
Cockburn, Sir Alexander 107, 113, 115, 116, 117, 118, 119–20
Coercion Bill 188, 351
Colenso, John 50
Common Law Act (1854) 79
Commonweal 189, 190, 191, 205, 211, 247; new 299, 301, 321
Comte, Auguste 91, 141
Condon (Manchester martyr) 26–7
Contagious Diseases Acts 122
contraception 98, 102, 104, 108, 111, 113–14, 115–18, 121–3, 342
Conway, Ellen 62, 92, 93, 109, 112, 129, 132, 183, 199
Conway, Moncure Daniel (1832–1907); South Place Chapel, 47–8, 64, 86, 134; relationship with Annie, 62, 69, 70–1, 85, 109, 112, 121, 125, 129,

INDEX

Conway, Moncure Daniel (*cont.*): 132, 181, 225–6, 243–4; religious doubt, 63; and Bradlaugh, 96–7; and Blavatsky, 230, 242, 358
Cook, Henry 102, 103, 276
Coulomb, M. and Mme 237–8, 239, 253
Cousins, James 300, 302
Cousins, Margaret 300–1
Cow Protection Society 270–1, 286
Cowen, Joseph 162, 347
Cox, Sergeant 222
Cracknell, Elizabeth 166, 175, 195
Criminal Law Amendment Act (1885), 198
Cromwell, Oliver 199
Cunninghame Graham, Robert 196, 204, 210, 250
Curzon of Kedleston, Marquess 286, 299
Custody of Infants Act (1873) 60, 129, 131

Dadhabhai Naoroji (1825–1917) 271, 286
Darling, Robert 196
Darwin, Charles 116–17, 342
Das, Babu Bhagavan 280, 289, 291–2, 295, 296, 330
Das, C. R. 321
Davidson, Thomas 177
Davies, Dr C. M. (1828–1910) 85–6, 88, 111
Davitt, Michael 145–7, 173, 189, 203, 204, 206, 255, 270, 271, 352
Deasy, Captain 26
de Brimont Brissac, Mina, Vicomtesse 92–3, 99
de la Warr, Muriel, Countess 293–4, 299, 308, 315, 318, 367
Dennis, Helen 283, 284
de Vaux, Mme Clothilde 91
Dialectical Society 85, 111, 139, 174–5, 222
Diggle, Revd 217, 218, 219
Dilke, Mrs Ashton 218, 219
Dillon, John 157
Disraeli, Benjamin, Earl of Beaconsfield 101, 147
Dodge, Mary 293, 299, 367
Drysdale, Charles 118, 121, 130
Drysdale, George 89–90, 115
Dufferin and Ava, Marquess of 270

Dunlop Smith, James 286–7, 288–9
Dwarkadas, Jamnadas and Kanji 316
Dyer, General 317

East London Observer 216
Elements of Social Science, The (Drysdale) 89–90, 94, 98, 104, 115, 150, 217, 251, 289
Ellis, Havelock 348
Ely, Talfourd 158
Emerson, Ralph Waldo 63
Engels, Friedrich 167, 170
Essays and Reviews 45, 47, 51
Everitt, Louisa 68, 94

Fabian Essays in Socialism 245, 255
Fabian Parliamentary League 181–2, 183, 213
Fabian Society 175, 177–8, 179, 180–1, 212, 213, 222, 254–5, 353
Fadeef, Elena 228
Family Herald 32–3
Fathers of the Church 15–16, 31–2, 51, 141
Fawcett, Henry 113, 114, 235
Fawcett, Millicent Garrett 113, 114, 117, 261
Fenian Brotherhood 25–6, 145–6, 200
Fern Hill, Charmouth, Dorset 9–10, 13
Folkestone, Kent 64–5
Foote, G. W. 244, 256, 259
Fowler, Alderman 155
Fox, Kate 223
Fox, Margaretta 223
Fox, W. J. 64
Frank, Isabel Campbell 167
Freethinker's Text Book, The (Besant) 130
freethought 49, 77–8, 85, 88–9, 95, 98, 101, 105, 108–9, 126, 142–4, 150, 152, 159, 235, 244, 259–60
Freethought Publishing Company 106, 107, 113, 168, 176, 183, 255–6
Fruits of Philosophy (Knowlton) 102–5, 107, 109, 113–19, 121, 127, 130–1, 150, 214, 237

Galsworthy, John 100
Gandhi, Mohandas Karamchand 257, 294, 299, 301, 306, 315, 316, 317, 319, 321, 327
Garrett, F. Edmund (1865–1907) 275, 361

INDEX 379

Garrett, Elizabeth 214
Geddes, Patrick (1854–1932), 174, 186
George, Henry 145, 165
Ghose, Motilal 270
Ghose, S. K. 234, 235
Giffard, Sir Hardinge 114–15, 119, 152
Gipsy Hill, Norwood, London 64, 65, 82
Gladstone, William Ewart 35, 87, 101, 126, 146, 151, 153–4, 157, 161, 194, 202, 326, 347
Gokhale, Gopal Krishna (1866–1915) 271, 286, 288, 295, 301, 303, 367
Gorky, Maxim 234
Grant Duff, Sir Mountstuart 238, 239
Granville, Earl 92

Haggard, Henry Rider 226, 264
Haldane, Viscount 297–8, 301, 320
Hall of Science 81, 88, 142–4, 156, 159–60, 162, 165, 168, 176, 183, 259, 280, 327
Hamilton, Lord George 280
Hardie, Keir 213
Hardinge of Penshurst, Baron 301
Harkness, Margaret 353
Harrow-on-the-Hill, London 5–6, 14
Hartman, Dr Franz 238
Hatherley, Baron (William Page Wood) 35, 39, 40, 46, 48, 57, 337
Headlam, Revd Stuart 143, 144, 147, 151, 160, 195, 196, 217, 218, 219, 350
Healaugh, Vicar of 62
Heard, Gerald 328
Heasly, Dr Philip 130
Hennell, Charles and Sara 49
Hetherington, Henry 77
Hewitt, Sir John 289
Hicklin, Regina v. 107
Higginson, Colonel Thomas Wentworth (1823–1911) 129
Hobart, Harry W. 207, 210, 211
Hodgson, Richard 239, 243
Holmes, Mr and Mrs Nelson 229
Holyoake, Austin 75, 76, 78, 102–4, 108, 109
Holyoake, George Jacob 22, 30, 70, 73, 76, 79, 102, 109–10, 111, 113, 127
Home, Daniel Dunglass (1833–86) 224, 229, 230, 231, 356
Home Rule League; in India, 271, 301–4, 306, 308, 310, 313, 316, 317, 319;
in Ireland, 271, 293
Hooker, Sir Joseph 159
Horniman, Benjamin 305
Hoskyns, Revd Edwin 217, 251
Howe, Julia Ward (1819–1910) 264, 360
Hume, Allan Octavian (1829–1912) 234–5, 238, 269, 357
Hume, Joseph 357
Huxley, Aldous 328
Huxley, T. H. 62, 157, 159, 346
Hyndman, Henry 161–2, 163, 164, 167, 173, 177, 180, 195, 247, 251, 329, 342, 344, 348

Iconoclast, see Bradlaugh, Charles
Impeachment of the House of Brunswick, The (Bradlaugh) 80
Ince, Mr 130, 136
Indian National Congress 134, 235, 256, 269, 271, 279, 295, 296, 301, 302–4, 307, 309, 311–12, 318, 319
Indian National Convention 320
Ireland 144, 146, 147, 154, 161, 162–3, 189, 203, 205, 253, 254, 262, 270, 293, 319
Irish Land League 145, 147, 191, 196
Ironside Circles (formerly Vigilance Circles) 200–1
Irrational Knot, The (Shaw) 174
Isis Unveiled (Blavatsky) 227, 241

James, Lord Justice 135
Jessel, Sir George 128–9, 131–2, 135, 138, 140, 158
Jesus, Shelley and Malthus (Bradlaugh) 80
Jinnah, Muhammad Ali (1876–1948) 306, 308, 321
Jones, Ernest 26
Jones, Reginald 58
Jowett, Benjamin 45
Judge, William Quan 230, 241, 258, 260, 262, 264–5, 272, 273, 275, 297, 362
Justice 163, 165, 172, 179, 200, 204, 205, 207, 209, 210, 214, 217, 218, 223, 247

Kalisch, Marcus 49
Keightley, Bertram 239, 240
Kelkar, N. C. 305
Kelly, Colonel 26, 146

Kenwood, Mrs 166
Khaparde, G. S. 286
Kitto, Revd 190
Knowlton, Charles 126, 143, 153, 158;
  Fruits of Philosophy, 102–5, 107, 109,
  113–19, 121, 127, 130–1, 150, 214,
  237
Krishnamurti, Jiddu 291, 293, 295,
  296–9, 304, 323–4, 325, 326, 327–8

Labouchère, Henry 153–4
Labour World 255
Lancet, The 118
land reform 85, 134, 144–5, 147, 154,
  203, 255, 271
Lansbury, George (1859–1940) 156,
  319, 321, 367
Lansdowne, Marquis of 271
Larkin (Manchester martyr) 26–7, 190
Larkin, James 300
Law, Harriet 70, 74, 127
Law Breakers and Law Makers (Besant)
  151
Law and Liberty League 195, 196,
  199, 200, 204, 211
*Law of Population: Its Consequences and
  Its bearing upon Human Conduct and
  Morals* (Besant) 121–2, 127, 130,
  138, 237, 252, 283
Leadbeater, Charles Webster; and Theosophical Society 281; relationship
  with Annie, 282, 289–91, 294, 296;
  reputation, 283–5, 323, 365; and
  Krishnamurti, 297–9, 325; Liberal
  Catholic Church, 323–4; death, 327;
  influence of, 328, 332
Levy, Joseph Hiam 139–40, 143, 162
Lewis, George Henry (1833–1911) 94,
  95, 106, 133, 136, 191
Liberal Catholic Church 322–3, 327
Liberal Social Union 69
Liddon, Henry 67
Lights and Shadows of Spiritualism
  (Home) 229, 230
Link 199, 203, 204, 205, 207, 208, 210,
  214, 220
Linnell, Alfred 195
Lloyd George, David 306, 307, 309
Local Government Act (1888) 246
Love Among the Artists (Shaw) 174
Lutyens, Edwin 293, 318, 322
Lutyens, Lady Emily 293, 299, 315,
  319, 322, 323–4, 325

Lutyens, Mary 343
Lyell, Sir Charles 62
Lyons, Lewis 217
Lytton, Earl of 293

Macdonald, Alexander 101
Macdonald, Sir Antony 279–80
MacDonald, Ramsay 320–1
Madras, India 269, 296
Maguire (Manchester martyr) 26–7
Malaviya, Pandit Mohun 296, 302–3
Malthus, Thomas Robert 113
Malthusian League 121, 251
Malthusianism 123; neo-, 80, 90, 104,
  107, 108, 111, 116, 150, 237, 251
Manchester martyrs 25–7, 146, 190
Mann, Emma 11, 13
Mann, Tom 213, 247
Marryat, Amy 6, 9, 11
Marryat, Charlotte 8–9
Marryat, Ellen 6–12, 13–14, 15, 23,
  31, 44, 51, 55, 82, 215, 268, 275,
  282, 331
Marryat, Captain Frederick 8–9, 217
Marryat, Joseph 7–8
Martineau, James 62
Martyrdom of Man, The (Reade) 226
Mary, Annie's maid 68, 84
Marx, Karl Heinrich 163, 167, 347, 348
Maskelyne, J. N. 224, 276
Matthew, Henry 192, 209–10
Maurice, Revd F. D. 38, 143
Mead, G. R. S. 290
Mearns, Revd Andrew 348
Meston, Sir James 302
Metropolitan Radical Federation 191–2,
  193, 194, 203, 204
Mill, John Stuart 113, 114
Miller, Florence Fenwick 214, 354
Milton, John 10
Minto, Earl of 286–7, 288–9, 314, 319
Modern Socialism (Besant) 178
Montagu, Edwin 294, 295, 299, 307,
  308, 309, 310, 312, 313, 316, 318,
  319
Montagu–Chelmsford Report 313–14,
  318–19, 321
Montefiore, Claude 217
Moore, Samuel 347
*Moral Physiology; or, a Brief and Plain
  Treatise on the Population Question*
  (Owen) 109, 126
Morison, Rosa 158

Morris, Constance Marion 100
Morris, May 186
Morris, Minnie 16, 18, 25
Morris, William 163, 177–8, 181, 184, 189, 190, 191, 193, 195, 196, 251, 283, 348
Moss, Arthur 219
Muir, John 49
Müller, Henrietta 214, 218, 282, 362
Müller, Max 231, 265–6
Mundella, A. J. 159

Nair, Dr T. M. 304, 365
Napoleon, Prince Jerome, 'Plonplon' 92–3
Narianiah, father of Krishnamurti 291, 296–8, 301, 364
National Agricultural Labourer's Union 43, 85
National Reformer 207, 233; Bradlaugh and, 47, 90, 93, 137, 139, 151, 162, 197; Annie and, 71, 76, 87, 91, 96, 99, 132, 136, 138, 163–4, 183, 188, 214, 217, 222–3, 245; Annie as Ajax, 82, 85, 153, 203; Watts and, 86, 104, 106; and *Fruits of Philosophy*, 107, 110, 111, 112; and *Law of Population*, 121–2; Aveling and, 142; Robertson and, 171; and Blavatsky, 236–7, 238, 240, 243
National Secular Society 70, 127, 145, 162, 188, 192, 203; Annie and, 71, 76, 87, 102, 183, 201, 244, 260, 331; Bradlaugh and, 74, 81, 256; Mme de Brimont and, 92; Aveling and, 142, 168–9; and Fabian Society, 179, 181; and Law and Liberty League, 200
National Sunday League 138
Neale, Cornelius 95
Nehru, Pandit Jawaharlal 296, 306, 327
Nehru, Pandit Motilal 279, 306, 321
Nesbit, E. 276
Nevinson, Henry (1856–1941) 299
New India 300, 301–2, 304, 305, 314, 319, 321
Nityananda, brother of Krishnamurti 291, 293, 296–9, 323, 324
Novikoff, Olga 202, 238, 239

Oatlands, St John's Wood, London 99, 100, 105, 124, 141
O'Brien, William 191, 192
O'Brien (Manchester martyr) 26–7, 190

Obscene Publications Act (1857) 107, 116
Occult Chemistry (Besant and Leadbeater) 282
Occult World, The (Sinnett) 234
O'Connor, Feargus 22–3
O'Connor, T. P. 75, 88–9, 155
Olcott, Henry Steele (1832–1907) 231–3, 362; and Theosophical Society, 224, 226, 227, 230, 238, 245, 253, 263, 283, 285; reputation, 229, 235; relationship with Annie, 237, 259, 262, 265–6, 267, 271–2, 327; and Blavatsky, 239, 249; and Judge, 273–5; and Leadbeater, 281; in Adyar, 277
Oliphant, Laurence 264
Olivier, Sydney 211, 245, 320–1
On the Deity of Jesus of Nazareth (Besant) 55–7
On the Nature and Existence of God (Besant) 87
Order of the Star 324–5
Origen 57
O'Shea, Katharine (Kitty) 100, 220, 340
Our Corner 173–4, 175, 178, 180, 183, 189, 199, 203, 205, 214, 220, 299
Owen, Robert 108–9, 178
Owen, Robert Dale 108, 109, 115, 126, 229
Owen, Rosamund Dale 178

Packer, Revd 77
Pal, Bepin Chandra 286, 299
Pall Mall Gazette 166, 192, 194, 195, 197–8, 239, 243
Pankhurst, Emmeline 196
Pankhurst, Dr Richard 195, 196
Paris 12–13, 247–8
Parker, John Henry 15
Parnell, Charles Stuart 100, 146, 181, 271
Pearson, Hesketh 185
Pease, Edward 180, 182, 222, 254, 255
Peel, Mr Speaker 176
Pentland, Baron 293, 300, 301, 304, 306, 307, 308, 318
Phelp, James 358
Place, Francis 108
Podmore, Frank 181, 222
Pole, David Graham (1877–1952) 315, 366
Principles of Political Economy (Mill) 113

INDEX

Prothero, Rowland 67
Pusey, Dr Edward 16, 50–2, 58, 282

Quin, Malcolm 87

Rai, Lala Lajpat 286
Ram Mohun Roy 358
Ramakrishna, Paramahansa 263
Ransom, Josephine 225–6, 228, 229, 233
Reade, William Winwood (1838–75) 226
Reading, Marquess of 319
Renan, Ernest 54
Review of Reviews (Stead) 261
Richardson, Dr 279
Richmond Terrace, Clapham, London 4, 16
Roberts, William Prowting 22–3, 25–6, 30, 42, 61–2, 98, 146, 173
Roberts, Mrs 22, 26
Robertson, F. W. 44
Robertson, John Mackinnon 76, 77, 146, 170–1, 173–4, 175, 186, 197, 199, 200, 211, 259, 349
Rogers, Thomas 79
Rothschild, Baron 206
Rowlatt, Mr Justice 315
Russell, G. W. 275
Russell, Lord John 111

St Leonards on Sea, Sussex 18, 25
Salt, Henry 142
Sanatama Dharma 278, 287
Sapru, Tej Bahadur (1875–1949) 320, 367
Saraswati, Swami Dayanand 232, 234, 235
School Boards 172–3, 215–20, 249, 251, 254, 262–3
Scott, Frank 196
Scott, Thomas (1808–78) 47–50; as publisher, 55, 56, 57, 69, 74, 86, 163; relationship with Annie, 60, 62, 64, 68, 70, 76, 82, 87, 134, 177, 225
Scott, Mrs 65
Secret Doctrine, The (Blavatsky) 239, 241–2, 243, 245, 253, 259, 331–2, 368
Seddon, James 196
Seeley, Sir John 62
Sen, Norendrath 361
Seymour, Digby 26
Sharples, Eliza 78
Shastri, Srinivasa (1869–1949) 320, 367

Shaw, George Bernard (1856–1950); and Aveling, 141–2; and Hyndman, 164; as orator, 165, 189, 193; as socialist, 166, 178, 350; serialization of novels, 174–5, 355; Social Democratic Federation, 177; as Fabian, 179, 182, 211, 222, 353; relationship with Annie, 180, 184–5, 187, 190, 195, 197, 199, 201, 213, 243, 252, 255, 281; Robertson and, 186; and Stead, 198; *Fabian Essays in Socialism*, 245–6; and Bradlaugh, 256–7; on Annie, 261–2, 331, 359
Shelley, Percy Bysshe 94, 135
Shuttleworth, Canon 201
Shyamaji, Krishnavarma 234
Sibsey, Lincolnshire 40, 41–3, 53–4
Sinn Fein 293, 319, 321
Sinnett, A. P. 224, 228, 230, 234, 235, 262, 285
Sling and the Stone, The (Voysey) 45, 49
Smith, W. H. 256
Snowden, Philip (1864–1937) 299, 320
Social Democratic Federation 161–2, 163, 168, 169–70, 177, 181, 182, 193, 212, 213, 215, 247, 250
Socialist League 177, 181, 189
South Place Chapel, Finsbury, London 62–3, 86, 97, 134, 181, 189
Southsea, Hampshire 28, 58
Sowerby, Mr 159
Stanley, Arthur Penrhyn, Dean 11, 16, 44–6, 66–8, 81, 85
Stanley, Lyulph 218
Stead, W. T. 198; on Annie's marriage, 27; relationship with Annie, 67, 187, 195, 196, 197–202, 240, 243, 250, 261, 275; on Annie and Bradlaugh, 125; as editor of *Pall Mall Gazette*, 166, 192, 194, 239; and Charrington, 211
Steiner, Rudolf 296
Stephens, Fitzjames 47
Stepniak, Sergius 179, 189
Stewart, G. A. 124
Stockwell Proprietary Grammar School, Clapham, London 19, 20
Story of the Amulet, The (Nesbit) 276
Strange, Thomas Lumisden 49

Tawney, R. H. 215
Taylor, Helen 162, 214, 218, 347, 354

INDEX 383

Taylor, Sedley 163
Tchaykowsky, Nicolai 179
Temple, Frederick 47
Theosophical Society; founding of, 224–6, 227; criticism of, 229–30, 267; Olcott and, 231; in India, 234–7, 263, 304; object of, 245; Esoteric section, 253, 258, 273, 275, 280, 282, 291, 311, 325; Annie and, 260, 262, 271, 282–3, 290, 322, 326, 327; Judge and, 265, 273–5; Cousins and, 300; Tilak and, 315; Krishnamurti and, 328
Theosophist, 236–7, 281
Thurlow, Baron 138, 344
Tilak, Bal Gangadur (1856–1920) 271, 278, 285, 297, 301, 303, 305, 314–15, 318, 319
Tillett, Ben 250
Times, The 128, 149–50, 156, 190, 194, 270, 272, 314, 327
Tingley, Katharine 297
Tower Hamlets, London 147, 214, 215, 217, 249, 263
trade unions 22, 43, 211–13, 250
Trafalgar Square, London 189–95, 203, 204, 205
Truck Amendment Act (1887) 205–6, 209
Truelove, Edward 71, 77, 90, 102, 126–7
Tyler, Sir Henry 159, 160
Tyndall, John 62, 87

Varley, Henry 153
Vaughan, Catherine 5, 11, 337
Vaughan, Dr Charles John (1816–97) 5, 18
Veil of Isis, The (Reade) 226–7
Vickery, Alice 118
Vivekananda, Swami 263, 264, 278, 279
Voice of the Silence, The (Blavatsky) 248
vo.' Hahn, General 228
Voysey, Revd Charles 39, 44–7, 49, 50, 55–6, 57–8, 60, 62, 67, 143, 225, 261

Wachtmeister, Countess 267, 268
Wadia, B. P. 299, 300, 304, 307, 313, 323
Walker, Edward, (W. D.?) 37–8, 50
Wallace, Alfred Russel 222, 227
Wallas, Graham 201, 211, 245–6
Warren, Sir Charles 191–2, 199, 204

Watson, James 77, 102–4, 108, 109
Watts, Charles 81, 84–5, 86, 101, 102–3, 105–6, 107, 108, 127
Watts, Kate 104, 105, 127
'W.D.' (Edward Walker?) 37–8, 50
Webb, Beatrice 55, 166, 212, 213, 221, 223, 225, 226, 257, 329
Webb, Sidney 177–8, 180, 211, 212, 245, 246–7, 254, 261, 350
Wedgwood, James 322
Wedgwood, Josiah (1873–1943) 320–1, 367, 368
Westminster Gazette 274
White Slavery in London (Besant) 207–8
Why did Gladstone Fall from Power? (Besant) 101
Why I am a Socialist (Besant) 178
Wilde, Oscar 283
Williams, John 209
Willingdon, Marquess of 308, 309, 318
Willson, Miss 267
Wilson, Charlotte 180, 184
Wilson, Thomas Woodrow 305, 306
Winterbotham, Lauriston 36, 39, 43
Wood, Alfred 2, 4, 68
Wood, Annie, *see* Besant, Annie
Wood, Benjamin 100
Wood, Emily; descent, 2, 3–4; relationship with Annie, 5–7, 14–15, 34, 44, 52, 59, 61, 64, 65, 66–7, 69; and religion, 16; and Frank Besant, 18–19, 21; and Annie's marriage, 24, 27; poverty, 35; illness, 66, 68; and Catherine Vaughan, 337
Wood, Henry Trueman (1845–1929); birth, 2; education, 4–7, 18, 25; relationship with Annie, 6, 59, 68, 95, 128; occupation, 35, 40, 264, 360
Wood, Maria 100, 220
Wood, Matthew (1768–1843) 1, 4, 340
Wood, Robert 4, 100
Wood, William Burton Persse 1–4, 66, 68
Wood, William Page, Lord Hatherley 35, 39, 40, 46, 48, 57, 337
Woodward, Mr 64–5
Wordsworth, Christopher 85
Wright, Francis 70

Yeats, William Butler 230, 231, 253, 273
Younghusband, Francis 287, 311